T0262026

Handbook of Weed Control

Handbook of Weed Control

Edited by **Jordan Smith**

New York

Published by Callisto Reference,
106 Park Avenue, Suite 200,
New York, NY 10016, USA
www.callistoreference.com

Handbook of Weed Control
Edited by Jordan Smith

International Standard Book Number: 978-1-63239-416-3 (Hardback)

Printed in the United States of America.

Contents

Preface

This book has been a concerted effort by a group of academicians, researchers and scientists, who have contributed their research works for the realization of the book. This book has materialized in the wake of emerging advancements and innovations in this field. Therefore, the need of the hour was to compile all the required researches and disseminate the knowledge to a broad spectrum of people comprising of students, researchers and specialists of the field.

The aim of this book is to educate the readers regarding the various techniques of weed control with the help of extensive information. Agricultural managers have been facing the problem of crop loss because of weeds ever since the first farming systems were developed. Over the course of the last century, significant progress has been made in order to minimize weed intrusion in crop settings with the help of efficient yet mostly non-sustainable weed control methodologies. With the commercial introduction of herbicides in the mid 1900's, developments in chemical weed control strategies have helped in effective suppression of a wide range of weed species for most agricultural practices. Presently, due to the need to design efficient sustainable weed management systems, research has been further pushing its boundaries regarding investigation on unified weed management options comprising of mechanical, cultural as well as chemical practices. The contributions made by authors in this book elucidate important topics of research which evaluate several options that can be utilized for establishing efficient and sustainable weed management systems for numerous areas of crop production.

At the end of the preface, I would like to thank the authors for their brilliant chapters and the publisher for guiding us all-through the making of the book till its final stage. Also, I would like to thank my family for providing the support and encouragement throughout my academic career and research projects.

Editor

Part 1

Integrated Cultural Weed Management

Intercropping of Pea and Spring Cereals for Weed Control in an Organic Farming System

Lina Šarūnaitė[1], Aušra Arlauskienė[2], Irena Deveikytė[1],
Stanislava Maikštėnienė[2] and Žydrė Kadžiulienė[1]
[1]*Institute of Agriculture, Lithuanian Research Centre for Agriculture and Forestry,*
[2]*Joniškėlis Experimental Station of the Lithuanian Research*
Centre for Agriculture and Forestry,
Lithuania

1. Introduction

Organic or sustainable management systems is focused on the creation of greater crop spatial and temporal diversification in crop rotation, and a reduction in the negative effects for food quality and environment, specifically a reduction in synthetic pesticide use (Lazauskas, 1990; Anderson, 2010). The relationship and competition beetween crop and weed populations is determined by the practical application of basic ecological principles in such management systems (Liebman & Davis, 2000; Singh et al., 2007). Crop diversification, which alters the composition of weed communities and influences their density, helps stabilize agricultural crop and weed communities (Barbery, 2002). Different seasonal types of agricultural crops (e.g. winter or spring crops) with different growth cycles and agronomic requirements provide unfavourable conditions for segetal plant life cycles. This prevents weed spread, germination, growth and seed ripening (Liebman & Dyck, 1993; Koocheki et al., 2009). In organic farming systems, an important role is assigned to a crop rotation (plant sequence diversification), catch crops and intercrops (Liebman & Davis, 2009; Anderson, 2010), and crop potential usage for suppressing and tolerating segetal plants (Liebman & Dyck, 1993).

Intercropping is the simultaneous production of more than one crop species in the same field (Willey & Rao, 1980). Intercrops can be combinations of two or more species, including both annuals and perennials or a mixture (Anil et al., 1998). When two or more crops are growing together, each must have adequate space to maximize synergism and minimize intercrop competition and decrease weed competition. Therefore, before implementing specific intercropping systems, it should be taken into account: spatial arrangement (Malezieux et al., 2009); plant density (Neumann et al., 2007; Andersen et al., 2007); maturity dates of the crops being grown (Anil et al., 1998); and plant architecture (Brisson et al., 2004).

One of the most commonly used intercropping mixtures is the legume/nonlegume (usually cereals) combination (Ofori & Stern, 1987; Anil et al., 1998; Hauggaard-Nilsen et al., 2008). Biologically fixed nitrogen (N_2) of legumes is the most common plant growth stimulating factor and improved crop competition with respect to weed species in organic or sustainable farming systems (Berry et al., 2002). Studies in the literature have demonstrated that grain legumes are weak suppressors of weeds, but mixing species in a cropping system becomes a way to improve the ability of the crop itself to suppress weeds (Lemerle et al., 2001; Mohler, 2001; Jensen et al., 2006). Therefore, intercropping of cereals and grain legumes: pea (*Pisum sativum* L. (Partim), lupin (*Lupinus angustifolius* L.), bean (*Vicia faba* L.), vetch (*Vicia sativa* L.) *et ctr* is a neglected theme in agricultural research. Weeds continue to play a major limiting role in agricultural production. The control of weeds using classical pesticides raises serious concerns about food safety and environmental quality, which have dictated the need for alternative weed management techniques.

The field experiments were carried out in 2007–2010 at the Institute of Agriculture (Dotnuva, loamy soil) and the Joniskelis Experimental Station (Joniskelis, clay loam soil,) of the Lithuanian Research Centre for Agriculture and Forestry. The aim of this study was to determine the effect of intercropping pea with spring cereals on crop competition, yield performance and weed control in organic farming conditions. The following trial design was used for intercrops and sole crops: 1) pea (cv. 'Pinochio', Ps, 2) pea/spring wheat (*Triticum aestivum* L. emend. Fiori et Paol., cv. 'Estrad'), PWi, 3) pea/spring barley (*Hordeum vulgare* L., cv. 'Aura'), PBi, 4) pea/oats (*Avena sativa* L., cv. 'Migla'), POi, 5) pea/spring triticale (× *Triticosecale* Wittm., cv. 'Nilex'), PTi, 6) spring wheat, Ws, 7) sprig barley, Bs, 8) oats, Os, 9) spring triticale, Ts. The experimental plots were laid out in a complete one–factor randomised block design in three replicates. Individual plot size was 2.5 × 12 m. The intercrop design was based on the proportional replacement principle, with mixed pea grain and spring cereals grain at the same depth in the same rows at relative frequencies (50:50 –a relative proportion of grain legume and spring cereals seeds). Wheat seeds rate were 5.5, barley 4.7, oat 6.0, triticale 4.5 and pea 1.0 mln seeds ha^{-1} for sole crop. Weeds were assessed twice: at stem elongation growth stage (BBCH 32–36) and at development of grain filling growth stage (BBCH 73). Mass of weeds and botanical composition was determined in 0.25 m^2 at 4 settled places of each treatment. The experimental data were processed by the analysis of variance and correlation-regression analysis methods using a software package "Selekcija". Weed number and mass data were transformed to $\sqrt{x+1}$.

2. Benefits of intercropping of cereal and grain legume

2.1 Yield and quality of intercrops

Cereal and legume intercropping systems are one of the important agronomic practices, wherein usually the productivity of the system as a whole is higher in comparison with that of their performance individually. Intercropping of cereals with grain legumes has been a common cropping system in rain-fed areas and especially in the Mediterranean countries (Anil et al., 1998; Lithourgidis et al., 2006). Grain legumes extensively used in intercropping with cereal include pea, vetch, lupin and bean (Hauggaard-Nielsen et al., 2001; Andersen et al., 2005; Ghaley et al., 2005; Berk et al., 2008). A number of different cereal crops have been proposed to be appropriate for intercropping with grain legumes such as barley oat, triticale, and wheat (Thomson et al., 1992; Berk et al., 2008). Intercropping advantages include improved soil conservation (Anil et al., 1998), yield stability (Hauggaard-Nielsen et al., 2003; Lithourgidis et al., 2006) and favours weed control (Banik et al., 2006). This leads to improved utilisation of environmental resources, light, water, and nutrients, in a multiple plant species community (Brisson et al., 2004; Corre-Hellou et al., 2007). The legume can provide N benefits to the nonlegume directly through mycorrizal links, root exudates, or decay of roots and nodules; or indirectly through a spring effect, where the legume fixes atmospheric dinitrogen (N_2), and thereby reducing competition for soil NO_3^- with the nonlegume (Anil et al., 1998). In intercropping the risk of nitrogen losses through leaching is substantially reduced in comparison to sole cropped pea (Neumann et al., 2007). Urbatzka et al. (2009) suggest, when pea is cultivated in a mixture with cereals, the N utilization effect was higher than in sole pea crop. In Danish and German experiments, the accumulation of phosphorous (P), potassium (K) and sulphur (S) was 20% higher in the intercrop (50:50) than in the respective sole crops (Hauggaard-Nielsen et al., 2009). The concentration of nitrogen is one of the most important criteria for grain quality evaluation. Pea intercropped with spring cereal increased the nitrogen concentration in intercrops compared with sole cereal (Ghaley et. al., 2005; Mariotti et al., 2006). Thus, better nutrition conditions are created in intercrops, therefore crops have a higher competitive ability against weeds.

Results from our study conducted in Dotnuva suggests that intercrops were less productive than sole pea crop (except for pea intercropped with wheat). However, the pea / barley and pea/triticale intercrops were slightly more productive than the sole cereal crops (Table 1). At Dotnuva, according to productivity, the dual-component intercrops were ranked in the following order: pea / wheat > pea / triticale; pea / barley > pea / oats. The data from the Joniskelis site show that in a heavy loam *Cambisol*, crop productivity was on average 20.5% higher, compared with that of crops grown in the Dotnuva site (Table 1).

Sole crops and intercrop	Crop component	Loamy soil (Dotnuva)			Clay loam soil (Joniskelis)		
		Grain yield (kg ha^{-1})	Nitrogen		Grain yield (kg ha^{-1})	Nitrogen	
			mg kg^{-1}	kg ha^{-1}		mg kg^{-1}	kg ha^{-1}
Ps	pea	2936.5	37.2	108.8	2896.6	33.6	83.8
P+SWi	pea	550.4	37.3	20.5	795.3	34.5	23.5
	wheat	2401.3	22.2	54.8	2473.2	20.0	42.3
	total	2951.9	24.9	75.3	3268.5	23.5	65.7
P+SBi	pea	565.7	36.9	20.6	649.0	33.1	18.2
	barley	2184.5	19.6	44.9	2386.4	18.7	38.4
	total	2750.1	24.2	65.5	3035.4	21.8	56.6
P+Oi	pea	445.2	37.0	16.6	432.9	33.3	12.2
	oat	2109.9	18.5	39.3	3837.5	17.4	57.1
	total	2555.4	22.4	55.9	4270.4	19.0	69.4
P+STi	pea	520.7	34.2	19.2	1240.8	33.4	35.5
	triticale	2214.6	26.6	48.9	2000.6	23.4	39.3
	total	2735.3	23.3	68.2	3241.4	27.2	74.8
SWs	wheat	3002.9	19.2	59.8	3387.9	18.7	53.9
SBs	barley	2583.3	16.6	46.4	2995.8	17.7	45.4
Os	oat	2897.0	17.6	48.3	3955.1	16.6	56.0
STs	triticale	2717.9	19.5	55.1	3220.5	21.7	59.9
LSD$_{05}$		743.12	5.70	20.06	611.8	6.32	9.42

Note. Sole crop: Ps –pea, SWs – spring wheat, SBs – spring barley, Os –oat, STs – spring triticale; intercrop: P+SWi – pea and spring wheat, P+SBi – pea and spring barley; P+Oi – pea and oat, P+STi – pea and triticale.

Table 1. Grain yield and nitrogen content of pea and spring cereals grown as sole crops and in dual-component intercrops data averaged over 2007-2010

Clay loam soils have high capillary water capacity, therefore plants are not so readily affected by lack of soil moisture (Maikštėnienė et al., 2006). The data from the Joniskelis site evidenced that all intercrops were more productive than sole pea crop. Moreover, the sole spring cereal was lower yielding than cereal intercropped with pea (except for pea intercropped with wheat). The rough structure of these soils was more favourable for cereals than for peas. At Joniskelis, according to productivity, intercrops were ranked in the following order: pea / oats > pea / wheat, pea / triticale > pea / barley. In dual-component intercrops with increasing productive density of cereals and their share in the yield, the total yield of the intercrops increased in Dotnuva (r = 0.650; $P<0.05$; r = 0.969; $P<0.01$, respectively) and in Joniskelis (r = 0.576; $P<0.05$; r = 0.916; $P<0.01$, respectively). Results obtained in various soils showed that when peas were grown mixed with oats or barley, their productivity was directly influenced only by cereals yield (r = 0.991; $P<0.01$; r = 0.971; $P<0.01$, respectively), whereas the productivity of peas grown in mixed crop with wheat or triticale was influenced by both components: yields of cereals (r = 0.825; $P<0.01$ and r = 0.984; $P<0.01$, respectively) and pea (r = 0.637; $P<0.05$ and r = 0.842; $P<0.01$, respectively) of the intercrop.

The accumulation of nitrogen in cereal grain is an indicator of different crop species competitive power. The findings from Dotnuva site showed that pea grown in sole crop accumulated 2.2% more nitrogen than pea intercropped with cereal. However, the grain nitrogen concentration of cereal intercropped with pea averaged 19.2% higher than that in sole cereal crop. The nitrogen concentration in pea grain was slightly lower in Joniskelis compared with Dotnuva. The amount of grain nitrogen did not differ between sole pea crops and pea intercrops. The grain nitrogen concentration of cereal intercropped with pea averaged 6.4% higher than that in sole crop. At both experimental sites, the highest grain nitrogen concentration was in spring wheat and triticale intercropped with pea.

In loamy soil (Dotnuva), sole pea produced a higher yield, therefore the nitrogen content was 29.8% higher compared to the corresponding data in clay loam soil (Joniskelis). The intercrops accumulated similar nitrogen concentrations in the total grain yield in both experimental sites. The amount of nitrogen in total grain yield of intercrops was greater by 26.3% in Dotnuva and by 23.8% in Joniskelis compared to the averaged amount of nitrogen in grain of sole cereal crop in corresponding experimental sites.

2.2 Intercropping for weed management

Weed management is a key issue in organic farming system (Bond & Grundy, 2001). Improvement of crop competition with weeds has been emphasised as the benefit of the increased sowing density of sole crops or intercropping (Auskalniene & Auskalnis, 2008; Liebman & Davis, 2000). Individual cereal species vary in their competitiveness against weeds. Weed suppression has been found to be greater in intercrops compared with sole crops, indicating synergism among crops within intercrops (Liebman & Dyck, 1993; Bulson et al., 1997; Szumigalski & van Acker, 2005; Deveikytė et al., 2008, 2009). In an organic farming weeds are controlled not only by direct means (manually or mechanacilly) and preventive measures (appropriate crop rotation, tillage, crop management) but also by increasing crop tolerance of weeds (choice of genotypes, sowing method, fertilization strategy) (Barbery, 2002; Anderson, 2010).

2.2.1 Weed species composition, germination time and conditions

According to literature, annual weeds, whose short vegetation period coincides with the cereal growth season, are most common in spring cereals (Rassmusen, 2002; Barbery, 2002). Weed seed germination is influenced by many factors such as field history and seed bank, soil properties, tillage practices and crop technologies. Annual weed seeds germinate at different times. *Thlaspi arvense* L. germinated at the earliest time, other weeds germinated when soil temperature had warmed up. *Fallopia convolvulus* (L.) A. Löve germinated a little later (Špokienė, 1995). In general, the intensive weed germination period is when soil warms up to 10-15 ° C. Therefore, the most intensive competitive interaction between weeds and crops occurs in the second half of May until mid June and late July to mid August in Eastern Europe (Špokienė, 1995).

Our results revealed that annuals dominated the weed flora composition (7-18 species) while there were fewer perennials (2-8 weed species). The most common annual weed species on fertile soils were: *Chenopodium album* L., *Veronica arvensis* L., *Stellaria media* (L.) Vill., *Galium aparine* L., *Fallopia convolvulus*, *Thlaspi arvense*, *Viola arvensis* Murray , *Lamium purpureum* L., *Polygonum aviculare* L., *Polygonum persicaria* L., *Fumaria officinalis* L., *Tripleurospermum perforatum* (Merat) M. Lainz. The dominance of these weed species can be explained by a very good adaptation to the existing soil and climatic conditions and soil tillage regime (Protasov, 1995). Such species are characterized by higher soil nutrient assimilation compared to agricultural crop plants. The following perennial weed species were identified: *Cirsium arvense* (L.) Scop., *Sonchus arvensis* L., *Taraxacum officinale* F.H. Wigg., *Equisetum arvense* L., *Tussilago farfara* L. Both experimental sites were similar in weed species and number (Table 2).

In a loamy soil (Dotnuva), plant diversity in crop rotation was higher (at cereal stem elongation growth stage BBCH 32-36) *Chenopodium album* was the dominant weed species from the 12-13 species identified. This species accounted for 61.7-77.2% of the total weeds documented. Whereas, in clay loam soil (Joniskelis), there were fewer (7-9) weed species of which the most frequent were *Stellaria media* (16.1-26.9 %), *Veronica arvensis* (9.8–16.8 %), *Galium aparine* (7.6-13.3%), *Chenopodium album* (7.3-12.9 %), *and Fallopia convolvulus* (6.9-9.9 %). Based on Špokienė and Povilionienė's (2003) findings, according to weed harmfulness reduction, the species can be ranked as follows: *Cirsium arvense* (10) > *Sonchus arvensis* (9) > *Taraxacum officinale (8)* > *Chenopodium album, Stellaria media (7)* > *Galium aparine, Fallopia convolvulus (6)*> *Polygonum* sp. (5)> *Thlapsi arvense (4)*. According to Lithuanian researchers' data, weed species such as *Viola arvensis, Veronica arvensis, and Lamium purpureum* are less harmful; however, the number of weed species in a crop (weed harmfulness threshold) is of great importance (Špokienė & Povilionienė, 2003).

Our research data (Joniskelis) revealed that weed germination was significantly lower in pea / barely intercrop, spring wheat and oat sole crops compared to the pea sole crop. At Dotnuva and Joniskelis experimental sites the weed number tended to decrease 3.6-19.5 % and 3.9-19.5 %, respectively. In Joniskelis, all intercrops and sole crops had good suppression of *Thlaspi arvense*. Slightly fewer weeds germinated in cereal sole crops compared to intercrops. The germination of *Galium aparine* was significantly lower, and the number of *Fallopia convolvulus* tended to decrease in Joniskelis' cereal sole crop. Different crops (intercrop and sole crop) had little effect on the variation of perennial weed number.

Further suppression of weeds depends on the crop's ability to impede weed growth. It is widely accepted that the competitive interaction between weeds and crops does not occur only at early stages of plant development (Lazauskas, 1990).

Species	Place	Sole crops and intercrop (BBCH 32–36)								
		Ps	P+SWi	P+SBi	P+Oi	P+STi	SWs	SBs	Os	STs
		Weed m²								
Viola arvensis	Dotnuva	2.3	1.7	1.9	2.3	1.9	2.6	2.2	1.6	1.7
	Joniskelis	5.8	3.5	3.3	2.8*	5.2	3.5	4.7	2.5*	5.8
Veronica arvensis	Dotnuva	1.1	0.6	0.6	0.3	1.2	0.6	0.4	0.4	0.7
	Joniskelis	8.7	8.8	7.7	5.5	10.0	8.7	10.0	6.8	7.3
Thlapsi arvense	Dotnuva	3.7	1.7	0.9	0.9	0.8	1.0	0.7*	1.6	0.4*
	Joniskelis	8.0	3.8*	2.8**	4.2	4.7	3.3**	3.3*	2.5**	3.3**
Galium aparine	Dotnuva	0	0	0	0	0	0	0	0	0
	Joniskelis	8.8	6.0	5.8*	7.0	6.7	5.0*	5.2*	4.0**	4.8*
Fallopia convolvulus	Dotnuva	1.9	1.1	1.6	1.3	0.7	2.1	1.3	2.0	1.2
	Joniskelis	5.3	4.2	4.7	5.5	6.3	3.5	3.0	3.8	5.5
Stellaria media	Dotnuva	2.7	1.9	2.2	3.2	2.3	2.4	2.4	1.8	2.2
	Joniskelis	10.7	12.2	11.0	12.5	9.3	10.5	14.7	9.5	17.0
Chenopodium album	Dotnuva	35.7	35.6	35.7	33.8	36.6	32.8	37.2	41.9	35.7
	Joniskelis	8.5	7.8	5.3	5.8	4.7*	4.2*	5.8	4.5*	4.8
Polygonum persicaria	Dotnuva	1.2	0.7	0.6	0.7	0.2**	0.1**	0.2**	0.7	0.3**
	Joniskelis	0	0	0	0	0	0	0	0	0
Polygonum aviculare	Dotnuva	1.0	0.1	1.4	0.6	0.7	0.7	0.6	0.8	1.7
	Joniskelis	0	0	0	0	0	0	0	0	0
Chaenorrhinum minus	Dotnuva	1.9	2.2	1.8	2.1	1.9	2.9	1.2	2.3	2.2
	Joniskelis	0	0	0	0	0	0	0	0	0
Cirsium arvense	Dotnuva	2.9	0.8	3.6	3.0	2.3	0.2	0.5	0.1	3.2
	Joniskelis	0.2	3.7*	1.3	2.7	3.0	2.5	1.7	0.5	4.2**
Sonchus arvensis	Dotnuva	1.8	0.7	0.1	0.9	1.0	0.3	1.0	1.4	0.4
	Joniskelis	0.7	0.3	0.5	1.5	2.2	0.7	0.3	0.5	0.3
Total number of weeds	Dotnuva	57.9	48.1	51.7	51.3	50.9	46.6	48.2	55.8	51.0
	Joniskelis	66.3	60.5	53.2*	55.9	63.7	51.7*	60.0	44.0**	63.2
Number of weeds species	Dotnuva	12	13	13	13	13	13	13	13	13
	Joniskelis	8	8	7	8	9	8	8	8	7

Note. *differences are statistically significant as compared to the control at P<0.05, **-at P<0.01
Sole crop: Ps –pea, SWs – spring wheat, SBs – spring barley, Os –oat, STs – spring triticale; intercrop: P+SWi – pea and spring wheat, P+SBi – pea and spring barley; P+Oi – pea and oat, P+STi – pea and triticale.

Table 2. Weed emergence and density of the most important species in sole crop and intercrop data averaged over 2007-2010

2.2.2 The competitive ability of pea intercropped with different spring cereal species

Intercropping advantages may be influenced by both plant density and relative frequency of the intercrop components (Subkowicz & Tendziagolska, 2005). The density of plants in intercrops varied between different experimental location, soil and cultivation conditions in our research. According to crop density data, pea plant accounted for 27.2% of barley intercrop and 29.7% of wheat intercrop at Dotnuva site. The greater density of pea was observed in intercrop with oat and triticale (35.2 and 34.7%, respectively). In Joniskelis, the number of pea plants was lower (20.3-24.6 %) in intercrops, except for pea intercropped with triticale (34.7%).

The highest productive density of pea in sole crop and intercrop was obtained in a loamy soil (Dotnuva) while a lower density was observed in clay loam soil (Joniskelis). Productive stem density of pea in crop structure was similar: 12.0-18.4 % (40-58 stems per m^2) in loamy soil, 10.2-20.4% (28-43 stems per m^2) in clay loam soil (Table 3). The more stable productive densities of intercrop were obtained in a loam soil (286-346 stems per m^2) compared to a clay loam soil (211-275 stems per m^2). This crop density in intercrop structure on a clay loam soil was determined by the specific properties of the soil (high clay content) and weather conditions. The weather conditions are essential on the formation of intercrop productivity and weed germination. They influence the optimal plant density and create the basis for competition between the components during crop germination period. The comparison between the different intercrops showed that the highest productive density was in pea intercropped with spring wheat (346 stems per m^2) and with barley (332 stems per m^2) in a loam soil, and peas with oats (275 stems per m^2) and with wheat (268 stems per m^2) in a clay loam soil.

Place	Crop component	Sole crops and intercrop (BBCH 73)								
		Ps	P+SWi	P+SBi	P+Oi	P+STi	SWs	SBs	Os	STs
		Productive stems per m^{-2}								
Dotnuva	pea	109	48	40	58	48				
	cereal		298	292	258	238	478	398	442	368
	total		346	332	316	286				
Joniskelis	pea	81	34	37	28	43				
	cereal		235	195	247	168	355	307	343	334
	total		268	231	275	211				

Note. Sole crop: Ps –pea, SWs – spring wheat, SBs – spring barley, Os –oat, STs – spring triticale; intercrop: P+SWi – pea and spring wheat, P+SBi – pea and spring barley; P+Oi – pea and oat, P+STi – pea and triticale.

Table 3. The productive density of sole crop and intercrop data averaged over 2007-2010

According to the literature, cereal has a stronger ability for weed suppression than pea (Andersen et al., 2007). German researchers note that crowding coefficients for semi-leafless pea cultivars were smaller than for conventional leafed types, therefore plant height of pea appears to be more important than plant leaf type for weed suppression (Rauber et al., 2001). The clay loam soil (Joniskelis) was more favourable for cereal growth: pea plants were shorter (13.2%), and cereals taller (2.6-4.9%, except for spring barley) compared with respective crops in a loamy soil (Dotnuva) (Table 4). The pea plants were 22.1-29.9 % shorter (Dotnuva) and 34.1-42.0% (Joniskelis) compared to oat, spring wheat and tricicale. The height of spring cereals ranked as follows: oat > triticale > wheat > barley. According to the study, the height of pea plants declined by 20.1-24.7% in higher density intercrops (Dotnuva), and in lower density intercrops (Joniskelis) by 11.0-25.0% compared to pea sole crops. The height of intercropped cereals was not significantly different than cereal sole crops. Pea plants intercropped with oat, in some cases with barely and triticale were taller than those of sole crops.

Place	Crop component	Sole crops and intercrop (BBCH 73)								
		Ps	P+SWi	P+SBi	P+Oi	P+STi	SWs	SBs	Os	STs
		Height of crop (cm)								
Dotnuva	pea	56.7	42.9	43.6	42.7	45.3				
	cereal		70.5	54.7	82.5	80.9	72.8	57.8	80.9	79.4
	weighted average		66.5	53.6	75.1	74.8				
Joniskelis	pea	49.2	36.9	38.2	39.7	43.8				
	cereal		72.4	58.7	89.4	79.6	74.7	56.6	84.9	82.8
	weighted average		67.9	55.5	84.4	72.2				

Note. Sole crop: Ps –pea, SWs – spring wheat, SBs – spring barley, Os –oat, STs – spring triticale; intercrop: P+SWi – pea and spring wheat, P+SBi – pea and spring barley; P+Oi – pea and oat, P+STi – pea and triticale.

Table 4. The plant height in sole crop and intercrop data averaged over 2007-2010

The weed suppression depended on the growth intensity of the crop aboveground biomass during the growing season. The mass per pea plant and per cereal stem at the beginning of cereal heading (BBCH 51) showed that the intercrops produced more biomass (0.18–1.05 g) compared to the cereal sole crops (Table 5). Comparison of different cereal species showed the lowest aboveground biomass per cereal stem was both in spring barley sole crop and intercropped with pea. Oat intercropped with pea accumulated the highest dry matter yield in the aboveground part. Here we identified the lowest aboveground mass per pea plant. The data of the aboveground mass suggested that pea grew slowly in intercrops until start of heading of cereals and poorly competed with cereals. During the experimental period, the aboveground mass was influenced by productive plant density but not by mass per stem. Peas produced more aboveground biomass in the second half of the vegetation period, in contrast to cereals, which already holds a dominant position in the first stages of growth. Weeds are suppressed for the durationof the vegetation period when the pea intercropped with cereal is established at appropriate densities. During the main crop growing period, when the development rate of the intercropped plant species do not coincide, favourable weather conditions for one or the other intercropped species can influence the degree of competition. The Joniskelis' experimental data indicated that the productive plant density in intercrops was lower for peas, which require higher nutrition area.

Indicators	Crop component	Sole crops and intercrop (BBCH 51)								
		Ps	P+SWi	P+SBi	P+Oi	P+STi	SWs	SBs	Os	STs
Dry matter of one stem (g)	pea	6.76	3.09	3.12	2.59	4.20				
	cereal		3.07	2.22	3.67	3.30	2.73	1.97	2.62	3.12
	weighted average		3.08	2.41	3.59	3.55				
Dry matter (g m^{-2})	pea	521.4	104.1	117.4	73.5	184.6				
	cereal		720.8	434.2	913.3	529.3	969.5	605.0	897.8	1010.9
	total		824.9	551.6	986.8	713.9				

Note. Sole crop: Ps –pea, SWs – spring wheat, SBs – spring barley, Os –oat, STs – spring triticale; intercrop: P+SWi – pea and spring wheat, P+SBi – pea and spring barley; P+Oi – pea and oat, P+STi – pea and triticale.

Table 5. The aboveground mass of crop during vegetation period in sole crop and intercrop (BBCH 51), Joniskelis 2007-2010 averaged data

The pea was suppressed in intercrops, where the productive density of pea stems was 40-58 (Dotnuva) and 28–43 stems m^{-2} (Joniskelis), the productive density of cereal was 238-298 and 168-247 stems m^{-2}, respectively. This indicates that the mass per pea plant in intercrops was 1.6-2.6 times lower compared to pea sole crop. Therefore, at Joniskelis site, the aboveground mass of crops during the growing season (BBCH 51) was lower for pea /wheat by 14.9%, pea/barley by 8.8%, and pea/triticale by 29.4% compared to the respective cereal sole crop. Only oat grown in intercrop produced more dry matter (9.9%) in aboveground mass compared to oat sole crop.

Indices allow researchers to quantify and express several attributes of plant competition, including competition intensity and importance, competitive effects and responses, and the outcome of competition (Weigelt & Jolliffe, 2003). An aggresivity value of zero indicates that component crops are equally competitive. If aggressivity value is higher than zero the species in the crop dominates, if this value is lower than zero the species is being chocked (Willey, Rao, 1980). Spring cereal has been dominant in intercrops due to the higher rate of aggression (Ac), the competitiveness ratio (CRc) in spring cereals. In most cases, oat was characterised as the strongest weed suppresser in intercropping system (Table 6).

The cultivation conditions were less favourable for crop growth in 2008 (in loamy soil) and 2009 (in clay loam soil), therefore weed density increased by up to 1.5-2 times until the harvesting period. The weakest competitive ability of cereal was obtained during 2008 and 2009. The study showed that the role of intercropped pea in weed suppression was limited. H. Hauggaard-Nielsen et al. (2008) indicate that a relative proportion of pea intercrop around 40-50% is needed in order to achieve a level of intraspecific competition.

2.2.3 Weed suppression in sole crops and intercrops

The ability of pea intercropped with cereal to suppress weed species was revealed only at the development of the grain at filling growth stage (BBCH 73) and during favorable crop growing conditions. The total number of weeds in intercrops was significantly reduced compared to pea sole crop at maturity stage (BBCH 73) at both experimental sites (Table 7).

The strongest weed suppression was observed in higher plant density intercrop and sole crop in Dotnuva. However, the number of weeds was 31.3-50.6% lower in intercrop compared to pea sole crop. In lower density crops (Joniskelis), the number of weeds in intercrop was 22.4-31.0% lower except for the oat sole crop and oat intercropped with pea. The oat displayed strong weed suppression capabilitieswith the number of weeds 72.5% lower in oat sole crop and 63.8% in oat/pea intercrop compared to pea sole crop. Comparison between cereal sole crops and intercrops showed a reduction in weed numbers in intercrop by an average 37.5%, in sole crops by 44.8% at Dotnuva, and by 36.3 and 39.1%, respectively in Joniskelis compared to pea sole crop. The number of weed species also significantly decreased except for pea/wheat (Dotnuva) and pea/triticale (Joniskelis) intercrops.

Intercrop	Year	Loamy soil (Dotnuva)		Clay loam soil (Joniskelis)	
		Ac	CRc	Ac	CRc
P+SWi	2007	1.06	4.75	0.56	3.37
	2008	0.80	2.99	0.40	2.89
	2009	1.10	5.22	1.09	2.13
	2010	-	-	0.37	2.63
P+SBi	2007	1.09	6.52	0.96	7.84
	2008	0.20	1.51	0.91	3.88
	2009	0.92	7.90	1.50	3.09
	2010			0.57	2.28
P+Oi	2007	1.25	4.89	1.22	15.63
	2008	1.17	2.90	1.15	5.45
	2009	1.41	9.89	1.26	4.67
	2010	-	-	0.85	5.25
P+STi	2007	0.78	3.65	0.38	2.44
	2008	0.70	4.05	1.52	1.33
	2009	1.99	7.92	-0.17	0.54
	2010	-	-	0.26	1.82

Note. Intercrop: P+SWi – pea and spring wheat, P+SBi – pea and spring barley; P+Oi – pea and oat, P+STi – pea and triticale.

Table 6. Plant aggressivety (Ac) and competition rate (CRc,) in organically grown intercrops

The number of weeds observed during the cereal maturity stage (BBCH 73) varied compared to the weed number in spring (BBCH 32-36) (Table 8). Thus, weed population dynamics was influenced not only by the crop suppression ability, but also by the total weediness of crop and weed species. In Joniskelis, in the lower density pea sole crop, the number of weeds increased by 4.9 m^{-2}, and in Dotnuva decreased by 3.1 m^{-2} during the period from emergence to cereal grain-filling growth stage. At Dotnuva, in the higher density crop, the total number of weeds decreased by 12.4-28.8 m^{-2} compared to the corresponding data in the spring. The variation of weed numbers during the growing season differed little between sole crop and intercrop (except oat sole crop and intercrop) with a decrease of 28.8 and 21.0 m^{-2}, respectively. At Joniskelis, more marked differences between crops were determined; however, the suppression of weeds was weaker compared to the Dotnuva data. According to the spring weed density, the lowest suppression of weeds was in pea / barley intercrop and wheat and barley sole crops. Pea intercropped with wheat or triticale and triticale sole crop exhibited similar weed suppression; the number of weeds per m² decreased by 9.8, 12.7 and 14.0, respectively. The best ability to suppress weeds was shown by oat sole crop and oat intercropped with pea with a decrease in weeds per m² by 24.4 and 30.1, respectively.

Species	Place	Sole crops and intercrops (BBCH 73)								
		Ps	P+SWi	P+SBi	P+Oi	P+STi	SWs	SBs	Os	STs
Viola arvensis	Dotnuva	1.3	1.4	0.7	0.4	1.0	0.6	1	0.0*	0.7
	Joniskelis	5.6	4.0	3.0	1.6**	4.2	4.1	5.4	1.2**	4.6
Veronica arvensis	Dotnuva	0.2	0.3	0.0	0.1	0.0	0.2	0.2	0.0	0.0
	Joniskelis	15.3	10.2	10.8	2.9**	8.2*	11.1	12.3	5.0**	8.6*
Thlaspi arvense	Dotnuva	1.1	0.2**	0.0**	0.0**	0.0**	0.0**	0.0**	0.0**	0.0**
	Joniskelis	3.7	1.3**	0.2**	0.0**	0.8**	0.5**	0.2**	0.0**	0.5**
Polygonum persicaria	Dotnuva	1.7	1.0	0.2**	1.1	0.7*	0.6*	0.4*	0.3**	0.1**
	Joniskelis	0.0	0.0	0.0	0.0	0.0	0.0	0.0	0.0	0.0
Polygonum aviculare	Dotnuva	2.0	0.7*	2.2	0.4*	0.9	0.2**	0.7	0.4*	0.7
	Joniskelis	0.0	0.0	0.0	0.0	0.0	0.0	0.0	0.0	0.0
Galium aparine	Dotnuva	0.0	0.0	0.0	0.0	0.0	0.0	0.0	0.0	0.0
	Joniskelis	4.4	2.2**	3.5	1.7**	3.4	3.1	2.5	1.3**	3.1
Fallopia convolvulus	Dotnuva	1.2	0.8	0.8	0.6	1.1	1.7	0.2	1.8	1.3
	Joniskelis	7.8	5.2	7.1	4.2*	6.6	5.3	5.9	4.2*	5.8
Stellaria media	Dotnuva	2.3	1.7	1.8	1.4	1.7	2.0	1.0	0.4*	1.0
	Joniskelis	11.3	8.4	9.4	3.3**	5.9**	6.0**	8.2	3.0**	5.4**
Chenopodium album	Dotnuva	37.6	27.0*	23.4**	22.0**	28.0*	22.9**	24.2**	21.9**	26.0**
	Joniskelis	7.2	7.0	6.1	5.1	9.3	6.1	8.1	1.7**	6.2
Sonchus arvensis	Dotnuva	2.2	0.3	0.7	0.9	0.7	0.3	0.4	1.4	0.2
	Joniskelis	6.1	3.4	5.3	1.4*	3.6	1.6*	3.9	0.5**	4.1
Cirsium arvense	Dotnuva	2.8	0.7	3.2	3.1	2.9	0.2	1.1	0.4	3.6
	Joniskelis	0.8	4.3	2.2	2.8	3.2	6.2*	3.0	0.7	4.8*
Total number of weeds	Dotnuva	54.7	35.7**	33.2**	30.3**	37.6**	29.4**	30.0**	27.0**	34.3**
	Joniskelis	71.3	50.7*	54.2*	25.8**	51.0*	49.4**	55.3*	19.6**	49.2**
Number of weeds species	Dotnuva	6	5	3**	4**	5*	4**	3**	3**	4**
	Joniskelis	9	7*	7*	5**	8	7*	8*	4**	8*

Note. *differences are statistically significant as compared to the control at P<0.05, **-at P<0.01
Sole crop: Ps –pea, SWs – spring wheat, SBs – spring barley, Os –oat, STs – spring triticale; intercrop: P+SWi – pea and spring wheat, P+SBi – pea and spring barley; P+Oi – pea and oat, P+STi – pea and triticale.

Table 7. Weed density (weed m⁻²) of the grain at filling growth stage (BBCH 73) of the most important species in sole crops and intercrops, data averaged over 2007-2010

Species	Place	Sole crops and intercrops (BBCH 73)								
		Ps	P+SWi	P+SBi	P+Oi	P+STi	SWs	SBs	Os	STs
Viola arvensis	Dotnuva	-1.0	-0.3	-1.2	-1.9	-0.9	-2.0	-1.2	-1.6	-1.0
	Joniskelis	-0.2	+0.5	-0.4	-1.3	-1.0	+0.6	+0.7	-1.3	-1.3
Veronica arvensis	Dotnuva	-0.9	-0.3	-0.6	-0.2	-1.2	-0.4	-0.2	-0.4	-0.7
	Joniskelis	+6.6	+1.4	+3.1	-2.6	-1.8	+2.4	+2.3	-1.8	+1.2
Thlaspi arvense	Dotnuva	-2.6	-1.5	-0.9	-0.9	-0.8	-1.0	-0.7	-1.6	-0.4
	Joniskelis	-4.3	-2.5	-2.7	-4.2	-3.8	-2.9	-3.2	-2.5	-2.8
Polygonum persicaria	Dotnuva	+0.5	+0.3	-0.4	+0.4	+0.5	+0.5	+0.2	-0.4	-0.2
	Joniskelis	-	-	-	-	-	-	-	-	-
Polygonum aviculare	Dotnuva	+1.0	+0.6	+0.8	-0.2	+0.3	-0.5	+0.1	-0.4	-1.0
	Joniskelis	-	-	-	-	-	-	-	-	-
Galium aparine	Dotnuva	-	-	-	-	-	-	-	-	-
	Joniskelis	-4.4	-3.8	-2.3	-5.3	-3.2	-1.9	-2.7	-2.6	-1.8
Fallopia convolvulus	Dotnuva	-0.7	-0.3	-0.7	-0.7	+0.5	-0.4	-1.1	-0.2	+0.1
	Joniskelis	+2.5	+1.0	+2.5	-1.3	+0.3	+1.8	+2.9	+0.4	+0.4
Stellaria media	Dotnuva	-0.4	-0.2	-0.4	-1.8	-0.6	-0.4	-1.4	-1.4	-1.2
	Joniskelis	+0.7	-3.8	-1.6	-9.3	-3.4	-4.5	-6.4	-6.5	-11.5
Chenopodium album	Dotnuva	+1.9	-8.6	-12.3	-11.8	-8.5	-9.9	-13.0	-20.0	-9.7
	Joniskelis	-1.3	-0.8	+0.7	-0.8	+4.6	+1.9	+2.3	-2.8	+1.4
Sonchus arvensis	Dotnuva	+0.4	-0.4	+0.6	0.0	-0.3	0.0	-0.6	0.0	-0.2
	Joniskelis	+5.4	+3.1	+4.8	-0.1	+1.4	+0.9	+3.6	0.0	+3.7
Cirsium arvense	Dotnuva	-0.1	-0.1	-0.3	+0.1	+0.6	0.0	+0.6	+0.3	+0.4
	Joniskelis	+0.7	+0.7	+0.8	+0.2	+0.2	+3.7	+1.3	+0.2	+0.7
Total number of weeds	Dotnuva	-3.1	-12.4	-18.2	-21.0	-12.9	-17.2	-18.2	-28.8	-16.7
	Joniskelis	+4.9	-9.8	+1.0	-30.1	-12.7	-2.3	-4.8	-24.4	-14.0

Note. Sole crop: Ps –pea, SWs – spring wheat, SBs – spring barley, Os –oat, STs – spring triticale; intercrop: P+SWi – pea and spring wheat, P+SBi – pea and spring barley; P+Oi – pea and oat, P+STi – pea and triticale

Table 8. The variation of weed numbers (weed m-2) of the most important species in sole crops and intercrops during growing season, data averaged over 2007-2010

At Joniskelis, the highest total mass of weeds was determined in pea sole crop and intercropped with triticale. At Dotnuva, the total dry matter (DM) of weeds in pea sole crop was 38.4% higher compared to the pea sole crop at Joniskelis (Table 9). At Dotnuva, *Chenopodium album, Cirsium arvense,* and *Sonchus arvensis* mass accounted for the largest share in the total weed mass. All cereal sole crops and intercrops significantly reduced weeds and the weed mass decreased by 72.0-90.7% compared to pea sole crop. At Joniskelis site, *Cirsium arvense* was spread unevenly in the experimental area; therefore, the total weed mass was substantially higher where this weed was present. The lowest total mass of weeds was determined in oat sole crop and pea intercropped with oat and barley.

The variation of weed total numbers and weight was determined by the response of different weed species to crop suppression. Weed species and their numbers at cereal maturity stage during maturity stage are presented in Tables 7, 8 and 9. The species *Viola arvensis, Veronica arvensis, Thlaspi arvense, Polygonum persicaria* and *Polygonum aviculare* are considered less harmful for agricultural crops (Špokienė & Povilionienė, 2003). The suppressing effect for *Viola arvensis, Veronica arvensis* was more pronounced in the lower density crop in Joniskelis. The number of these weed species slightly increased in the cereal sole crop compared to the pea intercropped with cereal. The number of these weeds significantly decreased in oat sole crop and intercropped with pea, and *Veronica arvensis* decreased even in triticale sole crop and intercropped with pea. Although the presence of *Thlaspi arvense* was low, the number of weeds was significantly reduced in all cereal sole crops and intercrops regardless of the site. At Dotnuva site, where *Polygonum persicaria* and *Polygonum aviculare* were found, the cereal sole crop suppressed these weeds slightly sronger compared to the pea intercrop. There were significantly fewer *Polygonum aviculare* plants in wheat, oat sole crops and intercropped with pea compared to the pea sole crop, whereas, *Polygonum persicaria* was suppressed by all cereal sole crops and some of their intercrops with pea.

The number of *Viola arvensis* was low at Dotnuva during cereal maturity stage compared to the findings in spring, (BBCH 32-36); the weed numbers were reduced. However, at Joniskelis, where the incidence of these weed species was higher, the number of weeds was reduced only in oat and triticale sole crops and intercropped with pea (by 1.3, 1.3 and 1.3, 1.0 weed m^{-2}, respectively). The number of *Viola arvensis* increased in pea/wheat and wheat, barley sole crop. At Dotnuva, the number of *Veronica arvensis* during the gropwing season was reduced in all crops 0.2-1.2 weed m^{-2}; however, at Joniskelis, the weed number increased in the majority of the crops. The number of *Viola arvensis* was reduced in oat sole crop (1.8 weed m^{-2}) and pea intercropped with oat (2.8 weeds m^{-2}), triticale (1.8 weed m^{-2}). The number of *Thlaspi arvense* was reduced in all crops (0.4–1.6 weed m^{-2}) at Dotnuva and at Joniskelis (2.5–4.3 weed m^{-2}). The number of *Polygonum persicaria* and *Polygonum aviculare* increased in the majority of crops. The number of weeds most consistently decreased in oat and triticale sole crops. *Viola arvensis, Veronica arvensis, Thlaspi arvense, Polygonum persicaria, Polygonum aviculare* are less harmful, the mass of the weeds was low and significantly decreased in the majority of crops compared to pea sole crop.

Galium aparine and *Fallopia convolvulus* are common in crops. Their numbers significantly decreased in oat sole crop and intercropped with pea; *Galium aparine* was also decreased in pea intercropped with wheat compared to pea sole crop. Cleare advantages of intercrops compared to sole crops were not detected against these two harmful species of weeds.

A strong suppressive effect of crops on *Galium aparine* was identified during cereal maturity stage when the number of weeds declined by 1.9–5.3 weed m^{-2} compared to the findings in spring. The advantages of intercrops were clear with oat, wheat and triticale intercrops reducing the number of *Galium aparine* by 5.3, 3.8 and 3.2 weed m^{-2}, respectively, than sole cereal crops by 2.6, 1.9 and 1.8 weed m^{-2}, respectively.The mass of *Galium aparine* decreased (22.9-96.1%) in all crops, except for intercropped wheat. Significantly lower mass of these weeds was in the oat intercrop, and oat and barley sole crop compared to pea sole crop.

At Dotnuva, the number of *Fallopia convolvulus* during the growing season decreased in the majority of crops, except for tricticale sole crop and intercropped with pea. At Joniskelis, the number of these weeds increased during the whole growing season compared to the respective number of weeds in spring. The number of weeds markedly increased in intercropped barley and pea and wheat and barley sole crops. The number of *Fallopia convolvulus* decreased only in intercropped oat compared to the findings in spring. The experimental crops had greater influence on *Fallopia convolvulus* mass rather than number. At both experiment sites, intercrops and cereal sole crops significantly decreased the mass of this weed (by 70.6-98.7% and 64.8-92.8%, respectively) except for intercrop and sole crop of wheat (Dotnuva) and intercropped triticale (Joniskelis) compared to pea sole crop.

The incidence of *Stellaria media* was high at Joniskelis. Significantly fewer *Stellaria media* plants were recorded in cereal sole crop (except for spring barley) compared to pea sole crop. The number of this weed was significantly reduced by the intercrops of oat and triticale. The reduction in *Stellaria media* numbers during the cereal maturity stage was marked (regardless of their abundance) compared to the number of these weeds in spring, except for pea sole crop at Joniskelis. At both sites, this weed species was more suppressed by cereal sole crop than the intercrop. The highest reduction of *Stellaria media* was determined in these crops: pea / oats, oats, triticale (Dotnuva and Joniskelis) and barley sole crop (Dotnuva) compared to the respective weed numbers in spring.At Dotnova, the mass of *Stellaria media* was low and the influence of the crops was not significant, except for oat sole crop. The incidence of this weed was high at Joniskelis where the influence of crops on the reduction of weed mass was significant (37.7–94.8%) compared to pea sole crop. The mass of *Stellaria media* was reduced by intercropped or sole oat. Also, this weed was suppressed by wheat, triticale sole crop and their intercrops with pea.

The incidence of *Chenopodium album* was high at Dotnuva (21.9-37.6 weeds m^{-2}); all crops significantly reduced the number of this weed species compared to pea sole crop. At Joniskelis, the number of *Chenopodium album* was reduced only by oat sole crop. The variation of *Chenopodium album* numbers during the growing season showed that these weeds were not as intensively suppressed as other weed species at Dotnuva. The number of *Chenopodium album* reduced in intercrop and sole crop was 8.5–20.0 weeds m^{-2} compared to the respective weed numbers in spring. The number of this weed species slightly increased in pea sole crop. However, at Joniskelis, the number of *Chenopodium album* incresed in the majority of sole crops and intercrops, where 4-9 times fewer weeds emerged in spring. The number of weeds slightly decreased in wheat and oat intercrops and pea sole crop, but the weed incidence decreased most in oat sole crop (2.8 weeds m^{-2}). The investigated crops at both experimental sites reduced the mass of *Chenopodium album* by 71.6-93.4% at Dotnuva and by 54.3-97.9% at Joniskelis compared to pea sole crop. The mass of this weed was lower in many cereal sole crops compared to intercrops.

Perennial weeds *Sonchus arvensis* and *Cirsium arvense* are more frequent on a clay loam soil, found at Joniskelis, compared to a loamy soil found at Dotnuva. At Joniskelis, *Sonchus arvensis* was more frequent in pea sole crop. The number of this weed significantly reduced in wheat, oat sole crops and intercropped with pea compared to pea sole crop. At Joniskelis, the number of *Cirsium arvense* decreased in all investigated crops, except for oat sole crop. Significantly higher numbers of this weed were found in spring wheat and triticale sole crop compared to pea sole crop. The trends of variation of this weed number were similar in Dotnuva. The crops were less suppressivefor perennial weeds than annual weed species observed in the experiment.

In spring, the number of *Sonchus arvensis* did not differ at either experimental site; however, variation of the weed numbers was noted. Consistent patterns of *Sonchus arvensis* variation were not determined in the higher density crops at Dotnuva. However, the number of weeds increased in all experimental crops, except for oat sole crop and intercropped with pea at Joniskelis. The influence of investigated crops on perennial weeds mass was not as marked as on annual weeds. At Joniskelis, the mass of *Sonchus arvensis* was significantly reduced in several crops including wheat and oat sole crops, and wheat and barley intercrops when compared to pea sole crop.

Species	Place	Sole crops and intercrops (BBCH 73)								
		Ps	P+SWi	P+SBi	P+Oi	P+STi	SWs	SBs	Os	STs
Viola arvensis	Dotnuva	0.16	0.16	0.05	0.06	0.14	0.03*	0.05	0.00*	0.08
	Joniskelis	0.75	0.20**	0.27**	0.10**	0.52	0.18**	0.32*	0.02**	0.44
Veronica arvensis	Dotnuva	0.02	0.02	0.00	0.01	0.00	0.02	0.01	0.00	0.00
	Joniskelis	2.02	0.88**	0.63**	0.12**	0.72**	0.44**	0.66**	0.10**	0.88**
Thlaspi arvense	Dotnuva	0.72	0.01*	0.00*	0.00*	0.00*	0.00*	0.00*	0.00*	0.00*
	Joniskelis	1.11	0.31*	0.0**	0.00**	0.42	0.02**	0.01**	0.00**	0.01**
Polygonum persicaria	Dotnuva	1.00	0.39**	0.02**	0.33**	0.06**	0.11**	0.06**	0.03**	0.04**
	Joniskelis	0.00	0.00	0.00	0.00	0.00	0.00	0.00	0.00	0.00
Polygonum aviculare	Dotnuva	2.16	0.44**	0.50**	0.14**	0.14**	0.03**	0.10**	0.11**	0.14**
	Joniskelis	0.00	0.00	0.00	0.00	0.00	0.00	0.00	0.00	0.00
Galium aparine	Dotnuva	0.00	0.00	0.00	0.00	0.00	0.00	0.00	0.00	0.00
	Joniskelis	3.36	3.87	1.47	0.22**	2.59	2.33	1.04*	0.13**	1.39
Fallopia convolvulus	Dotnuva	1.53	0.51	0.21*	0.09**	0.46*	0.86	0.02**	0.20*	0.37*
	Joniskelis	9.07	2.08**	3.19**	2.46**	6.55	1.27**	2.87**	0.65**	2.36**
Stellaria media	Dotnuva	0.71	0.47	1.04	0.21	0.52	0.54	0.20	0.06*	0.12
	Joniskelis	8.64	5.38*	3.56**	0.96**	2.85**	1.62**	3.79**	0.45**	2.02**
Chenopodium album	Dotnuva	37.15	10.54**	5.59**	6.25**	7.68**	5.03**	3.15**	2.44**	4.15**
	Joniskelis	7.94	2.41*	1.22**	1.08**	7.27	0.88**	1.63**	0.17**	3.63*
Sonchus arvensis	Dotnuva	4.51	0.26	0.18	4.00	0.49	0.41	0.50	1.82	0.18
	Joniskelis	2.97	0.19*	0.24*	1.58	3.33	0.40*	0.67	0.05**	0.82
Cirsium arvense	Dotnuva	8.01	1.54	2.48	1.59	6.32	0.06*	0.92	0.25*	4.78
	Joniskelis	0.66	19.37*	4.55	8.03	32.53*	30.30*	18.77	1.77	19.04*
Total mass of weeds	Dotnuva	56.89	14.61**	10.27**	12.72**	15.96**	7.34**	5.36**	5.27**	10.02**
	Joniskelis	41.10	35.64	17.18*	15.28**	60.44	38.84*	30.60	4.09**	33.55

Note. *differences are statistically significant as compared to the control at $P<0.05$, **-at $P<0.01$
Sole crop: Ps –pea, SWs – spring wheat, SBs – spring barley, Os –oat, STs – spring triticale; intercrop: P+SWi – pea and spring wheat, P+SBi – pea and spring barley; P+Oi – pea and oat, P+STi – pea and triticale

Table 9. The weed dry matter mass (DM g m^{-2}) of the most important species in sole crops and intercrops, data averaged over 2007-2010

At Joniskelis, the number of *Cirsium arvense* increased during the growing season in all investigated crops compared with its number in spring. At Dotnuva, an increase in the number of this weed was not consistent. The number of these weeds was slightly reduced by pea sole crop and intercropped with wheat and barley. More weeds germinated in cereal sole crop compared to intercrops at both expermental sites. The variation of *Cirsium arvense* mass differed between the experimental sites. At Dotnuva, *Cirsium arvense* mass was reduced in the wheat and oat sole crops of higher density but in other investigated crops, we established only a trend towards weed reduction. At Joniskelis, the mass of this perennial weed increased in all investigated crops, particularly in the wheat and triticale sole crops and their intercrops compared to pea sole crop.

Statistical data analysis showed that productive stem density had the greatest effect on weed suppression, while the effect of crop height and mass had a lesser affecting both soil conditions. In loamy soil (Dotnuva), the total number and mass of weeds were significantly related to the productive density (r= − 0.922, $P<0.01$, r= − 0.909, $P<0.01$) within the range 109–478 stems m^{-2}. In clay loam soil (Joniskelis), where productive crop density was lower (81–355 stems m^{-2}), the total number of weeds were significantly reduced by the height of crop (r= − 0.830, $P<0.01$).

Annual weed species had variable responses to the crop density, height and mass. At Joniskelis, the number of *Stellaria media* was significantly reduced with increasing productive density, height and mass of crops (r= − 0.685, $P<0.05$; r= − 0.952, $P<0.01$; r= − 0.816, $P < 0.01$, respectively) and the mass of this weed (respectively r= − 0.820, $P<0.01$; r= − 0.834, $P<0.01$; r= − 0.720, $P<0.05$). At Dotnuva, in the treatments with a lower *Stellaria media* incidence the weed mass was most markedly reduced by crop height (r= − 0.701, $P<0.05$). The investigated crops gave a good suppression of the following annual weeds as well: *Veronica arvensis, Thlaspi arvense, Polygonum aviculare, Fallopia convolvulus*. At Joniskelis, the number and mass of *Fallopia convolvulus* was significantly reduced as productive density of crops increased (r= − 0.759, $P<0,05$; r= − 0.930, $P<0.01$, respectively). The number of these climbing weeds was also significantly reduced by the height and mass of crops (r= − 0.818, $P<0.01$; r= − 0.799, $P<0.01$, respectively).

All crops competed well with *Veronica arvensis*. The number of this weed was significantly reduced by the height and mass of the crop (r= − 0.862, $P<0.01$; r= − 0.681, $P<0.05$, respectively), but weed mass was reduced by the productive density and height of crops (r= − 0.789, $P<0.05$; r= − 0.695, $P < 0.05$, respectively). At Dotnuva, a consistent pattern was not determined due to lower incidence of *Fallopia convolvulus* and *Veronica arvensis*. For *Thlaspi arvense*, the findings at the Dotnuva site were similar to those at Joniskelis. The number and mass of *Thlaspi arvense* were significantly reduced by the productive density of crops (r= − 0.823, $P<0.01$; r= − 0.821, $P<0.01$, Dotnuva and r= − 0.821, $P<0.01$; r= −0.889, $P<0.01$, respectively at Joniskelis). The data of suppression are less consistent for *Galium aparine* and *Chenopodium album* which are harmful weed species in this region. At Dotnuva, in denser crops, *Chenopodium album* numbers and mass were significantly reduced by the productive density of crops (r= − 0.867, $P<0.01$; r= − 0.873, $P<0.01$, respectively). At Joniskelis, in thinner crops, productive density of crops significantly reduced only weed mass (r= − 0.783, $P<0.05$).

Galium aparine is a climbing weed; therefore, the spread of this weed was negatively influenced by increasing productive stem numbers and crop height (r= − 0.671, $P<0.05$; r= − 0.670, $P<0.05$, respectively). The data of perennial weeds showed that *Sonchus arvensis* was suppressed more than *Cirsium arvense*. At Joniskelis, the spread of *Sonchus arvensis* depended on the density (r= − 0.719, $P < 0.05$), height (r= − 0.814, $P<0.01$) and mass (r= − 0.754, $P<0.01$) of crops. Also, the mass of *Sonchus arvensis* decreased due to increasing productive density of crops (r= − 0.731, $P<0.05$). At Dotnuva, *Sonchus arvensis* spread less;

therefore, the relationship was determined only between the number of this weed and productive density of crops (r= – 0.670, P<0.05). At Dotnuva, a strong relationship (r= – 0.856, P<0.01) was established between *Cirsium arvense* numbers and productive density of crops.

The relationship established between the total number and mass of weeds and intercrop competitive ability indicators (aggressivity - Ac; competition rate - CRc) showed that, with increasing competition rate of intercrops, weed incidence declined. This relationship was determined only at Joniskelis where the productive density was lower, the variation rate of CRc was higher, and weed species diversity and numbers were increased. With changing competition rates (0.54–15.63), weed number and mass declined by a linear inverse relationship. The correlation was medium (r= – 0.551, P<0.05; r= – 0.5031, P<0.05, respectively).

Researchers from five countries: Denmark, United Kingdom, France, Germany and Italy investigated the influence of pea and barley intercrop sown at different ratios – 45 peas and 150 barley plants m^{-2}, and 90 peas and 150 barley plants m^{-2} - on dry matter of weeds. The control of weeds was similar in sole barley and in intercrops, and no difference was established between the substitutive and the additive intercrops (Dibet et al., 2006). Researchers report the advantages of various intercropping managements such as pea with wheat against weeds (Szumigalski & van Acker, 2005), pea with barley (Hauggaard-Nielsen et al., 2006), and pea with oats (Rauber et al., 2001). Diversity of weeds was decreased in intercrops in comparison with sole crop (Gharineh & Moradi Telavat, 2009). Like cultivated plants, weeds obtain nutrients through root uptake from the soil solution. As a result, weeds and crops compete for space, nutrients, water and light. Both weeds and crop plants are similar in chemical composition; therefore, weeds can accumulate similar or even higher amounts of nutrients than crops (Lazauskas, 1990). Nitrogen increases weed and crop biomass (Kristensen et al., 2008). Peas use little nitrogen from the soil since they can fix atmospheric nitrogen for use. As a result, peas provide good conditions for weed growth, especially for high nitrogen demanding weed species. Dibet *et al.* (2006) reported a lower nitrogen concentration, 0.8 g m^{-2}, in weed mass due to competition between weed and cereal in intercrops compared to pea sole crop.

Statistical analyses of sole crop and intercrop grain yields and weed numbers and variation are presented in Figure 1.

Strong, inversely proportional relationships were established between grain yield and total weed number, and between grain yield and weed number variation during the growing season. This means that the number of weeds and their variation conversely affected crop yield. These relationships were determined only in lower density crops in clay loam soil (Joniskelis) when weed incidence markedly increased. The analysis of the statistical data suggested that an increase in the total weed number by one weed (regardless of the species) resulted in a grain yield reduction by 27.3 kg ha^{-1} (Figure 1a). The investigated crop competition characteristics describe the relationship between crop yields and weed number variation during the growing period. The grain yield changed 18.8 kg ha^{-1} by an inverse trend when changing one weed (Figure 1b). The relationship between grain yield and total weed mass was not significant (r= – 0.564, P>0.05). This relationship could determine perennial weed mass, which was especially high and the weeds were spread unevenly. It can be maintained that, in clay loam soil (Joniskelis), the majority of the investigated crops were not strongly competitive and, during maturity stage, there remained 71.3-49.2 weeds m⁻² in crops which had a negative impact on grain yield. Oat sole crop and intercropped with pea markedly differed from other investigated crops in that 19.6 and 25.8 weeds m⁻² remained during the maturity stage. At Dotnuva, this relationship was not determined. The number of weeds decreased by 54.7-27.0 m⁻² and such weed incidence had no significant negative effect on the crop productivity. This shows that sustainable plant communities are formed under organic farming conditions.

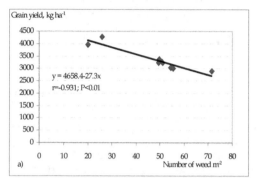 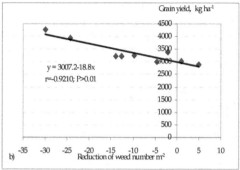

Fig. 1. The correlation between grain yield of sole crop, intercrop and total number of weeds (a), and their variation during grain-filling stage (BBCH 73) (b),Joniskelis 2007-2010

2.2.4 The effect of intercrops on subsequent crops in a crop rotation

During 2007-2010, studies were conducted at Joniskelis to assess the incidence of weeds in the intercropping system. The studies were set up during the transitional period from conventional to organic farming. The dynamics of weed germination (at crop growth stage BBCH 32-36) in the crop rotation is summarized in Table 10. In the spring of 2007, the highest number of weeds was in the pea sole crop. In various spring cereals and their intercrops with pea, weed density decreased by 1.7-35.7% compared to pea sole crop.

The averaged data suggested that the number of weeds in intercrops was slightly lower compared to cereal sole crops. After various sole cereal crops and intercrops with pea, as pre-crop to spring wheat (2008), the number of weeds in spring wheat was similar to that in the pre-crop (2007); however, in spring wheat grown after pea, weed numbers were the lowest. In spring wheat grown after various cereal sole crops and their intercrops with pea, the number of weeds increased by 3.6-69.4% compared with spring wheat grown after pea sole crop. In the third year (2009), under an organic cropping system, weed density significantly increased in sole cereal crops and their intercrops with pea. Weed germination was 1.8 times higher than that in 2007 or 2008. The number of weeds in pea sole crops increased by 50.7 m⁻² on average, with an increase of 26.5 m⁻² in intercrops and 12.5 m⁻² in cereal sole crop compared to the 2008 spring period. The number of weeds was significantly (P<0.01) influenced by crops. Like in 2007, the highest number of weeds was in pea sole crop. The number of weeds decreased by 17.1-53.3% in all other crops tested. Weed germination was significantly lower in oat intercrop and cereal sole crops. The variation of total weed numbers in 2010 was low in winter wheat grown after pea, cereal sole crops and intercrops compared to that in 2009. However, weed germination depended on different pre-crops. Weed density decreased by 23.3 m^{-2} in wheat grown after pea sole crop, but weed numbers increased on average by 3.3 and 8.6 m^{-2}, respectively, after intercrops and cereal sole crops compared to the 2009 spring period. Different pre-crops did not have any significant influence on weed density in winter wheat.

Rotation					Total number over crop rotation	Averaged number over the year
Intercrop and sole crop 2007		Spring wheat (SWs) 2008	Intercrop and sole crop 2009	Winter wheat (WWs) 2010		
crop	weed m^{-2}		crop	weed m^{-2}		
Ps	35.3	19.3	Ps	70.0	46.7	171.3	42.8
P+SWi	26.0	20.0	P+Oi	51.3*	62.0	159.3	39.8
P+SBi	22.7	32.7	P+STi	58.0	58.0	171.3	42.8
P+Oi	34.7	24.7	P+SWi	54.0	66.0	179.3	44.8
P+STi	26.0	34.7	P+SBi	54.7	45.3	160.7	40.2
SWs	27.3	27.3	Os	35.3**	47.3	137.3**	34.3
SBs	34.7	28.7	STs	52.7*	52.7	168.7	42.2
Os	24.0	28.0	SWs	32.7**	47.3	132.0**	33.0*
STs	27.3	32.7	SBs	46.0**	54.0	160.0	40.0
Mean	28.7	27.6	Mean	50.5	53.3	160.0	40.0
intercrop	27.4	28.0	intercrop	54.5	57.8	167.7	41.9
sole crop	28.3	29.2	sole crop	41.7	50.3	149.5	37.4

Note. *differences are statistically significant as compared to the control at $P<0.05$, **-at $P<0.01$
Sole crop: Ps –pea, SWs – spring wheat, SBs – spring barley, Os –oat, STs – spring triticale; intercrop: P+SWi – pea and spring wheat, P+SBi – pea and spring barley; P+Oi – pea and oat

Table 10. The dynamics of weed germination in the crop rotation with cereal sole crop and intercrop, 2007-2010, Joniskelis

Over a four-year period, the total number of weeds ranged from 132.0 to 179.3 m^{-2}. In the crop rotation with intercrops, the weed germination was 12.2% or 18.2 weeds m^{-2}, higher compared to the crop rotation with cereal sole crops. Significantly lower weed germination was seen in the four-course crop rotation, including oat sole crop: Os – SWs –SWs – WWs; SWs – SWs – Os – WWs, compared to the rotation including pea sole crop (Ps – SWs – Ps – WWs).

The number of weeds and their variation during cereal maturity stage (BBCH 73) are presented in Table 11. During the first experimental year (2007), the intercrops and sole crops were competitive and gave a good suppression of weeds. The number of weeds decreased by 12-28 m^{-2} compared to that in spring. According to the averaged data, the intercrops and sole crops did not differ markedly in their ability to suppress weeds. The number of weeds decreased by 20.9-79.1% in cereal sole crops and intercrops compared to pea sole crop. Significantly lower numbers of weeds were found in wheat and oat intercrops and oat sole crops during the cereal maturity stage. At the cereal maturity stage, the number of weeds was on average 4.3 times higher in spring wheat (2008) grown after various cereal sole crops and intercrops compared to the corresponding period in 2007. Although in 2008 weed emergence was similar to that in 2007, the number of weeds in spring wheat increased by an average 1.7 times compared to that during the spring period.

Averaged data showed that in the wheat crop grown after intercrops, the number of weeds increased by 20 m-2, by 17.9 m-2 after sole crops and by 24.7 m-2after pea, compared with the respective data in spring. Comparison of the effects of various pre-crops on weed incidence in spring wheat showed that the number of weeds declined in pea/oat, sole barely and oat crops (by 6.1, 19.8 and 15.2%, respectively), compared with pea sole pre-crop. However, these differences were not significant.

In 2009, the number of weeds further increased. In sole cereal crops and intercrops, during the cereal maturity stage, the number of weeds was 97.1 m-2, which was 2 times higherthan during the same period in 2008, and 1.9 times more than in spring (2009). Compared with the spring period, weed numbers in pea crop increased by 76 m-2, in intercrops by 49.3-54.0 m-2 (except for pea/oat crop) and in sole cereal crops by 53.3-77.3 m-2 (except for oat crop). Weed numbers declined in pea/oat and sole oat crops by 2.0 and 6.6 m-2 or 3.9 and 18.7% respectively, compared with the respective data in spring. All intercrops and cereal sole crops significantly decreased weed numbers by 15.5-80.3%, except for the sole triticale crop, compared with pea crop. Averaged data suggest that sole cereal crops suppressed weeds slightly more than intercrops. The lowest weed incidence was recorded in pea/oat and sole oat crops.

Rotation					Total weed number during crop rotation	Averaged number over the year
Intercrop and sole crop 2007		Spring wheat (SWs) 2008	Intercrop and sole crop 2009		Winter wheat (WWs) 2010	
Crop	weed m-2		crop	weed m-2		
Ps	17.7	44.0	Ps	146.0	91.3	299.0	74.8
P+SWi	8.7*	51.3	P+Oi	49.3**	66.7**	176.0**	44.0**
P+SBi	10.7	53.7	P+STi	112.0*	110.0	286.3	71.6
P+Oi	6.7**	41.3	P+SWi	103.3*	112.7	264.0*	66.0
P+STi	13.3	45.7	P+SBi	108.7*	88.7	256.3**	64.1
SWs	11.0	58.7	Os	28.7**	70.0	168.3**	42.1**
SBs	14.0	35.3	STs	106.0*	102.0	257.3**	64.3
Os	3.7**	37.3	SWs	96.7**	80.7	218.3**	54.6*
STs	10.0	45.0	SBs	123.3	88.7	267.0*	66.8
Mean	10.6	45.8	Mean	91.7	90.1	243.6	60.9
Intercrop	9.9	48	intercrop	93.3	94.5	245.7	61.4
sole crop	9.7	44.1	sole crop	88.7	85.4	227.8	56.9

Note. *differences are statistically significant as compared to the control at $P<0.05$, **-at $P<0.01$
Sole crop: Ps –pea, SWs – spring wheat, SBs – spring barley, Os –oat, STs – spring triticale; intercrop: P+SWi – pea and spring wheat, P+SBi – pea and spring barley; P+Oi – pea and oat

Table 11. The variation of weed numbers in the crop rotation with intercrops and sole crops at cereal maturity stage, 2007-2010, Joniskelis

In 2010, at the winter wheat maturity stage, weed numbers differed little from that in 2009. However, compared with the spring period, weed numbers increased an average of 1.7 times. Averaged data indicate that the greatest increase in weed numbers occurred in winter wheat grown after pea; a smaller increase occurred after intercrops and sole crops. Compared with pea, significantly lower weed numbers were determined in winter wheat crop grown after pea/oat intercrop.

Various crop rotations had significant effects ($P<0.01$) on the total weed number during cereal maturation stage. The highest number of weeds over a four-year period was established in the crop rotation with pea (Ps – SWs – Ps – WWs). Inclusion of semi-leafless pea in the crop rotation tended to increase crop weed incidence. In all other crop rotations with sole cereal crops or pea/cereal intercrops, the total weed incidence significantly declined by 10.7-43.7% (except for the crop rotation: P+SBi – SWs – P+STi - WWs), compared with the crop rotation with pea (Ps – SWs – Ps – WWs). Averaged over one year, significantly lower weed incidence was in the following crop rotations: P+SWi – SWs – P+Oi – WWs; SWs – SWs – Os – WWs and Os – SWs – SWs – WWs, compared with the crop rotation including pea.

Weed mass variation in different crops at cereal maturation stage (BBCH 73) is presented in Table 12. For the first year (2007) in intercrops and sole cereal crops, the weed incidence was low; consequently, their mass was low. Compared with pea crop, in all intercrops and sole cereal crops weed mass was significantly lower (61.1-97.3%). The lowest weed mass was recorded in oat and its intercrops with pea. The mass per weed varied in a similar way (except for pea/triticale crop). Averaged data suggest that higher total weed mass and mass per weed was in intercrops, compared with sole crops.

In the next year (2008), in spring wheat crop grown after different pre-crops, weed mass increased by 1.9 times. Different pre-crops exerted varying effects. When spring wheat had been grown after pea, the total weed mass declined by 2.0 times; after intercrops, it increased by 3.0 times and after sole cereals it increasedby 3.7 times, compared with respective weed mass in 2007. Pea/barley and pea/triticale intercrops tended to increase weed mass in spring wheat, compared with pea pre-crop. Other crops, as pre-crops, reduced weed mass. Averaged data indicate that the highest mass per weed was in spring wheat grown after pea sole crop; weed mass was lower after intercrops and it was the lowest after sole cereal crops. Pea/oat intercrop and sole spring wheat crop as pre-crops significantly reduced mass per weed compared with pea as pre-crop. Reduction of mass per weed decreased viability and number of mature seeds (Lazauskas, 1990; Liebman & Davis, 2000).

In the third year of the crop rotation, when growing various species of cereals and their intercrops with pea, the total weed number increased by an average of 3 times, compared with the average total weed mass in 2008, or by 5.6 times, compared with 2007. Many of the tested crops significantly reduced weed mass by 70.4-96.3% (except for pea/triticale and triticale crops) compared with sole pea crop. Significantly lower mass per weed was determined in sole cereal crops (except for triticale) and pea/barley crops, compared with pea crop.

In the final year of the experiment (2010), in the winter wheat crop, total weed mass increased by an average of 29.6% compared with that in 2009. After different pre-crops, total weed mass was variable. In winter wheat grown after pea, total weed mass declined by

34.3%, after intercrops and sole crops it increased by 23.7 and 94.3%, respectively, compared with the respective data in 2009. Significantly lower total weed mass and mass per weed in winter wheat was recorded when it was grown after pea/oat and sole oat crops.

Rotation										Total weed mass over crop rotation (g)
Intercrop and sole crop 2007			Spring wheat (SWs) 2008		Intercrop and sole crop 2009			Winter wheat (WWs), 2010		
	weed mass					weed mass				
crop	total (g m⁻²)	single weed (g)	total (g m⁻²)	single weed (g)	crop	Total (g m⁻²)	single weed (g)	total (g m⁻²)	single weed (g)	
Ps	23.65	1.338	12.11	0.275	Ps	67.27	0.461	44.17	0.484	147.2
P+SWi	1.77**	0.204**	10.02	0.195	P+Oi	12.44**	0.252	16.51**	0.248*	40.74**
P+SBi	3.21**	0.301**	13.58	0.253	P+STi	71.80	0.641	41.10	0.374	129.69
P+Oi	1.23**	0.184**	7.12*	0.172*	P+SWi	19.93**	0.193	60.33	0.535	88.61*
P+STi	9.19**	0.689	14.73	0.322	P+SBi	18.61**	0.171*	33.97	0.383	76.5**
SWs	2.71**	0.247**	6.66*	0.113*	Os	2.50**	0.087**	13.72**	0.196**	25.59**
SBs	2.57**	0.184**	7.63	0.216	STs	51.03	0.481	52.25	0.512	113.48
Os	0.63**	0.171**	8.80	0.236	SWs	9.05**	0.094*	37.06	0.459	55.54**
STs	2.91**	0.291**	9.22	0.205	SBs	15.01**	0.122*	47.78	0.539	74.92**
Mean	5.32	0.401	9.99	0.221	Mean	29.74	0.278	38.54	0.414	83.59
intercrop	3.85	0.345	11.36	0.236	intercrop	30.70	0.314	37.98	0.385	83.89
sole crop	2.21	0.223	8.08	0.193	sole crop	19.40	0.196	37.70	0.427	67.38

Note. *differences are statistically significant as compared to the control at $P<0.05$, **-at $P<0.01$
Sole crop: Ps –pea, SWs – spring wheat, SBs – spring barley, Os –oat, STs – spring triticale; intercrop: P+SWi – pea and spring wheat, P+SBi – pea and spring barley; P+Oi – pea and oat

Table 12. The variation of weed mass in the crop rotation with intercrops and sole crops during cereal maturity stage, 2007-2010, Joniskelis

Summarised data show that in cultivated heavy loam *Cambisol*, during the transition period from an input-intensive to an organic cropping system, weeds emerged more intensively in the third and fourth years of the crop rotation. Averaged data evidence that, in the crop rotations with sole cereal crops and intercrops, weed numbers annually increased. In the crop rotation with pea, the pea promoted weed emergence; pea as pre-crop effect on wheat reduced weed emergence. During the four year, significantly fewer weeds emerged in the crop rotation with sole oat crop.

During the cereal maturity stage, weed numbers and mass were more markedly influenced by sole cereal crops and their intercrops with pea compared with their effect as pre-crops. In the first year, compared with the spring period, weed numbers during the growing season markedly declined; over the following years, weeds were suppressed less and their numbers increased. Cereal sole crops and intercrops had a greater suppression of weeds during the growing season; therefore, their numbers per rotation (except for the crop rotation P+SBi – SWs – P+STi + WWs) and mass (also except for the crop rotation P+SBi – SWs – P+STi – WWs and SBi – SWs –STi – WWs) were significantly lower compared with the crop rotation with pea. Over the four years, during the cereal maturity stage in the crop rotations with intercrops, the total number of weeds was an average of 17.9 m⁻² higher and 55.3 m⁻² lower, and the total weed mass by 16.51 g m^{-2} higher and 63.31 g m^{-2} lower, compared with the crop rotations with sole cereals or pea. Averaged data indicate that for any one year, significantly fewer weeds were in the crop rotation including oat or its intercrop with pea. Literature provides data on allelopathy effects against weeds due to direct or indirect release of chemicals from live or dead plants (including microorganisms) (Bhadoria, 2011). The effect of sole oat crop against weeds was longer-lasting than that of pea/oat intercrop. This finding is consistent with other researchers' evidence suggesting that the sequence of oat/pea intercrop harvested for forage followed by winter wheat will suppress warm-season weeds during the 2-year interval (Anderson, 2010).

3. Conclusions

The weed suppression effect of intercrops verses sole crops markedly differed during the plant growing period. Competitive abilities of crops were determined by plant productive density, height, mass, index of aggressiveness of cereals (Ac), and competition rate (CRc). More stable productive densities of intercrops were obtained in a loam soil (286-346 stems m^{-2}) compared to a clay loam soil (211-275 stems m^{-2}). Productive stem density of pea in crop structure was similar. According to plant height, spring cereals ranked as follows: oat > triticale > wheat > barley. Pea plants were the shortest and their height and mass tended to decline in intercrops. In intercrops, cereals had greater influence on weed suppression than pea.

During the crop growing season, sole cereals and pea/cereal intercrops provided better weed suppression than pea (semileafless pea cultivars). At Dotnuva, in denser crop densities, the total weed numbers during the maturity stage declined by 12.4-28.8 weeds m^{-2} compared with that in spring;in pea crops, the reduction amounted to 3.1 weeds m^{-2}. At Joniskelis, in the crops with a lower population density, the effect on weeds was lower. At lower crop population densities, weed suppression differences between sole crops and intercrops were inappreciable. At both experimental sites, the best weed suppression was exhibited by sole oat crop and its intercrop with pea; total weed numbers during the maturation stage declined by 24.4–28.8 weeds m^{-2} and 21.0–30.1 weeds m^{-2}, respectively, compared with the spring period. At Dotnuva, all crops significantly reduced weed mass by 72.0-90.7%, compared with pea crop. At Joniskelis, due to higher and uneven incidence of *Cirsium arvense*, the variation of weed mass was less consistent. According to increasing total weed mass, the crops ranked as follows: cereals < intercrops < pea. The lowest weed mass was identified in sole oat crop.

The variation of total weed numbers and mass was influenced by weed species. With low incidence of *Viola arvensis*, *Veronica arvensis*, *Thlaspi arvense* and more abundant counts of *Galium aparine*, all crops tended to reduce weed numbers compared with the spring period. With higher incidence of *Viola arvensis*, *Veronica arvensis* and *Galium aparine* (Joniskelis), their number (for *Galium aparine* also mass) decreased most in pea / oat and oat crops, compared with pea crop. With higher incidence of *Fallopia convolvulus*, its numbers were reduced only by pea/oat crop, while other crops increased its number compared with the spring period. However, *Fallopia convolvulus*, *Viola arvensis*, and *Veronica arvensis* mass significantly declined compared with that in pea crop. Sole cereal crops gave a better suppression of *Stellaria media* compared with intercrops. When the incidence of this weed was high, all crops significantly reduced its mass, compared with pea crop. In spring, when the incidence of *Chenopodium album* was very high (32.8-41.9 m^{-2}) at the Dotnuva site, the number and mass of this weed significantly declined in intercrops and cereal crops compared with pea crop. When the incidence of this weed was lower (4.2-8.5 m^{-2}), only its mass declined more markedly compared with pea crop. At both experimental sites, sole cereal crops, particularly especially oat, reduced weed mass more appreciably than intercrops. Crops had the lowest suppressive effect on perennial weeds, *Sonchus arvensis and Cirsium arvense*. In many crops, the number and mass of these weeds increased. Slightly less sensitive to crop suppression, especially to oat and pea / oat intercrop, was *Sonchus arvensis*. An increase in crop productive density had a significant negative effect on the number and/or mass of many weed species. For many climbing weed species, *Galium aparine* and *Fallopia convolvulus*, an increase in crop height significantly reduced their density. Short-growing weeds *Veronica arvensis Stellaria media* responded negatively to many competitive properties of crops.

The greatest negative effect on crop grain yield (2896.6–4270.4 kg ha^{-1}) in a clay loam soil (Joniskelis) was exerted by weed numbers during crop maturation stage and its variation during the crop growing season. With a simultaneous increase in the number of these weeds (19.6-71.3 m^{-2} range), the yield of the crops tested statistically declined by 27.3 kg ha^{-1}. It was calculated that during the crop growing season, with one suppressed weed, grain yield increased by 18.8 kg ha^{-1}. In loamy soil (Dotnuva), the remaining number of weeds (27.0-54.7 weed m^{-2}) during cereal maturity stage did not have any significant effect on crop yield (2555.4-3002.9 kg ha^{-1}).

On a cultivated, heavy loam *Cambisol*, during the transition period from an intensive to an organic cropping system, the highest number of weeds emerged and persisted through the growing season in the third and fourth years of crop rotation. During the cereal maturity stage, sole cereal crops and their intercrops with peas had the greatest impact on weed numbers and mass, compared with their effect as pre-crops. Sole cereal crops and intercrops suppressed weeds during the growing season; therefore, in many crop rotations weed numbers and mass were significantly lower compared with a crop rotation with pea. Over a four-year period, during the maturity stage of cereals, in the crop rotations with intercrops, the total number of weeds was an average of 17.9 m^{-2} higher and 55.3 m^{-2} lower, and the total mass by 16.51 g m^{-2} higher and 63.31 g m^{-2} lower, compared with the respective crop rotations with sole cereal crops or pea. Averaged data showed, that during one year, significantly lower numbers of weeds were in the crop rotation with oat or its intercrop with pea.

4. Acknowledgment

The study has been supported by the Lithuanian Ministry of Agriculture and the Lithuanian Academy of Science.

5. References

Andersen, M. K.; Hauggaard-Nielsen, H.; Ambus, P. & Jensen, E. S. (2005) Biomass production, symbiotic nitrogen fixation and inorganic N use in dual and tri-component annual intercrops, *Plant and Soil*, Vol. 266, pp. 273–287, ISSN 0032-079X

Andersen, M. K.; Hauggaard-Nielsen, H.; Weiner, J. & Jensen, E. S. (2007) Competitive dynamics in two-component intercrops. *Journal of Applied Ecology*, Vol.44, No. 3, pp. 545–551, ISSN 0021-8901

Anderson, R.L.A. (2010) Rotation design to reduce weed density in organic farming. *Renewable Agriculture and Food Systems*, Vol. 25, No. 3, pp. 189–195, ISSN 1742-1705

Anil, L.; Park, R.; Phipps, R.H. & Miller, F.A. (1998) Temperate intercropping of cereals for forage: a review of the potential for growth and utilization with particular reference to the UK. *Grass and Forage Science*, Vol. 53, pp. 301-317, ISSN 1365-2494

Auskalniene, O. & Auskalnis, A. (2008) The influence of spring wheat plant density on weed suppression and grain yield. *Zemdirbyste=Agriculture*, Vol. 95, No. 3, pp. 5–10, ISSN 1392-3196

Arlauskienė, A. & Maikštėnienė S. (2008) Tarpinių pasėlių ir šiaudų naudojimo trąšai įtaka trumpaamžių piktžolių plitimui vasariniuose miežiuose. *Vagos*, Nr. 79 (32), pp. 13-23, ISSN 1648-116X

Banik, P; Midya, A.; Sarkar, B.K. & Ghose, S.S. (2006) Wheat and chickpea intercropping systems in an additive series experiment: Advantages and weed smothering. *Europea Journal of Agronomy*, Vol. 24, Iss. 4, pp. 325-332, ISSN 1161-0301

Barbery, P. (2002) Weed management in organic agriculture: Are we addressing the right issues. *Weed Research*, Vol 42, pp. 177-193, ISSN 0043-1737

Berk, A.; Bramm, A.; Böhm, H.; Aulrich, K. & Rühl, G. (2008) The nutritive value of lupins in sole cropping systems and mixed intercropping with spring cereals for grain production In: *Proceedings of the 12th International Lupin Conference: Lupins for Health and Wealth*, J.A. Palta and J.B. Berger (eds). 2008. '14-18 Sept. 2008, Fremantle, Western Australia. Canterbury, New Zealand. ISBN 0-86476-153-8

Berry, P.M.; Sylvester-Bradley, R.; Philipps, L.; Hatch, S.P.; Cuttle, F.W.; Gosling, P. (2002) Is the productivity of organic farms restricted by the supply of available nitrogen? *Soil Use and Management*, Vol. 18, pp. 248-255, ISSN 0266-0032

Bhadoria, P. B. S. (2011) Allelopathy: A Natural Way towards Weed Management. *American Journal of Experimental Agriculture*, Vol. 1, No. 1, pp. 7-20, ISSN 22310606

Bond, W. & Grundy, A.C. (2001) Non-chemical weed management in organic farming systems. *Weed Research*, Vol. 41, No. 5, pp. 383-405, ISSN 0043-1737

Brisson, N. ; Bussiere, F. ; Ozier-Lafontaine, H. ; Tournebize, R. & Sinoquet, H. (2004) Adaptation of the crop model STICS to intercropping. Theoretical basis and parameterisation. *Agronomie*, Vol. 24, pp. 1–9, ISSN 0249-5627

Bulson, H. A.; Snaydon, R. W. & Stopes, C. E. (1997) Effects of plant density on intercropped wheat and field beans in an organic farming system. *The Journal of Agricultural Science*, Vol. 128, pp. 59-71, ISSN 0021-8596

Corre-Hellou, G.; Brisson, N.; Launay, M; Fustec, J. & Crozat, I. (2007) Effect of root depth penetration on soil nitrogen competitive interactions and dry matter production in pea–barley intercrops given different soil nitrogen supplies. *Field Crops Research*, Vol. 103 (1), pp. 76-85, ISSN 0378-4290

Deveikyte I., Kadziuliene Z., Sarunaite L. (2009) Weed suppression ability of spring cereal crops and peas in pure and mixed stands. *Agronomy Research*, Vol. 7, No. 1, pp. 239–244, ISSN 1406-894X

Deveikyte, I.; Kadziuliene, Z.; Sarunaite, L. & Feiziene, D. (2008) Investigations of weed-suppressing ability of leguminous plant under organic agriculture conditions. *Vagos*, Vol.79, No. 32, pp. 43–48, ISSN 1648-116X

Dibet, A.; Hauggaard-Nielsen, H.; Kasyanova, E.; Ruske, R.; Gooding, M.; Pristeri, A.; Monti, M.; Dahlmann, Ch.; Fragstein, P.; Ambus, P.; Jensen, E. & Crozat, Y. (2006) Pea-barley intercropping for the control of weeds in European organic cropping systems. Poster at: Joint Organic Congress, Odense, Denmark, May 30-31, 2006.

Ghaley, B.B.; Hauggaard-Nielsen, H.; Høgh-Jensen, H. & Jensen, E.S. (2005) Intercropping of wheat and pea as influenced by nitrogen fertilization. *Nutrient Cycling in Agroecosystems*, Vol. 73, No. 3, pp. 201-212, ISSN 1385-1314

Gharineh, M.H. & Moradi-Telavat, M.R. (2009) Investigation of ecological relationship and density acceptance of canola in canola-field bean intercropping. *Asian Journal of Agricultural Research, Nr.* 3(1), pp.11–17, ISSN 18191894

Hauggaard-Nielsen, H.; Ambus, P. & Jensen, E. S. (2001) Interspecific competition, N use and interference with weeds in pea-barley intercropping. *Field Crops Research*, Vol. 70, pp. 101 – 109, ISSN 0378-4290

Hauggaard-Nielsen, H.; Ambus, P. & Jensen, E. S. (2003) The comparison of nitrogen use and leaching in sole cropped versus intercropped pea and barley. *Nutrient Cycling in Agroecosystems*, Vol. 65, pp. 289-300, ISSN 1385-1314

Hauggaard-Nielsen, H.; Jørnsgaard, B.; Kinane, J. & Jensen, E. S. (2008). Grain legume – cereal intercropping: The practical application of diversity, competition and facilitation in arable and organic cropping systems. *Renewable Agriculture and Food Systems*, Vol. 23, No. 1, pp. 3–12, ISSN 1742-1705

Hauggaard-Nielsen, H.; Knidsen, M. T.; Jørgensen, J. R. & Jensen, E. S. (2006) Intercropping wheat with pea for improved wheat baking quality Intercropping of cereals and grain legumes for increased production, weed control, improved product quality and prevention of N-losses in European organic farming systems, *Proceedings of the European Joint Organic Congress*, pp. 268–269, ISBN 87-991343-3-0, Odense, Denmark, May 30-31, 2006

Hauggaard-Nielsen, H.; Mundus, S.; Jensen, E.S. (2009) Nitrogen dynamics following grain legumes and subsequent catch crop and the effects on succeeding cereal crops. *Nutrient Cycling in Agroecosystems*, Vol. 84, No. 3, pp. 281-291, ISSN 1573-0867

Jensen, E.S.; Ambus, N.; Bellostas, N.; Boisen, S.; Brisson, N.; Corre-Holou, G.; Crosat, Y. ; Dahlman, C. ; Dibet, A.; Fragstein, F.; Gooding, M. ; Hauggaard-Nielsen, H.; Kasyanova, E.; Launay, M. & Pristeri, A. (2006) Intercropping of cereals and grain legumes for increased production, weed control, improved product quality and prevention of N-losses in European organic farming systems, *Proceedings of the European Joint Organic Congress*, pp. 180–181, ISBN 87-991343-3-0, Odense, Denmark, May 30-31, 2006

Koocheki, A.; Nassiri, M.; Alomoradi, L.; Ghorbani, R. (2009) Effect of cropping systems and crop rotations on weeds. *Agronomy for Sustainable Development*, Vol. 29, pp. 401-408, ISSN 1774-0746

Kristensen, L.; Olsen, J. & Weiner, J. (2008) Crop density, sowing pattern, and nitrogen fertilization effects on weed suppression and yield in spring wheat. *Weed Science*, Vol. 56, No. 1, pp. 97–102, ISSN 0043-1745

Lazauskas, P. (1990). *Agrotechnika prieš piktžoles*, Mokslas, ISBN 5-420-00206-X Vilnius, Lietuva

Lemerle, D.; Gill, G. S.; Murphy, C.E.; Walker, S R.; Cousens, R. D.; Mokhtari, S.; Peltzer, S. J.; Coleman, R. & Luckett, D.J.(2001) Genetic improvement and agronomy for enhanced wheat competitiveness with weeds. *Australian Journal Agricultural Research*, Vol. 52, No. 5, pp. 527–548, ISSN 0004-9409

Liebman, M.; Davis, A. S. (2009). Managing weeds in organic farming systems: An Ecological Approach. *Organic Farming: The Ecological System* In: Francis, C. (Ed.) 173–195, American Society of Agronomy, ISBN 978-0-89118-173-6 Madison, Wisconsin, USA

Liebman, M. & Davis, A.S. (2000) Integration of soil, crop and weed management in low-external-input farming systems. *Weed Research*, Vol. 40, No. 1, pp. 27-47, ISSN 0043-1737

Liebman, M. & Dyck, E, (1993) Crop rotation and intercropping strategies for weed management. *Ecological Applications*, Vol. 3, No. 1, pp. 92-122 ISSN 1051-0761

Lithourgidis, A.S.; Vasilakoglou, I.B.; Dhima, K.V.; Dordas, C.A. & Yiakoulaki, M.D. (2006) Forage yield and quality of common vetch mixtures with oat and triticale in two seeding ratios. *Field Crops Research*, Vol. 99, pp. 106-113, ISSN 0378-4290

Malézieux, E.; Crozat, Y; Dupraz, C.; Laurans, M.; Makowski, D.; Ozier-Lafontaine, H.; Rapidel, B.; de Tourdonnet, S. & Valantin-Morison, M. (2009). Mixing plant species in cropping systems: concepts, tools and models. A review. *Agronomy for Sustainable Development*, Vol. 29, Nr.1, pp. 43-62, ISSN 1774-0746

Maiksteniene, S., Velykis, A., Arlauskiene, A. & Satkus, A. (2006). Javų stelbiamosios galios prieš sunkiuose priemoliuose plintančias piktžoles tyrimai. *Vagos*, Vol. 72, No. 25, pp. 24–33, ISSN 1648-116X

Mariotti, M.; Masoni, A.; Ercoli, L. & Arduini, I. (2006) Forage potential of winter cereal/legume intercrops in organic farming. *Italian Journal of Agronomy*, Vol. 3, pp. 403–412, ISSN 1125-4718

Mohler, C. L. (2001) Enhancing the competitive ability of crops. In: *Ecological Management of Agricultural Weeds*, Matt Liebman, M.; Mohler, C. L.; Staver, C. P., pp. 1231–1269, Cambridge University Press, ISBN 978-052-1560-68-9, Cambridge, UK

Neumann, A.; Schmidke, K. & Rauber, R. (2007). Effects of crop density and tillage system on grain yield and N uptake from soil and atmosphere of sole intercropped pea and oat. *Field Crops Research*, Vol., 100, No. 2-3, pp. 285–293, ISSN 0378-4290

Ofori, F. & Stern, W.R. (1987) Cereal-legume intercropping system. *Advances in Agronomy*, Vol. 41, pp. 41-90, ISSN 0065-2113

Rasmussen, K. 2002. Influence of liquid manure application method on weed control in spring cereals. *Weed Research*, Vol., 42, No. 4, pp. 287-298, ISSN 0043-1737

Rauber, R.; Schmidtke, K. & Kimpel-Freund, H. (2001). The performance of pea (*Pisum sativum* L.) and its role in determining yield advantages in mixed stands of pea and oat (*Avena sativa* L.). *Journal of Agronomy and Crop Science*, Vol. 187, No. 2, pp. 137–144, ISSN 0931-2250

Protasov, N. (1995). The tendency of change of weed phytocenosis in the North-Eastern part of Byelorussia. In: *Proceedings of the International conference: Weed Control in the Changing Situation of Farming in the Baltic Region*, Lithuanian Academy of Agriculture, Trečiokas, K.; Liakas, V. (Eds.), pp. 203-210, Kaunas, Lithuania

Špokienė, N. (1995) Vyraujančių trumpaamžių piktžolių dygimas dirvoje. In: *Proceedings of the International conference: Weed Control in the Changing Situation of Farming in the Baltic Region*, Lithuanian Academy of Agriculture, Trečiokas, K.; Liakas, V. (Eds.), pp. 267-275, Kaunas, Lithuania

Špokienė, N. & Povilionienė E. (2003) *Piktžolės*. Lietuvos žemės ūkio universitetas, ISBN 9955-552-03-4, Kaunas, Lietuva

Singh, S.; Ladha J.K.; Gupta, R.K.; Bhushan L.; Rao, A.N.; Sivaprasad, B & Singh, P.P. (2007) Evaluation of mulching, intercropping with *Sesbania* and herbicide use for weed management in dry-seeded rice (*Oryza sativa* L.). *Crop Protection*, Vol. 26, pp. 518-524, ISSN 0261-2194

Subkowicz, P. & Tendziagolska, E. (2005) Competition and productivityin mixture of oats and wheat. *Journal of Agronomy and Crop Science*, Vol. 191, No. 5, pp. 377–385, ISSN 0931-2250

Szumigalski, A. & van Acker, R. (2005) Weed suppression and crop production in annual intercrops. *Weed Science*, Vol. 53, No. 6, pp. 813-825, ISSN 0043-1745

Thomson, D.J.; Stout, D.G. & Moore, T. (1992) Forage production by four annual cropping sequences emphasizing barley irrigation in southern interior British Columbia. *Canadian Journal of Plant Science*, Vol. 72, pp. 181-15-85, ISSN 0008-4220

Urbatzka, P.; Graβ, R.; Haase, T.; Schüler, C. & Heβ, J. (2009) Fate of legume-derived nitrogen in monocultures and mixtures with cereals. *Agriculture, Ecosystems and Environment*, Vol. 132, pp. 116–125, ISSN 0167-8809

Weigelt, A. & Jolliffe, P. (2003) Indices of competition, *Journal of Ecology*, Vol. 91, pp. 707–720, ISSN 1365-2745

Willey, R.W. & Rao, M.R. (1980) A competitive ratio for quantifying competition between intercrops. *Experimental Agriculture*, Vol. 16, pp. 117–125, ISSN 0014-4797

2

Mulches for Weed Management in Vegetable Production

Timothy Coolong
Department of Horticulture,
University of Kentucky,
USA

1. Introduction

The practice of applying mulches for the production of vegetables is thousands of years old (Lightfoot, 1994; Rowe-Dutton, 1957). Typically mulching involves placing a layer of material on the soil around the crop of interest to modify the growing environment to improve crop productivity. The primary purpose for using mulches is for weed suppression in the crop to be grown. Mulches typically function by blocking light or creating environmental conditions which can prevent germination or suppress weed growth shortly after germination. However, numerous other benefits are often obtained including: increased earliness, moisture conservation, temperature regulation of the root zone and above-ground growing environment, reduced nutrient leaching, altered insect and disease pressures, and, in some instances, reduced soil compaction or improved soil organic matter (Lamont, 2005; Lamont, 1993; Ngouajio and McGiffen, 2004; Rowe-Dutton, 1957). The use of mulches typically results in higher yields and quality in vegetable crops enhancing profitability for the grower.

1.1 History of mulching in vegetable systems

A wide variety of mulches have been utilized throughout history. Lithic-mulches, which include pebbles and gravel as well as volcanic ash, may be some of the earliest documented mulches used in vegetable production. Depending on the site and crop grown, lithic-mulches could take the form of mounds around individual plants, long rows or ridges of larger stones, or vast areas where an entire production site is covered in pebbles or volcanic ash (Lightfoot, 1994). Although primarily used in areas with scarce moisture, lithic-mulches also modulate fluctuations in soil temperatures as well as reduce weeds (Lightfoot, 1994). Some of the earliest documented sites where lithic-mulches were used date to 200 B.C. and are found in the Negev desert of Israel (Kedar, 1957). These mulches may have been used with grapevines or olive trees; though, it is unclear if they were used for vegetable production (Mayerson, 1959 as cited in Lightfoot, 1994). The Maori people of New Zealand used gravel mulches in fields between the years of 1200 and 1800 AD to grow sweetpotatoes [*Ipomoea batatas* (L.) Lam.] and maize (*Zea mays* L.) (Lightfoot, 1994; Rigg and Bruce, 1923). In a practice dating back several hundred years, growers in the Lanzhou area of China have used river pebbles at a depth of 7-10 cm for the production of melons (*Cucumis sp.)*

(Lightfoot, 1994; Rowe-Dutton, 1957). Pieces of slate were also used as mulches under melons in England nearly 200 years ago; however, this was likely a way to keep fruit dry as well as warm the plants rather than for weed control (Williams 1824 as cited in Rowe-Dutton, 1957).

Dust mulching is another practice that persists to this day; although, it is not a true mulching technique since no materials are applied to the soil. Dust mulching is the practice of repeatedly and shallowly cultivating the soil surrounding the crop to create a pulverized (dust) layer of soil (James, 1945). A theory, though proven to be incorrect, is that by creating a finely textured layer of soil at the surface, capillarity in the soil is "broken" and the movement of water out of the soil via evaporation is reduced (James, 1945; Ladewig, 1951). It is generally accepted that the primary benefit from dust mulching comes from the destruction of weeds around the crop and not a reduction in evaporation at the soil surface (Rowe-Dutton, 1957).

Organic-based mulches such as plant waste, straw, sawdust, and manure have also been used to a great extent for vegetable production. Traditionally, organic mulches have consisted of materials which are locally plentiful. Organic-based mulches can be as diverse as the region in which they are used. For instance, banana (*Musa* sp.) leaves and water hyacinth [*Eichhornia crassipes* (Mart.) Solms] have been used for mulching tomato (*Solanum lycopersicum* L.) in Bangladesh (Kayum et al., 2008), while cane (*Saccharum officinarum* L.) bagasse (sugarcane stalks) have been used in Hawaii (Gilbert, 1956) and sawdust in Pennsylvania (Isenberg and Odland, 1950). When applied as a thick layer, organic-based mulches can effectively suppress weeds and increase soil moisture levels (Diaz-Perez et al., 2004). However, research dating to the late 19th century has shown variable results of organic mulches on yield. In areas with warm temperatures and limitations on water for plant growth, straw mulches have been found to positively affect growth and yields of several cucurbits (Emerson, 1903). However, when used in cool climates, the addition of straw mulch, while beneficial for controlling weeds, has been shown to retard the growth of warm season vegetables and decrease yields (Rowe-Dutton, 1957). Nonetheless, organic mulches remain popular due to their low cost and ready availability.

1.2 Paper-based mulches

Paper-based mulches represent some of the earliest mulching systems developed for fruit and vegetable production. Paper was an ideal mulch because it could be transported long distances and easily applied from a roll in the field. Paper-based mulches were extensively used in Hawaii in sugarcane production in the early 20th century. In sugarcane production, lightweight-tar or asphalt-impregnated paper mulches were placed over rows of seed cane and held to the ground with soil. Newly emerged cane shoots were sharp enough to pierce the mulch and continue to grow, while weeds were unable to penetrate the paper and died shortly after germination (Stewart et al., 1926). Using a system remarkably similar to modern plastic mulch, Stewart et al. (1926) evaluated asphalt-impregnated paper for in-row weed control for pineapple [*Ananas comosus* (L.) Merr]. In that trial, the mulches were unrolled over beds using a tractor mounted mulch layer, much like those used today. In addition to reducing weed pressure on the crop, paper mulches generally increased the soil temperature by several degrees Fahrenheit on sunny days, with little effect on cloudy or rainy days (Stewart et al., 1926). Soil moisture and nitrate levels were also generally greater under mulches compared to bare ground leading to enhanced pineapple production (Stewart et al., 1926).

Paper-based mulches were utilized through much of the early 20th century with positive results. Thompson and Platenius (1931) reported positive results with paper mulches in several vegetable crops including pepper (*Capsicum* sp.), tomato, and muskmelon (*Cucumis melo* L.). Paper mulches controlled weeds, increased soil temperatures and moisture levels, resulting in greater yields (Thompson and Platenius, 1931). By altering the growing environment, black-paper mulches affected root distribution of several vegetable crops when compared to bare-ground production (Knavel and Mohr, 1967). The use of paper mulches, particularly those impregnated with asphalt or tar, suppressed weeds, conserved moisture, and warmed the soil, increasing yields in most warm-season crops. However, issues with cost and durability led to the development of alternative mulches.

1.3 Other mulches

In addition to paper and organic mulches several other substances were evaluated through the early and mid 20th century. Aluminum foil was shown to be an effective mulch, increasing yields (Burgis, 1950). The reflective nature of aluminum foil mulch actually cooled soil, while affecting insect predation and the spread of some insect-transmitted diseases (Adlerz and Everett, 1968; Burgis, 1950; Wolfenbarger and Moore, 1968). However, while effective and more durable than paper, aluminum foil mulches were never implemented on a large scale due to high costs.

Petroleum-based spray mulches were also evaluated as an in-row band for cucurbits grown in Florida (Nettles, 1963). These spray mulches functioned as an effective pre-emergent herbicide and warmed the soil. Early and total season yields were significantly greater for petroleum mulches compared to a bare-ground control (Nettles, 1963; Takatori et al., 1963). Despite success with cucurbits, petroleum spray mulches were found to be no more effective than non-mulched controls for potato production (Hensel, 1968).

2. Polyethylene mulch

Invented in its solid form in 1935 by British chemists Fawcett and Gibson, and first made into a sheet form in 1938, polyethylene has changed vegetable production around the world (Lamont et al., 1993; Lamont, 1996; Partington, 1970; Wright, 1968). Much of the pioneering research using low density polyethylene (LDPE) mulch was conducted by Dr. Emery Emmert in the 1950s at the University of Kentucky. In his earliest research, Emmert utilized 0.0015 gauge (1.5 mil-thick) black and black-aluminum pigmented plastic sheets. Transplanted tomatoes and direct-seeded pole beans (*Phaseolus* sp.) were some of the first crops tested with plastic mulches (Emmert, 1956; Emmert, 1957). Irrigation was achieved by cutting furrows in the ground next to the crop, covering with plastic, and cutting holes in the plastic for the water to penetrate the plant bed. In the earliest trials with plastic mulches, Emmert found similar results as previous researchers observed with paper mulches. Weed control and yields, particularly early in the season, were significantly better in treatments grown using the plastic mulch compared to a non-mulched control. In some treatments, Emmert reported an increase in yield of more than 200 bushels/acre (5000 kg/ha) for pole beans grown on plastic compared to a bare-ground control (Emmert, 1957). Although expensive, Emmert estimated that if the plastic material lasted four years in a field, the annual cost would be approximately $12-$16 per acre per year (Emmert, 1957).

Much early research evaluated the effect of mulches on yields and microclimate. Soil temperatures were generally higher under black and clear plastic mulches than non-mulched controls (Army and Hudspeth, 1960; Clarkson and Frazier, 1957; Harris, 1965; Nettles, 1963; Oebker and Hopen, 1974; Takatori et al., 1964). Moisture and nitrate levels were generally greater under plastic mulches (Clarkson, 1960; Harris, 1965). This led to earlier (7-14 day) and greater yields in most crops tested (Clarkson and Frazier, 1957). Interestingly, much of the earliest research with plastic mulches indicated that they altered the soil-root zone microclimate in a similar manner as previously reported for asphalt-impregnated paper mulches in the 1920s and 1930s. However, unlike early paper mulches, the plastic-mulch production system has become the dominant mulching tactic for vegetable production.

2.1 Equipment for the plastic-mulch production system

Early plastic mulches were placed in the field by hand; however, to increase efficiency, specialized equipment was developed. Initial land preparation is similar for bare-ground and plastic-mulch production systems. Soil is ploughed and disked until a fine tilth is achieved. A piece of equipment which can form a raised bed and lay plastic mulch in a single operation is pulled through the field to form the planting bed. When using a raised-bed plastic-mulch system, rows must be spaced further apart in order to accommodate the bed shaping equipment than would be necessary in a flat-bed system. Therefore, raised-bed plastic-mulch rows are typically spaced on 1.7 to 2.2 m centers. Raised beds are often preferred with plastic mulches because they warm quicker than flat beds and offer superior drainage (Lamont, 1996; Tarara, 2000). Herbicides which must be incorporated with tillage may be applied to the soil prior to bed formation or under the mulch while it is being laid in the field. Chemical fumigants are often knifed into the soil under plastic mulches during this process as well (Hartz et al., 1993). Fumigation is an important component of many plastic-mulch production systems. Fumigants have the ability to kill weed seeds, which may potentially germinate, as well as control soil pathogenic fungi, bacteria, and nematodes (Goring, 1962; Wilhelm and Paulus, 1980). Drip irrigation tubing is placed under plastic mulch during the same process. Early research with plastic mulches was conducted using overhead irrigation or furrow irrigation (Emmert, 1957); however, with the introduction of drip irrigation in the 1970s, the vast majority of plastic-mulch production now utilizes this method (Hartz, 1996). The combination of drip irrigation with plastic mulch has significantly increased irrigation water use efficiency in vegetable production (Howell, 2001).

After the plastic is laid in the field, transplants can be placed by hand or using a mechanized transplanter. Plastic mulches must fit tightly against the soil; not only to obtain the maximum benefit of heat transfer from mulch to soil; but also because warm air, when trapped under the mulch, can escape through the holes where transplants are placed, desiccating and damaging the crop (Lamont, 2005). Due to the increased productivity of plastic mulches, in-row spacing of plants is often less compared to bare-ground production systems. Crops which may normally be planted in a single row fashion when grown without mulches are often planted in double rows with plastic mulches (Lamont, 1991). Plant populations per unit area may also be increased in plastic-mulch production systems.

At the end of the growing season, plastic mulches must be removed from the field; though in warmer climates mulches are often double or triple cropped (Hanna and Adams, 1989). Double cropping plastic mulch decreases input costs for growers; however, weed pressures are often increased during the second crop as pre-emergent herbicides have dissipated. Although additional herbicides may be applied to spaces between rows; in-row weeds, growing through the planting holes of the previous crop, can be difficult to control (Waterer et al., 2008). In regions with shorter growing seasons, most plastic mulch is removed after one crop, though double-cropping mulches that have been left in fields over a winter have been evaluated (Waterer et al., 2008). To remove plastic mulch from fields, a specialized piece of equipment (mulch lifter) is required. A mulch lifter is a device which undercuts and lifts plastic mulch out of the soil at which time it can be collected and disposed.

2.2 Characteristics of plastic mulches

The earliest plastic mulches evaluated were 1.5 mil-thick and black (Emmert, 1957). There are now arrays of mulches available. The most common mulches are 1.0 or 1.25 mil-thick and are sold on a 1.2 m-wide roll, though widths of 0.9 – 1.5 m are also produced. Mulches that are thinner than 1.0 mil are easily punctured by weeds. Most degradable plastic mulches are 0.5-0.75 mil-thick, which allows for quicker decomposition. Rolls of mulch commonly range from 730 – 1830 m in length. Mulches may be smooth or embossed. Mulches that are embossed tend to resist excessive expansion and contraction which can cause mulches to become loose from raised beds (Lamont, 1993).

2.3 Colored mulches

The most popular plastic mulch world-wide is black, though white-on-black and clear mulches are also used (Schales, 1990). Other colors that that have been evaluated include: blue, green, red, yellow, brown, white, and silver (Brault et al., 2002; Gough, 2001; Hanna, 2000; Ngouajio and Ernest, 2004). Different colored mulches have multiple effects on the crops being grown. The optical properties of various colored mulches can influence soil and air temperatures around the crop as well as impact weed growth under the mulch. Moreover, in some cases, colored mulches can alter insect behaviour, which can directly (insect feeding) and indirectly (vectoring diseases) affect crop growth. Colored mulches can be separated into those that do not discriminate between different wavelengths of light transmitted and those that selectively prevent transmission of photosynthetically active radiation (PAR) (400-700 nm) (Ngouajio and Ernest, 2004; Tarara, 2000). Mulches that selectively filter out light in the PAR range are called infrared transmitting (IRT) mulches. In addition to restricting light of the PAR range, IRT mulches tend to transmit high percentages of light at longer wavelengths (>900 nm). By selectively filtering light in the PAR range and transmitting longer wavelength light energy, IRT mulches allow for greater soil warming while reducing light available for weed growth.

2.3.1 Non-IRT colored mulches effects on light, temperature, and weed growth

The most common non-IRT mulches are black, clear, white-on-black, and reflective silver. A myriad of other colors exist including: yellow, blue, red, and green (Figure 1). These colored mulches comprise a very small portion of the total mulch utilized. Although benefits have been obtained from colored mulches, particularly red in tomatoes (Decoteau et al., 1989), some allow excessive light transmittance, resulting in unacceptable weed growth. The potential for weed growth and higher costs associated with colored plastic mulches has limited their use.

Fig. 1. Muskmelons being grown on blue, brown, red, and white-on-black mulches[1].

As would be expected, clear mulches transmit the most shortwave radiation (84%) and absorb the least (5%) (Ham et al., 1993). Clear mulches also reflect a high percentage (88%) of long-wave radiation. Clear-plastic mulches increase soil temperatures from 4.4 – 7.8 ºC when measured at a depth of 5 cm below the soil surface (Lamont et al., 1993). However, the ability of clear mulch to heat the soil also depends on how it is applied. As noted, clear mulches largely transmit shortwave radiation and reflect long-wave radiation. When clear mulches are loosely applied, long-wave radiation emitted from the soil becomes trapped under the plastic creating a greenhouse-type environment (Ham et al., 1993; Lamont, 1993; Liakatas et al., 1986). However, if the clear mulch is placed tightly on the soil surface, then less convective heating occurs and soil temperature increases may not be as large as expected (Ham and Kluitenberg, 1994; Ham et al., 1993). Diurnal temperature fluctuations are also greater in clear plastic that has not been held tightly to the soil compared to those that have (Tarara, 2000). It has also been reported that the warming effects of clear mulches compared to other colors are substantially reduced in overcast or cloudy environments with less solar radiation (Johnson and Fennimore, 2005).

Clear plastics are utilized for soil solarization. This is the process by which light energy from the sun is trapped, heating the soil enough to cause thermal degradation of bacterial, nematode, fungal, or weed pests (Katan, 1981b; Katan and DeVay, 1991). Soil is prepared for the crop of interest and then solarized for a period of time prior to planting. Disturbing the soil after solarization reduces weed control. When soils are disturbed after solarization, weed seeds that were deep in the soil and unaffected by the treatment, can be brought to the surface to germinate. Clear plastic is the best choice for solarization due to superior heating ability. Reports from California show soil temperatures, measured at a depth of 5 cm from the surface, reaching 60 ºC under clear plastic (Katan, 1981a; Katan, 1981b). Plastics are applied more loosely for solarization than they are when mulching plant beds. This may explain the higher temperatures observed in solarization trials than when using clear plastic as a mulch. Solarization has been documented to control a variety of weed pests in many crops (Basavaraju and Nanjappa, 1999; Katan and DeVay, 1991; Law et al., 2008; Megueni et

---

[1] Photos courtesy of Dr. John Strang, University of Kentucky, Department of Horticulture.

al.; Standifer et al., 1984). However, to properly solarize soil, clear plastic must be exposed to high light and temperatures for a fairly long period of time; therefore, its use is limited in cooler climates (Katan and DeVay, 1991).

Clear plastic functions well for soil solarization, but its use as a mulch is limited. Higher yields have been reported for crops such as strawberries (*Fragaria* sp.) when using clear plastic in combination with soil fumigation with methyl bromide and chloropicrin (Johnson and Fennimore, 2005). However, due to the methyl bromide phase-out and the absence of suitable replacements (Locascio et al., 1997), the ability to control weeds under clear-plastic mulches has limited their use. In non-fumigated soils, clear mulches only controlled 64% of weeds compared to black mulches (Johnson and Fennimore, 2005). Clear plastic is generally unsuitable as a mulch unless supplemental herbicides or fumigants are applied to control weeds (Lamont, 2005).

Black plastic is the predominate mulch utilized in vegetable production today. Much of this popularity is due to a lower cost per acre compared to other mulches. However, black-plastic mulch also effectively warms the soil, improving early crop production and eliminates most in-row weed growth. Unlike clear mulches, black plastic absorbs nearly all shortwave radiation to heat the soil (Ham et al., 1993). By absorbing radiation, black-plastic mulch heats the soil through conduction. A tightly formed plant bed where the mulch makes consistent contact with the soil is necessary for optimal soil warming (Lamont, 1993; Tarara, 2000). By absorbing nearly all shortwave radiation, the surface temperatures of black plastic mulches can reach 55 °C (Tarara, 2000). Soil temperatures 10 cm under the mulch may increase 3-5 °C (Ham et al., 1993). Once crop canopies develop, shading of the mulches increases, and soil temperatures under mulches often decrease compared to bare-ground treatments. Though weed seeds may germinate under black-plastic mulch, subsequent weed growth is limited, with the notable exception of yellow and purple nutsedges (*Cyperus* spp.) (Patterson, 1998). Therefore, black plastic is the mulch of choice for early season vegetable production.

White-on-black and silver-reflective plastic mulches are less popular than black plastic, but still serve an important role in vegetable production and weed management. During periods when soil temperatures are elevated, warming the soil with black-plastic mulch can actually harm plants and reduce yields. To avoid damaging the crop, but still provide in-row weed control, white and silver reflective mulches were developed. White mulches were largely ineffective for weed control, without the use of fumigants or herbicides, because they transmitted too much light. Ngouajio and Ernest (2004) reported that white mulches transmitted 48% of solar radiation. This level of light transmission led to substantial weed growth under white mulch. Trials where black mulches were painted white demonstrated benefits of a reflective mulch where weeds could be controlled (Decoteau et al., 1988). White and black-colored mulches are now coextruded forming white-on-black mulch. This mulch is popular because it combines the weed control properties of black mulches (Johnson and Fennimore, 2005) with the soil cooling properties of white-reflective mulch. Ham et al. (1993) reported that white-on-black and silver mulches reflect 48%and 39% of shortwave radiation, respectively. The reflection of shortwave radiation can result in slightly lower root-zone temperatures in reflective mulches compared to bare soil (Diaz-Perez, 2010; Diaz-Perez et al., 2005; Ham et al., 1993; Tarara, 2000).

White-on-black and silver mulches reflect significantly more light into the plant canopy than black mulches, though this decreases as the canopy expands. The upwardly reflected light from white or silver mulches decreases the ratio of red to far-red light compared to black mulches (Decotcau, 2007; Decoteau et al., 1988). The alteration of the light microenvironment is thought to lead to greater leaf areas, shorter internodes, and greater branching in plants grown on reflective mulches compared to black plastic (Decotcau, 2007; Decoteau et al., 1988; Diaz-Perez, 2010). However, the impact of the optical characteristics of the reflective mulches is limited at certain heights above the bed, and wanes as the plant canopy forms (Lamont, 2005). It is also difficult to isolate differences in light effects on plant growth from root-zone temperatures when comparing different colored mulches (Diaz-Perez and Batal, 2002). Light reflective mulches have also been suggested to influence insect predation on vegetable crops as well (Brown and Brown, 1992; Caldwell and Clarke, 1999; Csizinszky et al., 1995; Funderburk, 2009; Lu, 1990).

2.3.2 The effect of IRT mulches on temperature, light, and weed growth

IRT mulches allow transmission of light outside of the PAR spectrum. By transmitting infrared radiation, but excluding PAR, IRT mulches combine the soil-warming benefits of clear plastic mulches with the weed control of black plastic mulch. IRT mulches are most commonly manufactured in green and brown colors. Ngouajio and Ernest (2004) reported that IRT-green and IRT-brown mulches transmitted 42% and 26% of light, respectively, between the wavelengths of 400 and 1100 nm. This was compared to just 1% in black and 2% in white-on-black mulches, respectively. However, the green and brown-IRT mulches transmitted 16% and 6% of PAR (400-700 nm), respectively. Ham et al. (1993) reported 37% of total short-wave light (300-1100 nm) transmitted for an IRT mulch, while Johnson and Fennimore (2005) reported 10.6% and 10.9% transmittance of PAR (400-700 nm) for green and brown IRT mulches, respectively. This selective transmittance of light allows IRT mulches to provide similar weed control as black-plastic mulches (Johnson and Fennimore, 2005; Ngouajio and Ernest, 2004).

The soil warming properties of IRT mulches are reported to be more similar to clear plastic mulches (Lamont, 1993). However, effects of IRT mulches on soil temperatures may vary. Ngouajio and Ernest (2004) reported heat accumulation in growing degree days (base 10 °C) in IRT mulches was similar to black plastic mulch and better than white and white-on-black mulches. In that trial, clear mulches were not included for comparison. Johnson and Fennimore (2005), using a degree-hour model for heat accumulation, reported that IRT brown and green mulches accumulated 5200 and 6300 degree hours, respectively, while clear and black-plastic mulches accumulated 11000 and 4400 degree hours, respectively. This suggests that IRT mulches provide soil warming abilities between clear and black-plastic mulches. However, in the same trial, the authors reported that black-plastic mulch accumulated more degree hours than clear and IRT mulches in a cooler, cloudier location. Ham et al. (1993) trialled several mulches and reported that IRT mulch had similar soil warming characteristics as clear-plastic mulch. However, both IRT and clear-plastic mulches failed to warm the soil as much as a black plastic mulch (Ham et al., 1993). Therefore, while IRT mulches may control weeds as well as black-plastic mulch, the relative soil warming abilities of IRT mulches compared to black plastic may vary based on local climate.

2.4 Polyethylene mulches influence the root zone affecting weeds and crops

Numerous studies show that vegetables grown with plastic mulches typically out yield those grown on bare ground, even with complete weed control for the bare-ground plots. (Table 1). It has been well documented that plastic mulches reduce evaporation, nutrient leaching, and soil compaction in the plant bed (Lamont, 2005). However, the impact of plastic mulch on root architecture and root-zone temperatures are particularly notable; especially as the yield benefits of black plastic mulch are often greater in the spring than in the fall after soil has warmed (Table 1).

Treatmemt	Total Yield [mean ± s.e (kg·ha⁻¹)]			
	Spring		Fall	
Black Plastic	37905 ± 1492	a^z	27214 ± 953	a
Bare ground hand-weeded	19693 ± 1352	b	21843 ± 1214	b
Bare ground non-weeded	10524 ± 722	c	17330 ± 1866	b

[z] Treatments within a column not followed by the same letter are different by Duncan's Multiple Range Test $P<0.05$

Table 1. Yields of summer squash (*Cucurbita* sp.) under black plastic mulch and bare-ground treatments grown in summer and fall [adapted from (Coolong, 2010)].

Knavel and Mohr (1967) reported summer squash, tomato, and pepper plants had significantly more and longer roots when grown with plastic mulches compared to unmulched controls. However in graphic representations, roots under plastic mulches were also significantly shallower and spread out over the surface of the bed compared to bare-ground plots (Knavel and Mohr, 1967). Other trials have reported that plastic mulches influenced adventitious root development, but overall root architecture remained similar compared to bare-ground production (Gough, 2001).

Significant research has been conducted evaluating the impact of mulch type and color on root-zone temperature and subsequent yield impacts (Diaz-Perez, 2009; Diaz-Perez, 2010; Diaz-Perez and Batal, 2002; Diaz-Perez et al., 2005). Generally, black plastic mulch is preferred for spring plantings as a method to warm the root zone and increase yields (Diaz-Perez, 2009; Diaz-Perez, 2010; Diaz-Perez et al., 2005). However, during summer, soil temperatures under black plastic mulches may be greater than 30 °C (Ham et al., 1993; Tarara, 2000; Tindall et al., 1991). Vegetable growth and yield has been shown to respond quadratically to root-zone temperature, increasing up to a point then rapidly decreasing (Coolong and Randle, 2006; Diaz-Perez, 2010; Tindall et al., 1990). Depending on the crop grown, the critical root-zone temperature for maximum yield and growth may be several degrees cooler than is present under the black plastic mulch. Diaz-Perez and Batal (2002) reported an increase of 5-fruit per plant for tomatoes grown in black plastic mulch compared to bare-ground. However, in the same trial, plants grown on reflective gray and silver mulches, which reduced root-zone temperatures compared to black plastic, had an additional 6-7-fruit per plant compared to the black-plastic mulch treatment. Because of high root-zone temperatures, reflective mulches are encouraged for summer-planted crops. To double-crop black plastic mulches during the summer, a system was developed which utilized a photodegradable black mulch placed over a white non-degradable mulch

(Graham et al., 1995). The black mulch warmed the soil in the spring and then degraded, exposing the white mulch used for a second planting. This system was effective in reducing soil temperatures late in the summer; however, a co-extrusion process has not been commercialized for developing such a system.

2.4.1 Mulch type influences weed morphology

Two common weeds that are not controlled by black plastic mulches are purple and yellow nutsedge. These are two of the most problematic weeds for vegetable production in the Southern U.S. (Webster and MacDonald, 2001). Unlike most weeds, both yellow and purple nutsedge have the ability to pierce plastic mulches (Figure 2) and successfully compete with crops (William, 1976; William and Warren, 1975).

Fig. 2. Yellow nutsedge penetrating white-on-black mulch.

Traditionally, growers have relied on fumigation with methyl bromide to control nutsedge when using plastic mulches. However, as methyl bromide use has been phased out with the exception of some critical-use exemptions, the management of yellow and purple nutsedge under plastic mulches has become a pressing issue (Webster, 2005). Interestingly, some research has demonstrated that controlling yellow and purple nutsedge may depend on the light transmittance of mulches used. Purple nutsedge shoot and tuber growth was shown to be greater under white-on-black mulch compared to IRT mulch when grown under sunlight in a greenhouse (Patterson, 1998). However, in the same trial, all mulches failed to prevent nutsedge shoot emergence when treatments were conducted in total darkness in growth chambers. This suggests that the transmission of light may alter the ability of purple nutsedge to penetrate mulches. Chase et al. (1998) reported similar results when evaluating yellow and purple nutsedge. In that trial, yellow and purple nutsedges penetrated black mulches to a greater extent than clear and IRT mulches. All mulches controlled yellow nutsedge to a greater degree than purple nutsedge. The authors theorized that nutsedge rhizomes have a sharp tip that will penetrate opaque mulches. However, upon exposure to light, photomorphogenic initiation of leaf expansion occurs and the leaves do not have the ability to penetrate the plastic mulches as well as the rhizome (Chase et al., 1998). Although nutsedges will sprout under clear or IRT mulches, they rarely penetrate through the film. Webster (2005) reported similar results, also noting the greater relative ability of purple compared to yellow nutsedge to overcome any plastic mulch. Over time, this may result in a shift in the weed population from yellow to purple nutsedge in mulched vegetable cropping systems (Webster, 2005).

3. Waste issues and mulches

Although plastic mulches provide excellent in-row weed control and enhance productivity of many vegetable crops, waste is a significant issue. It is estimated that world-wide plastic film use is 700,000 tons per year (Espi et al., 2006) with more than 140,000 tons used annually in the U.S. (Shogren, 2001). Most of these mulches end up in landfills or are burned (Hemphill Jr, 1993; Kyrikou and Briassoulis, 2007). As landfill space becomes limited and concerns rise about discarding plastic, which may potentially contain pesticide residues, disposal of mulch films has become a significant issue for farmers. Recycling is not typically an option as used mulches contain dirt and debris from production fields that must first be removed prior to the recycling process. At this time, processes to remove dirt from mulches are too expensive. An alternative to recycling that has been pilot-tested was to compress used-plastic mulches into dense pellets and using them as a fuel source. These pellets have been effectively co-fired with coal in trials conducted at Pennsylvania State University (Lawrence et al., 2010). Plastic mulches are petroleum-based products and contain roughly the same energy content as fuel oil on a weight basis (Hemphill Jr, 1993). In addition to environmental concerns, the costs for removal and disposal of plastic mulches are approximatley $250/ha (Waterer, 2010).

3.1 Degradable mulch films

Economic and environmental concerns have spurred interest in degradable mulch films. Designing degradable mulches with properties similar to LDPE is challenging. The degradable mulch must be flexible, lightweight, prevent light transmittance, and degrade in a timely manner after harvest. Exposure to light, temperature, and moisture can influence degradation (Kyrikou and Briassoulis, 2007). Normalizing degradation rates between growing regions with vastly different climates is a challenge as well. Some crops will quickly form a canopy shading mulches and thus delaying degradation; while others do not. Developing a mulch that will degrade on-demand at a competitive cost is a challenge.

Degradable plastic mulches are often labeled as biodegradable. However, to be considered biodegradable, a polymer must be completely converted by microorganisms to water, minerals, carbon dioxide and biomass (Kyrikou and Briassoulis, 2007). Some mulches that have been marketed as biopolymers do not biodegrade, but fragment, leaving synthetic polymers in the environment in microscopic fragments. Starch-polymer blends fragment as the starch co-polymers degrade, with the synthetic co-polymer remaining in the field (Halley et al., 2001). It is debated whether the synthetic polymers which are left in the field biodegrade. Nonetheless, a variety of mulches have been developed that are reported to completely degrade.

Halley et al. (2001) developed a mulch film using modified-starch polymers that performed as well as a conventional polyethylene mulch for pepper production. This mulch withstood 14 weeks of water exposure and remained largely stable during crop production. However, just two weeks after the mulch was plowed into the soil, it was visually undetectable, with composting trials indicating that the mulch completely degraded to carbon dioxide and water after 45 days (Halley et al., 2001). Waterer (2010) tested clear, black, and wavelength-selective starch-based mulches. In this trial, clear and wavelength-selective starch-based mulches degraded quickly in the field. The clear starch-based mulch broke down completely within 8 weeks of application. Although weed growth occurred in the clear and wavelength selective-starch mulches, the yields of the crops trialled (zucchini, cantaloupe, pepper, eggplant, and corn) were not significantly different between mulch types (starch-based polyethylene) of a given color (clear, wavelength-selective, black). The black-colored starch mulch remained intact for the entire growing season (Waterer, 2010). This trial was conducted at a northern latitude (Saskatoon, Saskatchewan, CA) with a short cool growing season. Different results may be expected in warmer environments.

Another commonly utilized polymer for degradable mulches is polybutylene adipate-co-terephthalate (PBAT). PBAT is reportedly a fully biodegradable polymer that has similar physical characteristics as traditional LDPE mulches, although, PBAT mulches are typically slightly thinner and tear easier than common LDPE mulches (Kijchavengkul et al., 2008a; Witt et al., 2001). During the time PBAT mulches are set out in the field for crop production they begin photodegrading with a period of intensive biodegradation after crop removal and subsequent plowing into the ground. However, the absolute biodegradability of the PBAT mulches has been questioned due to cross-linking that can occur between benzene rings contained in the PBAT polymer (Kijchavengkul et al., 2008a; Kijchavengkul et al., 2008b). Typically, white or green PBAT mulches have been found to degrade quicker than black mulches, often breaking apart while the crop is in the field (Moreno and Moreno, 2008; Ngouajio et al., 2008). White-colored PBAT films can contain titanium oxide, which may catalyze photodegradation leading to premature breakdown (Gesenhues, 2000; Kijchavengkul et al., 2008a). Black-colored PBAT mulches typically last longer and it is proposed that the carbon black added to the PBAT film absorbs light energy, reducing photodegradation (Kijchavengkul et al., 2008a; Schnabel, 1981).

Trials of PBAT mulches indicate that white-colored PBAT mulches have lower yields compared to black-polyethylene mulches, but black-PBAT mulches usually perform as well as traditional polyethylene mulches (Miles et al., 2006; Moreno and Moreno, 2008; Ngouajio et al., 2008). Usually white-PBAT mulches break down prematurely allowing weeds to grow, affecting crop yields. Interestingly, soil temperatures under black-PBAT mulches are often lower than under black-polyethylene mulches (Moreno and Moreno, 2008; Ngouajio et al., 2008). Although promising, PBAT and starch-based mulches are not used on a large scale at this time.

3.2 Degradable paper mulches

Nearly 100 years after the use of paper mulches was documented in Hawaii, they are again being evaluated for use in vegetable production (Stewart et al., 1926). Paper is a renewable resource that readily biodegrades. Newspaper-based mulches represent an available and cost effective resource and have been frequently trialled; though they often deteriorate rapidly under field conditions, reducing effectiveness (Shogren, 2001). Shredded newspapers have been successfully used as a weed suppressing mulch in organic high-tunnel cucumber production (Sanchez et al., 2008). A high-tunnel environment (no wind or rain) is conducive for using newspaper mulches. Traditionally, paper mulches degrade quickly under field conditions and may tear when using traditional mulch-laying and planting equipment (Coolong, 2010). To improve the durability of paper-based mulches, several trials have utilized mulches with polyethylene, wax, or vegetable oil coatings used to slow degradation of paper mulches in the field (Shogren, 1999; Shogren and David, 2006;

Shogren and Hochmuth, 2004; Vandenberg and Tiessen, 1972). Coating 30-40 lb (14-18 kg) kraft paper with vegetable oils will retard degradation by repelling water and also by filling voids in the cellulose fibers of paper, preventing microorganism infiltration (Shogren, 1999). When oils are applied to the kraft-paper mulches, field-life can be increased to 14 weeks; giving adequate weed control and yields comparable to black-plastic mulch (Shogren, 1999; Shogren and David, 2006). Coolong (2010) reported adequate weed control and yields comparable to black plastic mulches for 40-lb kraft paper coated with a thin layer of clear polyethylene. Mating a thin degradable coating to paper mulch may be a potential solution to the premature degradation of paper mulches; however, the weight and subsequent shipping costs for paper-based mulches at the present time precludes them from widespread use.

4. Summary and conclusions

Mulches have been used for centuries for weed control in vegetable crops. Despite the development of a range of herbicides available, mulches still continue to play a significant role in the production of vegetable crops. The introduction of polyethylene mulches in the 1950s significantly altered the way mulches were utilized. When combined with tillage techniques and herbicides, plastic mulches allow vegetable growers to maintain nearly weed-free fields. The ability of plastic mulches to alter crop microclimate can also lead to improved earliness, quality, and yields (Lamont, 2005). Plastic mulches are now an indispensible part of the modern vegetable production system. However, as concerns regarding the environmental impact of the disposal of mulches increase, alternatives are being sought. Paper-based mulches are degradable and made of a renewable resource, but are bulky and costly to produce. Organic mulches such as straw improve soil health by increasing organic matter and improving soil structure. However, they do not provide the same soil warming benefits as polyethylene mulches. This may limit their use in certain crops or cooler climates. As technologies improve, a completely degradable mulch film made from natural polymers may replace traditional polyethylene mulches. However, until that time polyethylene plastic will remain the most widely used mulch for the production of warm-season vegetable crops.

5. References

Adlerz, W. and Everett, P. 1968. Aluminum foil and white polyethylene mulches to repel aphids and control watermelon mosaic. *Journal of Economic Entomology*. 61:1276-1279.

Army, T. and Hudspeth, E. 1960. Alteration of the microclimate of the seed zone. *Agronomy Journal*. 52:17-22.

Basavaraju, H.K. and Nanjappa, H.V. 1999. Weed dynamics in chilli-maize cropping sequence as influenced by soil solarization. *Indian Journal of Weed Science*. 31:183-186.

Brault, D., Stewart, K.A., and Jenni, S. 2002. Optical properties of paper and polyethylene mulches used for weed control in lettuce. *HortScience*. 37:87-91.

Brown, S.L. and Brown, J.E. 1992. Effect of plastic mulch color and insecticides on thrips populations and damage to tomato. *HortTechnology*. 2:208-210.

Burgis, D.S. 1950. Mulching vegetable crops with aluminum foil. *Proceedings of the Florida State Horticultural Society.* 63:141-144.

Caldwell, J.S. and Clarke, P. 1999. Repulsion of cucumber beetles in cucumber and squash using aluminum-coated plastic mulch. *HortTechnology.* 9:247-250.

Chase, C.A., Sinclair, T.R., Shilling, D.G., Gilreath, J.P., and Locascio, S.J. 1998. Light effects on rhizome morphogenesis in nutsedges (*Cyperus* spp.): Implications for control by soil solarization. *Weed Science.*575-580.

Clarkson, V. 1960. Effect of black polyethylene mulch on soil and microclimate temperature and nitrate level. *Agronomy Journal.* 52:307-309.

Clarkson, V.A. and Frazier, W.A. 1957. Plastic mulches for horticultural crops. *Bulletin of the Oregon Agricultural Experiment Station.* 562.

Coolong, T. 2010. Performance of paper mulches using a mechanical plastic layer and water wheel transplanter for the production of summer squash. *HortTechnology.* 20:319-324.

Coolong, T.W. and Randle, W.M. 2006. The influence of root zone temperature on growth and flavour precursors in *Allium cepa* l. *Journal of Horticultural Science & Biotechnology.* 81:199-204.

Csizinszky, A.A., Schuster, D.J., and Kring, J.B. 1995. Color mulches influence yield and insect pest populations in tomatoes. *Journal of the American Society for Horticultural Science.* 120:778-784.

Decotcau, D.R. 2007. Leaf area distribution of tomato plants as influenced by polyethylene mulch surface color. *HortTechnology.* 17:341-345.

Decoteau, D.R., Kasperbauer, M.J., Daniels, D.D., and Hunt, P.G. 1988. Plastic mulch color effects on reflected light and tomato plant-growth. *Scientia Horticulturae.* 34:169-175.

Decoteau, D.R., Kasperbauer, M.J., and Hunt, P.G. 1989. Mulch surface color affects yield of fresh-market tomatoes. *Journal of the American Society for Horticultural Science.* 114:216-219.

Diaz-Perez, J.C. 2009. Root zone temperature, plant growth and yield of broccoli [*Brassica oleracea* (plenck) var. *Italica*] as affected by plastic film mulches. *Scientia Horticulturae.* 123:156-163.

Diaz-Perez, J.C. 2010. Bell pepper (*Capsicum annum* l.) grown on plastic film mulches: Effects on crop microenvironment, physiological attributes, and fruit yield. *HortScience.* 45:1196-1204.

Diaz-Perez, J.C. and Batal, K.D. 2002. Soil-plant-water relationships-colored plastic film mulches affect tomato growth and yield via changes in root-zone temperature. *Journal of the American Society for Horticultural Science.* 127:127-135.

Diaz-Perez, J.C., Randle, W.M., Boyhan, G., Walcott, R.W., Giddings, D., Bertrand, D., Sanders, H.F., and Gitaitis, R.D. 2004. Effects of mulch and irrigation system on sweet onion: 1. Bolting, plant growth, and bulb yield and quality. *Journal of the American Society for Horticultural Science.* 129:218-224.

Diaz-Perez, J.C., Phatak, S.C., Giddings, D., Bertrand, D., and Mills, H.A. 2005. Root zone temperature, plant growth, and fruit yield of tomatillo as affected by plastic film mulch. *HortScience.* 40:1312-1319.

Emerson, R.A. 1903. *Experiments in mulching garden vegetables.* University of Nebraska, Agricultural Experiment Station of Nebraska.

Emmert, E.M. 1956. Polyethylene mulch looks good for early vegetables. *Market Growers' Journal*. 85:18-19.

Emmert, E.M. 1957. Black polyethylene for mulching vegetables. *Proceedings. American Society for Horticultural Science*. 69:464-469.

Espi, E., Salmeron, A., Fontecha, A., García, Y., and Real, A. 2006. Plastic films for agricultural applications. *Journal of plastic film and sheeting*. 22:85.

Funderburk, J. 2009. Management of the western flower thrips (thysanoptera: Thripidae) in fruiting vegetables. *Florida Entomologist*. 92:1-6.

Gesenhues, U. 2000. Influence of titanium dioxide pigments on the photodegradation of poly vinyl chloride. *Polymer Degradation and Stability*. 68:185-196.

Gilbert, J.C. 1956. Soil mulches of local materials. *Hawaii Farm Science*. 4:4-5.

Goring, C. 1962. Theory and principles of soil fumigation. *Advances in Pest Control Research* 5:47-84.

Gough, R.E. 2001. Color of plastic mulch affects lateral root development but not root system architecture in pepper. *HortScience*. 36:66-68.

Graham, H.A.H., Decoteau, D.R., and Linvill, D.E. 1995. Development of a polyethylene mulch system that changes color in the field. *HortScience*. 30:265-269.

Halley, P., Rutgers, R., Coombs, S., Kettels, J., Gralton, J., Christie, G., Jenkins, M., Beh, H., Griffin, K., and Jayasekara, R. 2001. Developing biodegradable mulch films from starch based polymers. *Starch*. 53:362-367.

Ham, J.M., Kluitenberg, G., and Lamont, W. 1993. Optical properties of plastic mulches affect the field temperature regime. *Journal of the American Society for Horticultural Science* 118:188-188.

Ham, J.M. and Kluitenberg, G. 1994. Modeling the effect of mulch optical properties and mulch-soil contact resistance on soil heating under plastic mulch culture. *Agricultural and Forest Meteorology*. 71:403-424.

Hanna, H.Y. 2000. Double-cropping muskmelons with nematode-resistant tomatoes increases yield, but mulch color has no effect. *HortScience*. 35:1213-1214.

Hanna, H.Y. and Adams, A.J. 1989. Doublecropping tomatoes and cucumbers. *Louisiana Agriculture*. 33:22-23.

Harris, R. 1965. Polyethylene covers and mulches for corn and bean production in northern regions. Proceedings of the American Society for Horticultural Science 87:288-294.

Hartz, T. 1996. Water management in drip-irrigated vegetable production. *HortTechnology*. 6:165-167.

Hartz, T., DeVay, J., and Elmore, C. 1993. Solarization is an effective soil disinfestation technique for strawberry production. *HortScience*. 28:104-106.

Hemphill Jr, D. 1993. Agricultural plastics as solid waste: What are the options for disposal? *HortTechnology*. 3:70-73.

Hensel, D. 1968. Response of potatoes to mulching at different planting and harvesting dates. *Proceedings of the Florida State Horticultural Society*. 81:153-158.

Howell, T.A. 2001. Enhancing water use efficiency in irrigated agriculture. *Agronomy Journal*. 93:281-289.

Isenberg, F.M. and Odland, M.L. 1950. Comparative effects of various organic mulches and clean cultivation on yields of certain vegetable crops. *Progress Report. Pennsylvania Agricultural Experiment Station*. 35.

James, E. 1945. Effect of certain cultural practices on moisture conservation on a piedmont soil. *Journal of the American Society of Agronomy* 37:945-952.

Johnson, M.S. and Fennimore, S.A. 2005. Weed and crop response to colored plastic mulches in strawberry production. *HortScience*. 40:1371-1375.

Katan, J. 1981a. Solar heating (solarization) of soil for control of soilborne pests. *Annual Review of Phytopathology*. 19:211-236.

Katan, J. 1981b. Solar heating (solarization) of soil for control of soilborne pests. *Annual Review of Phytopathology*. 19:211-236.

Katan, J. and DeVay, J.E. 1991. *Soil solarization*. CRC Press. Boca Raton, Fl.

Kayum, M., Asaduzzaman, M., and Haque, M. 2008. Effects of indigenous mulches on growth and yield of tomato. *Journal of Agriculture & Rural Development*. 6:1-6.

Kedar, Y. 1957. Ancient agriculture at Shivtah in the Negev. *Israel Exploration Journal*. 7:178-189.

Kijchavengkul, T., Auras, R., Rubino, M., Ngouajio, M., and Fernandez, R.T. 2008a. Assessment of aliphatic-aromatic copolyester biodegradable mulch films. Part I: Field study. *Chemosphere*. 71:942-953.

Kijchavengkul, T., Auras, R., Rubino, M., Ngouajio, M., and Fernandez, R.T. 2008b. Assessment of aliphatic-aromatic copolyester biodegradable mulch films. Part II: Laboratory simulated conditions. *Chemosphere*. 71:1607-1616.

Knavel, D.E. and Mohr, H.C. 1967. Distribution of roots of 4 different vegetables under paper and polyethylene mulches. *Proceedings of the American Society for Horticultural Science*. 91:589-597.

Kyrikou, I. and Briassoulis, D. 2007. Biodegradation of agricultural plastic films: A critical review. *Journal of Polymers and the Environment*. 15:125-150.

Ladewig, J.E. 1951. Soil conservation in Queensland: Soil conservation in horticultural areas. *Queensland Agricultural Journal*. 73:1-18.

Lamont, W.J., Jr. 1991. The use of plastic mulches for vegetable production. *Extension Bulletin –Asian and Pacific Region, Food & Fertilizer Technology Center*.

Lamont, W.J., Jr. 1993. Plastic mulches for the production of vegetable crops. *HortTechnology*. 3:35-39.

Lamont, W.J., Jr. 1996. What are the components of a plasticulture vegetable system? *HortTechnology*. 6:150-154.

Lamont, W.J. 2005. Plastics: Modifying the microclimate for the production of vegetable crops. *HortTechnology*. 15:477-481.

Lamont, W.J., Hensley, D.L., Wiest, S., and Gaussoin, R.E. 1993. Relay-intercropping muskmelons with Scotch pine Christmas trees using plastic mulch and drip irrigation. *HortScience*. 28:177-178.

Law, D., Bhavsar, V., Snyder, J., Mullen, M., and Williams, M. 2008. Evaluating solarization and cultivated fallow for johnsongrass (*Sorghum halepense*) control and nitrogen cycling on an organic farm. *Biological Agriculture and Horticulture*. 26:175-191.

Lawrence, M., Garthe, J., and Buckmaster, D. 2010. Producing solid fuel from non-recyclable agricultural plastics. *Applied Engineering in Agriculture*. 26:217-223.

Liakatas, A., Clark, J., and Monteith, J. 1986. Measurements of the heat balance under plastic mulches. Part i. Radiation balance and soil heat flux. *Agricultural and Forest Meteorology*. 36:227-239.

Lightfoot, D.R. 1994. Morphology and ecology of lithic-mulch agriculture. *Geographical Review*.172-185.

Locascio, S.J., Gilreath, J.P., Dickson, D.W., Kucharek, T.A., Jones, J.P., and Noling, J.W. 1997. Fumigant alternatives to methyl bromide for polyethylene-mulched tomato. *HortScience*. 32:1208-1211.

Lu, F.M. 1990. Color preference and using silver mulches to control the onion thrips, *Thrips-tabaci* Lindeman. *Chinese Journal of Entomology*. 10:337-342.

Megueni, C., Ngakou, A., Mabong, M.R., and Abdoulaye. Weed control and yield improvement of *Solanum tuberosum* (l.) in Adamawa region (Cameroon) by soil solarization and chicken manure. *Journal of Applied Biosciences*. 42:2867-2875.

Miles, C., Klingler, E., Nelson, L., Smith, T., and Cross, C. 2006. *Alternatives to plastic mulch in vegetable production systems*. 12 Aug. 2011.
http://vegetables.wsu.edu/MulchReport06.pdf

Moreno, M.M. and Moreno, A. 2008. Effect of different biodegradable and polyethylene mulches on soil properties and production in a tomato crop. *Scientia Horticulturae*. 116:256-263.

Nettles, V. 1963. Planting and mulching studies with cucurbits. *Proceedings of the Florida State Horticultural Society*. 76:178-182.

Ngouajio, M. and Ernest, J. 2004. Light transmission through colored polyethylene mulches affected weed population. *HortScience*. 39:1302-1304.

Ngouajio, M. and McGiffen, M.E. 2004. Sustainable Vegetable production: effects of cropping systems on weed and insect population dynamics. Acta Hort:638:77-83.

Ngouajio, M., Auras, R., Fernandez, R.T., Rubino, M., Counts, J.W., Jr., and Kijchavengkul, T. 2008. Field performance of aliphatic-aromatic copolyester biodegradable mulch films in a fresh market tomato production system. *HortTechnology*. 18:605-610.

Oebker, N.F. and Hopen, H.J. 1974. Micro climate modification and the vegetable crop ecosystem. *HortScience*. 9:564-568.

Partington, J.R. 1970. *A history of chemistry*. Macmillan, London, UK.

Patterson, D.T. 1998. Suppression of purple nutsedge (*Cyperus rotundus*) with polyethylene film mulch. *Weed Technology*.275-280.

Rigg, T. and Bruce, J. 1923. The Maori gravel soil of Waimea West, Nelson, New Zealand. *Journal of The Polynesian Society*. 32:85-93.

Rowe-Dutton, P. 1957. *The mulching of vegetables*. Commonwealth Bureau of Horticulture and Plantation Crops, Bucks, UK.

Sanchez, E., Lamont, W.J., Jr., and Orzolele, M.D. 2008. Newspaper mulches for suppressing weeds for organic high-tunnel cucumber production. *HortTechnology*. 18:154-157.

Schales, F. 1990. Agricultural plastics use in the United States. *Proceedings of the 11th International Congress of Plastics in Agriculture*. 54-56.

Schnabel, W. 1981 *Polymer degradation, principles and practical applications*. Hanser Publishing. New York, NY.

Shogren, R.L. 1999. Preparation and characterization of a biodegradable mulch: Paper coated with polymerized vegetable oils. *Journal of Applied Polymer Science*. 73:2159-2167.

Shogren, R.L. 2001. Biodegradable mulches from renewable resources. *Journal of Sustainable Agriculture*. 16:33-47.

Shogren, R.L. and Hochmuth, R.C. 2004. Field evaluation of watermelon grown on paper-polymerized vegetable oil mulches. *HortScience.* 39:1588-1591.

Shogren, R.L. and David, M. 2006. Biodegradable paper/polymerized vegetable oil mulches for tomato and pepper production. *Journal of Applied Horticulture.* 8:12-14.

Standifer, L.C., Wilson, P.W., and Porchesorbet, R. 1984. Effects of solarization on soil weed seed populations. *Weed Science.* 32:569-573.

Stewart, G., Thomas, E., and Horner, J. 1926. Some effects of mulching paper on Hawaiian soils. *Soil Science.* 22:35-39.

Takatori, F.H., Lippert, L.F., and Whiting, F.L. 1963. Petroleum mulch aids germination and stand establishment in preliminary vegetable crop studies. *California Agriculture.* 17:2-3.

Takatori, F.H., Lippert, L.F., and Whiting, F.L. 1964. The effect of petroleum mulch and polyethylene films on soil temperature and plant growth. *Proceedings of the American Society for Horticultural Science.* 85:532-540.

Tarara, J.M. 2000. Microclimate modification with plastic mulch. *HortScience.* 35:169-180.

Thompson, H. and Platenius, H. 1931. Results of paper mulch experiments with vegetable crops. *Proceedings of the American Society for Horticultural Science.* 28:305-309.

Tindall, J.A., Mills, H., and Radcliffe, D. 1990. The effect of root zone temperature on nutrient uptake of tomato. *Journal of Plant Nutrition.* 13:939-956.

Tindall, J.A., Beverly, R.B., and Radcliffe, D.E. 1991. Mulch effect on soil properties and tomato growth using microirrigation. *Agronomy Journal.* 83:1028-1034.

Vandenberg, J. and Tiessen, H. 1972. Influence of wax coated and polyethylene coated paper mulch on growth and flowering of tomato. *HortScience.* 7:464-465.

Waterer, D. 2010. Evaluation of biodegradable mulches for production of warm-season vegetable crops. *Canadian Journal of Plant Science.* 90:737-743.

Waterer, D., Hrycan, W., and Simms, T. 2008. Potential to double-crop plastic mulch. *Canadian Journal of Plant Science.* 88:187-193.

Webster, T.M. 2005. Mulch type affects growth and tuber production of yellow nutsedge (*Cyperus esculentus*) and purple nutsedge (*Cyperus rotundus*). *Weed Science.* 53:834-838.

Webster, T.M. and MacDonald, G.E. 2001. A survey of weeds in various crops in Georgia. *Weed Technology.* 15:771-790.

Wilhelm, S. and Paulus, A.O. 1980. How soil fumigation benefits theCalifornia strawberry industry. *Plant Disease.* 64:264-270.

William, R. 1976. Purple nutsedge: Tropical scourge. *HortScience.* 11:357-364.

William, R. and Warren, G. 1975. Competition between purple nutsedge and vegetables. *Weed Science.*317-323.

Witt, U., Einig, T., Yamamoto, M., Kleeberg, I., Deckwer, W.D., and Müller, R.J. 2001. Biodegradation of aliphatic-aromatic copolyesters: Evaluation of the final biodegradability and ecotoxicological impact of degradation intermediates. *Chemosphere.* 44:289-299.

Wolfenbarger, D. and Moore, W. 1968. Insect abundances on tomatoes and squash mulched with aluminum and plastic sheetings1. *Journal of Economic Entomology.* 61:34-36.

Wright, J. 1968. Production of polyethylene film. *Proceedings of the National Agricultural Plastics Congress* 8:72-79.

3

Living Mulch as a Tool to Control Weeds in Agroecosystems: A Review

G.R. Mohammadi

Department of Crop Production and Breeding,
Faculty of Agriculture and Natural Resources, Razi University, Kermanshah,
Iran

1. Introduction

Weeds are a serious constraint to increased production in crops due to reduced yield and economic returns. Weed problems are particularly problematic in row crops as a result of widely spaced crop rows. Weed control in most agroecosystems is highly dependent on conventional cultivation and herbicide applications. Conventional interrow cultivation represents an additional cost for the producer due to the consumption of fossil fuels (Lybecker et al. 1988) and is also associated with increased soil erosion as soil particles are more susceptible to displacement after tillage (Dabney et al. 1993; Fuller et al. 1995). Moreover, ground and surface water pollution by pesticides are causes for concern (Hallberg 1989), and herbicides used in crops have been among the pesticides most frequently detected in these waters (National Research Council 1989). Improving water quality and decreasing herbicide carry over is one of the more important environmental issues for farmers and agriculture researchers (Stoller et al. 1993). Herbicide-resistant weed ecotypes are being discovered more frequently, due to increased herbicide applications and subsequent selection, is also posing a serious threat to agricultural production (Holt and LeBaron 1990).

Increased interest in sustainable agricultural systems has led to significant developments in cropping practices over the past decade (Thiessen-Martenes et al., 2001). Interest in alternative and sustainable agricultural production systems that require fewer production inputs is growing (Calkins and Swanson 1995). The current emphasis on reduced pesticide use has led to increased interest in alternative weed management methods (Bellinder et al. 1994). In sustainable agriculture, an alternative method to chemical and mechanical weed control in crops is the use of living mulches. Living mulches are cover crops that are planted between the rows of a main crop such as corn (*Zea mays* L.), soybean (*Glycine max* L.), etc., and are maintained as a living ground cover during the growing season of the main crop. Although living mulches are sometimes referred to as cover crops, they grow at least part of the time simultaneously with the crop.

In addition to providing adequate cover to reduce soil erosion (Wall et al. 1991) and increase soil water infiltration (Bruce et al. 1992), legume living mulches improve soil nutrient status through addition of organic nitrogen (N) (Holderbaum et al. 1990; Brown et al., 1993) via fixed atmospheric nitrogen which improves soil physical properties (McVay et al. 1989; Latif et al. 1992). Incorporating legume living mulches can also increase the yield of the succeeding crop (Bollero and Bullock 1994; Decker et al. 1994). Leguminous living mulches have the potential to reduce dependence on fossil fuels and reduce negative environmental effects of crop production systems. Some functions these living mulches can perform are (1) fixing atmospheric N that is made available to main crop, (2) protecting soil from erosion during the main crop growing season, (3) improving soil quality, (4) reducing evaporation and increasing infiltration during the main crop growing season and (5) suppressing weeds (SAN, 1998). Improvement of soil organic matter and production of forage for animal feed are other potential uses of living mulches.

2. Necessity to develop alternative weed control methods in agroecosystems

Weeds are one of the major problems in crop production around the world, and we are trending toward controlling these weeds with herbicides, which comes with an increased environmental impact. At present, in most agroecosystems, weed control highly depends on chemical and mechanical practices that are very expensive, hazardous for the environment and, consequently, unsustainable. For example, currently about 95 % of the soybean acreage in the state of Minnesota in the United States (U.S.) is treated with herbicides with about 6 million kg of herbicides applied annually (Minnesota Agricultural Statistics 2001). Herbicide costs account for 35% of the variable cost of production. Overall, in the U. S. alone, 75% of crop production is based on herbicide input (Duke 1999).

However, herbicide-based control has failed to achieve long-term weed seedbank management (Mortensen et al. 2000; Weber and Gut 2005). Even with herbicides, weeds remain prominent in croplands and producers still lose considerable crop yield due to weeds (Bridges 1994). Furthermore, herbicide resistance is forcing producers to use more expensive management tactics, thereby increasing production costs. Public concern over safety has also caused a reassessment of toxicological and environmental impacts of synthetic herbicides. Therefore, because synthetic herbicides represent a significant expense and environmental concern and cannot be used by those wishing to be certified as organic producers, many producers seek alternative weed control strategies.

3. Benefits of living mulch systems

Living mulches have the potential to form an important component in agroecosystems and can be a useful tool for weed suppression in sustainable agricultural systems (Teasdale 1996; Bond and Grundy 2001; Kruidhof et al. 2008) including many useful advantages such as: improvement of soil structure (Harris et al., 1966), regulation of soil water content (Hoyt and Hargrove 1986), enhancement of soil organic matter, carbon dynamics and microbiological function (Steenwerth and Belina 2008), reducing soil erosion (Malik et al., 2000), soil enrichment by nitrogen fixation (Sainju et al. 2001), insectarium for many beneficial arthropod species (Grafton-Cardwell et al. 1999), and enhancement of populations of soil macrofauna (Blanchart et al. 2006). Living mulches also have the potential to suppress weed growth (De Haan et al. 1994), increase soil water infiltration (Bruce et al. 1992), decrease soil erosion (Cripps and Bates 1993), contribute N to the main crop (Corak et al. 1991) and reduce economic risk (Hanson et al. 1993).

The symbiotic relationship of legume living mulches with rhizobia bacteria allows them to use N from the atmosphere. Of major interest is whether some of the fixed N will be available to a cereal grown simultaneously with the legume. If this is the case, living mulches of legumes could reduce the need for fertilizer N. The primary mechanism of N transfer from a legume to a nonlegume is decomposition of leaves, roots, and stems of the legume (Fujita et al. 1992). However, observations in cereal–legume intercrops have confirmed that N is also excreted from legume roots and leached from leaves, thus becoming available to the cereal immediately (Fujita et al. 1992).

More efficient use of environmental resources is another important benefit of a living mulch system. For example, a legume such as crownvetch (*Coronilla varia* L.) with different root architecture than corn might absorb relatively immobile potassium (K) from deep zones in the soil that would not be accessed by corn (Vandermeer, 1990). Corn requires less soil nutrients (Richie et al. 1993) and light as it matures late in the growing season. In the presence of kura clover (*Trifolium ambiguum* M. Bieb.) as a living mulch, nutrients and light penetrating through the maturing corn canopy may be utilized by the living mulch rather than fostering weed growth (Zemenchik et al. 2000).

3.1 Weed suppression

Living mulches are crops grown simultaneously with the main crop that can suppress weed growth significantly without reducing main crop yield through an ability to grow fast or because they are planted at a high density (De Haan et al., 1994). Living mulches can suppress weed growth by competing for light (Teasdale 1993), water and nutrients (Mayer and Hartwig 1986), and through the production of allelopathic compounds (White et al. 1989) which may ultimately result in reduced herbicide applications. Many studies have confirmed the weed suppressing ability of living mulches in different cropping systems.

There is wide agreement in the literature that a vigorous living mulch will suppress weeds growing at the same time as the living mulch (Stivers-Young 1998; Akobundu et al. 2000; Creamer and Baldwin 2000; Blackshaw et al. 2001; Favero et al. 2001; Grimmer and Masiunas 2004; Peachey et al. 2004; Brennan and Smith 2005). In one study (Echtenkamp and Moomaw 1989), chewing fescue or red fescue *(Festuca rubra* L.*)* and ladino clover *(Trifolium repens* L.*)* were effective living mulches for controlling weed growth. Reductions in weed infestations have been reported with sunn hemp (*Crotalaria juncea* L.) as a living mulch in citrus and avocado (*Persea americana* Mill.) (Linares et al. 2008; Severino and Christoffoleti 2004). According to Mohammadi (2010) weed dry weight was reduced by 34 and 50.9% when hairy vetch (*Vicia villosa* Roth) was interseeded in corn at 25 and 50 kg ha^{-1}, respectively.

Moynihan et al. (1996) also reported a 65% reduction in fall weed biomass compared with non-living mulch control following a grain barley (*Hordeum vulgare* L.) and medic (*Medicago* sp.) intercrop. Velvetbean (*Mucuna pruriens* L.) suppressed the radical growth of the local weeds alegria (*Amaranthus hypochondriacus* L.) by 66% and barnyardgrass (*Echinochloa crusgalli* L.) by 26.5% (Caamal-Maldonado et al. 2001). In another research, a subterranean clover (*Trifolium subterraneum* L.) living mulch reduced weed biomass and increased soybean yield by 91 % relative to weedy control plots (Ilnicki and Enache 1992). The list of weed suppression intensity by different living mulch species are shown in Table 1.

Living mulch species	Percentage weed suppression*	Reference
Red clover, hairy vetch	75	Palada et al. (1982)
Subterranean clover	53-94	Enache and Ilnicki (1990)
Hairy vetch	70-90	Oliver et al. (1992)
Subterranean clover	91	Ilnicki and Enache (1992)
Black mucuna (*Mucuna pruriens* L.), smooth rattlebox (*Crotalaria pallida* L.)	95-99	Skora Neto (1993)
Jack bean (*Canaualia ensiformis* L.), pigeon pea (*Cajanus cajan* L.)	71-90	
Cowpea (*Vigna unguiculata* L.)	29-48	
Hairy vetch	96	Hoffman et al. (1993)
Yellow mustard (*Sinapis alba* L.)	80	De Haan et al. (1994)
Annual medics (*Medicago* spp.)	65	Moynihan et al. (1996)
Annual medics (*Medicago* spp.)	41-69	De Haan et al. (1997)
Subterranean clover, white clover	45-51	Brandsaeter et al. (1998)
Velvetbean	68	Caamal-Maldonado et al. (2001)
Hairy vetch	79	Reddy and Koger (2004)
Alfalfa	34.2-56.9	Ghosheh et al. (2004)
Rye	37-76	Brainard and Bellinder (2004)
Persian clover, white clover, berseem clover, hairy vetch, alfalfa, and black alfalfa	60.1-80.5	Mohammadi (2009)
Hairy vetch	34-50.9	Mohammadi (2010)

*Percentage suppression relative to a control without living mulch.

Table 1. Suppression of weeds by different living mulch species.

4. Factors determining the success of a living mulch system to suppress weeds

4.1 Living mulch species

Interseeding of a crop and living mulches have not always resulted in a positive gain (Nordquist and Wicks 1974; De Haan et al. 1997). Consequently, the success of these kinds of living mulch-crop systems is largely determined by the selection of the most appropriate species and, additionally, by the design of an optimal management strategy for the intercrop. Living mulches differ in their ability to establish well in an interseeding situation. For example, Exner and Cruse (1993) found that alfalfa (*Medicago sativa* L.) and sweet clover (*Melilotus officinalis* L.) usually established better and produced more cover than either red clover (*Trifolium pratense* L.) or alsike clover (*T. hybridum* L.) when interseeded under corn. The competitive ability against weeds is also another important characteristic determining the suitability of a plant species as a living mulch. In a study on six leguminous species (Persian clover, *Trifolium resupinatum* L.; white clover, *T. repens* L.; berseem clover, *T. alexandrinum* L.; hairy vetch; alfalfa; and black alfalfa, *M. lupulina* L.), Mohammadi (2009) found that the highest corn (as the main crop) plant traits including yield, yield components, height, leaf area index and leaf nitrogen content and the lowest weed dry weight were obtained from the plots interseeded with hairy vetch as compared with the other living mulch species (Table 2). Corn yield was increased 79% and weed dry weight was reduced 80.5% when the plots were interseeded with hairy vetch as compared with full season weedy conditions.

Treatment	Yield (g m^{-2})	Ear per plant	Seed per ear	100-seed weight (g)	Height (cm)	Leaf area index	Leaf nitrogen content (%)	Weed dry weight (g m^{-2})
Weed free control	1282.89 a	1.20 a	760.10 a	32.34 a	273.41 a	5.10 a	2.73 ab	0.00 e
Hairy vetch	1188.78 ab	1.10 b	725.30 ab	30.94 ab	254.83 ab	4.82 ab	2.88 a	32.13 d
Berseem clover	1084.31 bc	1.05 bc	689.08 bc	29.86 b	252.99 b	4.52 abc	2.57 bc	49.23 bcd
Persian clover	1063.37 c	1.05 bc	683.60 bc	26.91 c	247.50 b	4.44 abc	2.74 ab	53.65 bc
White clover	999.52 cd	1.00 cd	671.10 bc	26.84 c	248.00 b	4.60 abc	2.23 de	46.33 cd
Black alfalfa	945.16 d	1.03 cd	653.50 c	26.78 c	251.50 b	3.96 c	2.17 e	65.73 b
Alfalfa	938.08 d	1.02 cd	645.05 c	27.16 c	251.75 b	4.19 bc	2.41 cd	62.15 bc
Weedy	664.03 e	0.98 d	519.40 d	23.49 d	205.42 c	3.13 d	1.89 f	164.78 a
LSD (0.05)	109.32	0.07	63.71	2.27	20.41	0.74	0.24	17.21

Similar letters at each column indicate the non significant difference at the 0.05 level of probability.

Table 2. Means comparison of corn plant traits and weed dry weight under different living mulch treatments (from Mohammadi 2009).

Living mulches should be species that establish more rapidly than weeds and whose peak period of growth coincides with that of early weed emergence but does not coincide with that of the crop. Ideally the living mulch should suppress weed growth during the critical period for weed establishment, i.e., the period when emerging weeds will cause a loss in crop yield (Buhler et al. 2001). Beard (1973) recommended chewing fescue as a good living mulch because it adapts to the shady conditions under corn and soybean. This grass is also well adapted to dry and poor soils.

Total biomass production and nitrogen fixation are the main factors determining the suitability of leguminous species for improvement of soil fertility, but if used as a component crop in intercropping systems, competitive ability is another obvious criterion. Morphological growth characteristics, such as early relative growth rate of leaf area and earliness of height development, have been identified to determine competition in intercropping systems (Kropff and van Laar 1993).

Different phenological characteristics and growth patterns were observed among living mulches species ranging from the short-lived species *Mucuna pruriens*, which germinated quickly and covered the ground surface rapidly (LAI=1 at GDD=476°Cd), to the long-lived species *Aeschynomene histrix*, which is slow to establish and only reached a canopy LAI of 1 at around 800°Cd. These characteristics make *M. pruriens* a relatively strong competitor, which may explain its use against the perennial grass *Imperata cylindrica* in maize-based systems in Africa and North Honduras (Versteeg and Koudopon 1990; Akobundo 1993; Triomphe 1996). Based on early growth characteristics, *Crotalaria juncea*, *Cajanus cajan* and *M. pruriens* can be considered as species with a higher competitive ability than *Calopogonium mucunoides*, *Stylosanthes hamata* and *A. histrix*. This can be explained by the combination of high initial growth rates for height and leaf area development. Additionally, the high final height of *C. juncea* and *C. cajan* may confer higher competitiveness throughout the growing season (Akanvou et al. 2001).

According to De Haan et al. (1994) medics used as living mulches in row crops should be small, prostrate, and early maturing. Because of their prostrate growth habit, short life span, and good seedling vigour, medics have potential as living mulches. A living mulch should control weeds, have a relatively short growing season, provide a constant N supply, and give minimal competition to the main crop for water, light, and nutrients (De Haan et al. 1994).

In general, ideal living mulches for weed suppression should have the following characteristics:

1. Ability to provide a complete ground cover of dense vegetation.
2. Rapid establishment and growth that develops a canopy faster than weeds.
3. Selectivity between suppression of weeds and the associated crop (Teasdale 2003).

Usually, living mulches that establish an early leaf canopy cover are most competitive with weeds.

4.2 Living mulch planting rate and time

The success of a living mulch system also depends on appropriate management. Both time and rate of living mulch interseeding can be important factors determining the success of a crop-living mulch system. These factors are critical to reduce living mulch competition with the main crop for environmental resources while allowing the mulch to grow and cover the soil surface sufficiently to reap potential benefits such as weed suppression.

For example, interseeding rye (*Secale* sp.) or small-grain living mulches tended to provide higher levels of weed suppression when interseeded at or near planting of the main crop (Rajalahti et al. 1999; Brainard and Bellinder 2004). In another study, Mohammadi (2010) observed that the plant traits of corn and weed dry weight were not significantly influenced by hairy vetch (as a living mulch) planting times (simultaneous with corn planting or 10 days after corn emergence), but increased hairy vetch planting rate from 0 to 50 kg ha[-1] improved corn yield (by 11%) and reduced weed dry weight (by 50.9%). It was hypothesized that as living mulch density is increased, canopy closure would occur more rapidly, decreasing the amount of photosynthetically active radiation (PAR) available beneath the canopy. This would result in a concomitant decrease in weed biomass until an optimum living mulch density is achieved, beyond which, no further decrease in weed biomass could be obtained. Generally, the biomass produced by a living mulch highly depends on its planting rate. Moreover, there is often a negative correlation between living mulch and weed biomass (Akemo et al. 2000; Ross et al. 2001; Sheaffer et al. 2002). Meschede et al. (2007) expressed that the biomass accumulation by the living mulches was inversely proportional to the weed biomass. Mohammadi (2010) also reported that increasing the hairy vetch dry weight led to the reduction of weed dry weight produced. As for every 1.18 g m^{-2} hairy vetch dry weight produced, 1 g m^{-2} weed dry weight was reduced (Fig. 1).

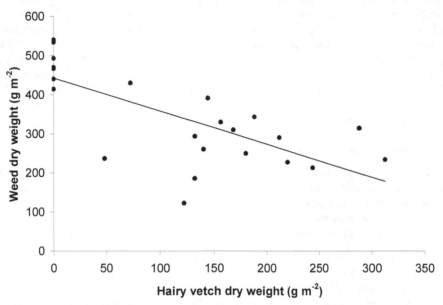

Fig. 1. Relationship between hairy vetch dry weight and weed dry weight loss in field corn as obtained using linear regression model, y = 442.81 - 0.8485x, R^2 = 0.52. The points indicate the individual weed-infested plot values (n = 24) (from Mohammadi 2010).

However, Akobundu et al. (2000) found that development of early ground cover was more important than the quantity of dry matter produced for suppression of cogongrass by velvetbean as a living mulch.

Sowing time of a living mulch is also a very important factor for controlling the weed flora. Some weed species will germinate faster than the living mulch, while some of them will germinate simultaneously with mulch species and others will germinate after the living mulch. Species that germinate after the living mulch cannot grow well since the mulch species shades and mechanically blocks growth of these weed species (Kitis et al. 2011). This can lead to the alteration of weed flora in cropping systems.

5. Mechanisms by which living mulches can suppress weeds

There are a number of mechanisms by which living mulches can suppress weeds such as: their competition for light (Teasdale 1993; Teasdale and Mohler 1993), moisture and nutrient availability (Mayer and Hartwig 1986,); stimulating microorganisms; shading; changes in physical factors of soil such as pH, water holding capacity, temperature and aeration; and the release of allelochemicals (Leather 1983; Liebel and Worsham 1983; Putnam and DeFrank 1983; Weston et al. 1989; Yenish et al. 1995; Liebman and Davis 2000). Overall, weed suppression is thought to be based on alleopathic properties, physical impedance of germination and seedling growth, and competition for light, water, and nutrients (Teasdale, 1993; Teasdale and Mohler, 1993).

Because weed and living mulch plants compete for the same resources, weeds can be suppressed by the introduction of living mulches into cropping systems. In other words, including a living mulch in a cropping system can contribute to weed suppression by occupying the niche that would normally be filled by weeds (Teasdale 1998). Once established, living mulches can rapidly occupy the open space between the rows of the main crop and use the light, water, and nutritional resources that would otherwise be available to weeds. This can result in the inhibition of weed seed germination and reduction in the growth and development of weed seedlings. Therefore, weeds attempting to establish along with a living mulch would be in competition for resources and may not develop sufficiently. Moreover, physical impediments to weed seedlings is another mechanism by which living mulches suppress weeds (Facelli and Pickett 1991; Teasdale 1996; Teasdale and Mohler 1993).

If a living mulch becomes established before the emergence of weeds, then the presence of green vegetation covering the soil creates a radiation environment that is unfavorable for weed germination, emergence, and growth. Moreover, a more diverse biological and physical environment at the surface of soils such as that associated with living mulches offers opportunities for regulating and minimizing weed populations (Teasdale 2003).

Weed seed germination can be negatively affected by quality and quantity of light and the smaller amplitude of soil temperature fluctuation that result from the presence of living mulches (Gallagher et al. 1999; Teasdale 1998). Germination of weed seeds may be inhibited by complete light interception (Phatak 1992) by the living mulch or by secretion of allelochemicals (White et al. 1989; Overland 1966). A delay in emergence of weeds because of the presence of living mulches can also adversely affect weed seed production. Moreover, the presence of living mulches leads to greater seed mortality of weeds by favoring predators (Cromar et al. 1999). Teasdale (1998) also suggested that living-mulch suppression of weeds occurs through resource competition, promoting conditions that are unfavorable for germination and establishment, retaining living mulch residues as ground cover, and by means of allelopathy.

Water competition is another mechanism by which living mulches suppress weeds. Plants exposed to water stress for a limited time (i.e., several hours) respond by a reduction in the transpiration rate through a lowering of the leaf water potential and closing of stomata. Stomatal closing will affect the rate of leaf photosynthesis, which influences the growth and yield. However, under prolonged moisture stress (i.e., days to weeks), whole plant photosynthesis is reduced with a possibility of permanent damage to the photosynthetic apparatus (Nissanka et al. 1997). The severity of this damage will affect total dry matter accumulation and allocation among various organs of the plant. However, since most crop-living mulch systems are sufficiently supported by water and nutrients, it seems that light is the most important resource for competition between living mulches and weeds.

5.1 Light

Plants grown together frequently compete primarily for solar radiation (Redfearn et al. 1999). Two components of light affect the outcome of competition: quantity and quality. The quantitative component of light (i.e., intensity and amount intercepted by a plant) determines canopy photosynthesis, whereas light quality is a driving variable of plant morphology. Both aspects of light are changed in a crop–weed competition situation when compared to the sole crop or weed canopy. Most crops and weeds attain their maximum photosynthetic rates at high levels of irradiance. In a mixed crop–weed community, mutual shading of leaves causes reduction of available photosynthetic photon flux density (PPFD), which results in reduction of photosynthetic rates (Rajcan and Swanton 2001).

In general, one of the important factors of weed suppression mechanisms of living mulch is light interception. Because plants need light to develop and living mulches are blocking sunlight reaching the weeds, weed species, especially decumbent weeds, cannot get enough light for germination and growth. Kruidhof et al. (2008) reported that weed suppression is positively correlated to early light interception by the living mulch and is sustained by the strong negative correlation between cumulative light interception and weed biomass. Similarly, according to Steinmaus et al. (2008), weed suppression was linked to light interception by the mulch cover for most weed species. Caamal-Maldonado et al. (2001) also found that canopy closure of velvetbean decreased the amount of light reaching the soil and inhibited weed growth. They reported that smooth pigweed (*Amaranthus hybridus* L.) and spiny amaranth (*Amaranthus spinosus* L.), among other weeds, were well controlled by a velvetbean living mulch.

Several studies have shown that the presence and nearness of the other vegetation influences the far red/red (FR/R) ratio received by a plant (Ballare et al. 1987; Kasperbauer 1987; Smith et al. 1990), where the FR/R ratio received by plants in a dense canopy was higher than the FR/R ratio in a sparse canopy. In fact, weeds that grow underneath or within a canopy are not only exposed to a reduced amount of PPFD, but they also receive a different quality of light than the plants grown in full sunlight. Light within the lower canopy is enriched in FR radiation (730–740 nm). This is caused by selective absorption of red light (660–670 nm) by photosynthetic pigments and FR light reflectance from and transmittance by green leaves. Chlorophyll preferentially absorbs R light and reflects FR light, thereby decreasing R/FR as sunlight moves through plant canopies. In turn, the pool of R light-absorbing phytochrome decreases relative to that of the FR light-absorbing pool, creating a signal transduction pathway that leads to an altered growth response (Rajcan and Swanton 2001). This causes the FR/R ratio of the light in the lower portion of the canopy to be higher than the FR/R ratio of the incoming light above the canopy. This may lead one to speculate that weeds growing with living mulches would have a lower root/shoot ratio than the living mulch-free condition, which would be a major disadvantage for a plant later in the season when competition for below-ground resources (such as water) may be more limiting (Rajcan and Swanton 2001). Moreover, decreased tillering may be another morphological change in weed grass species growing in this condition (Davis and Simmons 1994).

Living mulches can also change phenological development of weeds. Because FR is a determinant of photoperiod, within a dense crop canopy (FR enriched), long-day weed species may have accelerated phenological development, whereas short-day weeds (i.e., pigweed) will take longer to complete their life cycle (Huang et al., 2000). Branching and tillering are also influenced by FR light (Begonia et al. 1988; Davis and Simmons 1994; Ghersa et al. 1994; McLachlan et al. 1993). Thus, the competitive ability of weed species would be also affected by light quality. Weed seed germination is also influenced by living mulches. It is known that light can break weed seed dormancy and stimulate germination (Hartmann and Nezadal, 1990). Therefore, including living mulches in cropping systems can prevent weed seed germination by shading the soil surface.

In general, the common important traits that determine competition for light between plants are inherent to the species. Amongst these traits are growth rate and architecture of the canopy (Davis and Garcia 1983; Kropff and van Laar 1993).

5.2 Allelopathy

The term allelopathy was first introduced by Hans Molisch in 1937 and refers to chemical interactions among plants, including those mediated by microorganisms. Allelopathy can be defined as an important mechanism of plant interference mediated by the addition of plant-produced secondary products to the soil rhizosphere (Weston 2005).

In certain cropping situations, allelopathy may have the potential to be integrated into a weed management plan in order to reduce the use of synthetic herbicides as well as provide other added benefits from the allelopathic crop. Allelopathy could potentially be used for weed control by producing and releasing allelochemicals from leaves, flowers, seeds, stems, and roots of living or decomposing plant materials (Weston 1996). Allelopathic compounds can be released into the soil by a variety of mechanisms that include decomposition of residues, root exudation, and volatilization (Weston 2005). They can be broadly classified into plant phenolics and terpenoids, which show great chemical diversity and are involved in a number of metabolic and ecological processes (Sung et al. 2010). These naturally produced secondary compounds can have chemical structures as complex as synthetic herbicides; they can also have the same wide range of selectivity and control for weeds (Westra 2010).

Allelopathy is another mechanism by which living mulches may suppress weeds (Fujii 1999). However, this is difficult to separate experimentally from mechanisms relating to competition for growth resources. In some situations, the allelopathic properties of living mulches can be used to control weeds. For example, the allelopathic properties of winter rye (*Secale cereale* L.), ryegrasses (*Lolium spp*), and subterranean clover (*Trifolium subterraneum* L.) can be used to control weeds in sweet corn (*Zea mays* var "rugosa") and snap beans (*Phaseolus vulgaris* L.) (De Gregorio and Ashley 1986). Root exudation produces allelopathic compounds that are actively secreted directly into the soil rhizosphere by living root systems. The allelochemicals then move through the soil by diffusion and come into contact neighboring plants. This creates a radius effect, where proximity to the allelopathic species results in greater concentrations of the allelochemical, which, in turn, typically decreases the growth of neighboring plants (Westra 2010).

Usually, using allelopathic species as a living mulch can provide normal weed suppression traits seen for mulch, as well as slowly releasing allelochemicals from their biomass which provide further weed suppression especially for weed seedling control. Therefore, if allelopathic living mulches could be incorporated in certain cropping systems to provide weed suppression, this could reduce dependency on synthetic herbicides that are potentially hazardous to our environment.

6. Hairy vetch as a good living mulch

Hairy vetch is a well-known living mulch in the U.S. and Europe. It provides a number of advantages in agroecosystems. Benefits include: nitrogen fixation, quick addition of biomass, prevention of soil erosion and promotion of soil porosity, amelioration of microclimate, and, primarily, weed suppression owing to its allelopathic effects (Fujii 2001). It is extensively used to suppress weeds in different cropping systems. Johnson et al. (1993) observed that hairy vetch mulch completely inhibits the weeds under a no-tillage system. According to Fujii (2001), complete weed control can be achieved by direct application of the hairy vetch to the rice (*Oryza sativa* L.) paddy fields. He suggested that hairy vetch is a promising legume in abandoned paddy fields, grasslands and orchards in the central and southern parts of Japan and its inhibitory effect toward weeds was similar to that of herbicide applications.

Oliver et al. (1992) also reported that a hairy vetch (*Vicia villosa* Roth ssp. villosa) living mulch established into soybean reduced morningglory (*Ipomoea lacunosa* L.) and spotted spurge (*Euphorbia maculata* L.) biomass by about 90% and large crabgrass (*Digitaria ischaemum* Schreb) biomass by about 70 % compared to weedy controls. When grown with the hairy vetch living mulch, soybean had yields that were comparable to a conventional production system using herbicides. In another study, Mohammadi (2010) suggested that interseeding of hairy vetch as a living mulch can be used as a beneficial method to control weeds in corn fields without causing any reduction in corn yield. In other research, best results were obtained from vetch species such as hairy vetch among living mulch species for weed control because of competitive ability, high biomass, densely growing habit, and allelopathic features of these species (Moonen and Barberi 2002; Batool and Hamid 2006; Nakatsubo et al. 2008; Mohammadi 2009).

Overall, hairy vetch is a legume living mulch that suppresses weed emergence and supplies nitrogen for sustainable cropping systems (Ngouajio and Mennan 2005; Choi and Daimon 2008) and it can be proposed as a promising candidate for an integrated weed management program.

7. Living mulch vs. cover crop residue

Living mulches are generally considered to be more competitive with weeds than cover crop residue because they are actively growing and can compete efficiently for water, nutrients, and light. Dead cover crop residue does not suppress weeds as consistently as living mulches (Teasdale and Daughtry 1993; Reddy and Koger 2004). Once weed seedlings become established, cover crop residue will usually have a negligible impact on weed growth and seed production or may even stimulate these processes through conservation of soil moisture and release of nutrients (Teasdale and Daughtry 1993; Haramoto and Gallandt 2005). A living mulch competes with emerging and growing weeds for essential resources and inhibits emergence and growth more than cover crop residue does (Teasdale and Daughtry 1993; Reddy and Koger 2004).

In one study, a chemically stunted stand of crownvetch gave better weed control than dead rye mulch (Hartwig 1989). Teasdale and Daughtry (1993) found that weed suppression by live hairy vetch was more than that by paraquat desiccated cover crop residues. Therefore, weed control can be maximized by keeping hairy vetch live for a longer period rather than killing / desiccating. Living plant tissue of wheat (*Triticum* sp.), crimson clover (*Trifolium incarnatum* L.), subterranean clover and rye inhibited the emergence of weeds like ivyleaf morning glory (*Ipomoea hederacea* L.) and redroot pigweed (*Amaranthus retroflexus* L.) (Lehman and Blum 1997). However, if these were used after desiccation with glyphosate, only wheat and crimson clover were inhibitory. Likewise, subterranean clover cover crops, when used as living mulch under field conditions, can efficiently control weeds such as fall panicum (*Panicum dichotomiflorusm* Michx) and ivyleaf morning glory without affecting the yield of corn (Enache and Ilnicki 1990; Ilnicki and Enache 1992).

Several requirements for breaking dormancy and promoting germination of weed seeds in soils (light with a high red-to-far red ratio and high daily soil temperature amplitude) are reduced more by living mulches than by desiccated residue (Teasdale and Daughtry 1993). A living mulch absorbs red light and will reduce the red/far-red ratio sufficiently to inhibit phytochrome-mediated seed germination, whereas cover crop residue has a minimal effect on this ratio (Teasdale and Daughtry 1993).

Enache and Ilnicki (1990) reported that weed biomass was reduced 53 to 94 percent by subterranean clover living mulch whereas weed biomass in desiccated rye mulch ranged from an 11 percent decrease to a 76 percent increase compared to a no-mulch control. In another study, a live hairy vetch cover crop was more effective than a desiccated cover crop in suppressing weed emergence during the first four weeks and throughout the season (Teasdale et al. 1991). In addition, if growth suppression is sufficient, a living mulch can inhibit weed seed production (Brainard and Bellinder 2004; Brennan and Smith 2005). Weed seed predation at the soil surface was also higher when living mulch vegetation was present (Davis and Liebman 2003; Gallandt et al. 2005), suggesting a role for living mulches in enhancing weed seed mortality.

Generally, it can be concluded that living mulches will suppress weeds more completely and at more phases of the weed life cycle than will cover crop residue. The inhibitory effect of typical cover crop residue or living mulch on weeds at various life cycle stages has been shown in Table 3.

8. Competition between living mulch and main crop

Although living mulches can efficiently suppress weeds, they may compete for nutrients and water with the main crop (Echtenkamp and Moomaw 1989) which can reduce yields. For example, Elkins et al. (1983) examined the use of tall fescue (*Festuca arundinacea* Schreb), smooth bromegrass (*Bromus inermis* Leyss), and orchargrass (*Dactylis glomerata* L.) as living mulches. They found corn yield was reduced 5% to 10% at the end of the harvest. Regnier and Janke (1990) indicated that the majority of previously conducted studies showed that the species, when selected as living mulches do not suppress weeds selectively, but suppress the crop as well; therefore, living mulches must be managed carefully to reduce their competition with the crop. In that regard, Jeranyama et al. (1998) found a reduction of 13 to 18% in grain yield when corn was intercropped with legumes. Norquidst and Wicks (1974) found corn dry matter yield to be reduced by up to 47% and grain yield by up to 31% when alfalfa was interseeded at the time of corn establishment. Hoffman et al. (1993) observed a corn reduction of over 76% in corn grown with untreated hairy vetch.

Weed life cycle stage	Cover crop residue	Living mulch
Germination	Moderate	High
Emergence/establishment	Moderate	High
Growth	Low	High
Seed production	Low	Moderate
Seed survival	None?[a]	Moderate?[a]
Perennial structure survival	None?[a]	Low-moderate?[a,b]

[a] More research is needed to provide definitive estimates of cover crop effects on these processes.
[b] When living mulches are combined with other practices such as soil disturbance or mowing, perennial structure survival may be more effectively reduced.

Table 3. Potential impact of typical cover crop residue or living mulch on inhibition of weeds at various life cycle stages (from Teasdale et al. 2007).

Typically, a living mulch that is competitive enough to suppress weeds will also suppress crop growth and yield. Much of the research with living mulches has focused on documenting and alleviating this problem (Liebman and Staver 2001; Teasdale, 1998). Many studies in the North Central U.S. on legume interseeding in established corn stands report grain yield losses that are attributed to moisture stress (Kurtz et al. 1952; Pendleton et al. 1957), N deficiency (Scott et al. 1987; Triplett 1962), and reduced corn populations associated with wider row spacing (Schaller and Larson 1955; Stringfield and Thatcher 1951). Marks (1993) also suggested that reduced growth of the main crop may be due to competition for water or some other limited resource, or the mulch may be having an allelopathic effect.

De Haan et al. (1997) used burr medic (*Medicago polymorpha* L.) and snail medic [*Medicago scutellata* (L.) Mill.] as living mulches in corn and found that, although both medics suppressed weeds, corn and medics competed strongly for resources. Consequently, medic living mulches significantly reduced corn grain yields. The reduction was due to competition for nutrients or moisture when medic and corn were planted at the same time. Yield loss in transplanted cabbage due to competition with the living mulch for light or moisture was also recorded by Bottenberg et al. (1997).

When the growth of a living mulch is not restricted, or when soil moisture is inadequate, even a relatively vigorous crop like potato may suffer competition and loss of yield (Rajalahti and Bellinder 1996). Generally, without irrigation, it becomes more challenging to implement a living mulch system. However, there are successful examples of annual or biennial living mulches established after emergence of the main crop, which gives the main crop a competitive advantage (Scott et al. 1987; Wall et al. 1991). If living mulches are established before or after the main crop is planted, competition of the living mulch for water may reduce crop yields (Echtenkamp and Moomaw 1989; Eberlein et al. 1992; Masiunas et al. 1997; Teasdale et al. 2000). Thus, it can be concluded that living mulches can severely compete with the main crop for water which is particularly problematic during a dry period. In one study, corn yields were not negatively affected by competition from the crownvetch, birdsfoot trefoil (*Lotus corniculatus* L.), and flatpea (*Lathyrus sylvestris* L.) living mulches in years with adequate precipitation. However, in a year with very low rainfall in July and August, crownvetch and birdsfoot trefoil reduced corn yields (Duiker and Hartwig 2004).

In general, although legume living mulches compete weakly with cereals for light, N, phosphorus (P), and K, they can compete strongly for water. If water stress is eliminated by irrigation, living mulches of legumes rarely reduce and sometimes increase main crop yields (Grubinger and Minotti 1990; Fischer and Burrill 1993; Costello 1994).

9. The ways to prevent or reduce the competition between living mulch and main crop

A serious problem in living mulch cropping systems is reduced main crop yield because of competition. Management of living mulches becomes critical to reduce competition with the main crop for resources while allowing the mulch to grow sufficiently to reap potential benefits. Different ways have been suggested to overcome this problem in such cropping systems. One of them is the selection of suitable living mulch species and the others have been employed to suppress the living mulch, such as tillage, mowing, and herbicides (Grubinger and Minotti 1990; Fischer and Burrill 1993; Costello 1994; Martin et al. 1999; Zemenchik et al. 2000).

9.1 Selection of suitable species

It is important to make the correct choice of a living mulch (Ingels et al. 1994). According to Ilnicki and Enache (1992), to avoid competition with the main crop in the subterranean clover cropping system, it is essential to use species and cultivars which have a low canopy height and terminate vegetative growth early in the summer. Greater potential benefits might be expected from living mulches with a very different active growth period than the main crop. For example, kura clover does not produce abundant dry matter during dry periods of the growing season and should therefore compete less than other perennial legumes with corn for limited resources, especially water (Zemenchik et al. 2000). Kura clover and corn in the living mulch systems were more compatible after tasselling because the species differed greatly in stature and corn had sequestered much of the resources necessary to complete its life cycle. This ecological differentiation is a necessary condition for coexistence according to the competitive exclusion principle (Hardin 1960).

Another approach suggested by Ilnicki and Enache (1992) was to use winter annual legumes, e.g., subterranean clover, as a living mulch. Winter annual legumes sown in late summer grow vegetatively during autumn, become dormant in winter, and resume vegetative growth the following spring. Later in the spring or early summer the plant flowers, senesces, and dies. Because of this unique life cycle, a main crop transplanted into the senescencing mulch would be able to use all available water and nutrients.

Moynihan et al. (1996) reported that black medic was found to be the least competitive medic species when it was intercropped with barley as a living mulch. In another study, black medic did not significantly reduce corn yields compared with the medic and weed-free check, whereas all other species caused significant yield reductions. Therefore, it is suited to this system because it does not grow aggressively early in the year when it could reduce corn yield.

Newenhouse and Dana (1989) also evaluated different grass living mulches for strawberries (*Fragaria* sp.) and found perennial ryegrass (*Lolium perenne* L.) was best because it covered the ground quickly but did not spread into the crop rows. In raspberries (*Rubus* sp.), a white clover (*Trifolium repens* L.) living mulch did not affect the crop but perennial ryegrass reduced berry yield (Freyman 1989).

According to Akanvou et al. (2001), slow-growing species with longer duration such as *Stylosanthes hamata* and *Aeschynomene histrix* are expected to be less competitive and therefore appropriate for early establishment in rice-legume intercropping systems. Intercropping research has shown that most legumes do not compete strongly with cereals for light, N, P, and K, whereas they compete equally for water (Ofori and Stern 1987; Vandermeer 1990). The low stature of most legumes and their horizontally positioned leaves reduce competition for light with tall, erect cereals. Since many legumes are C_3 crops with low light saturation points and low temperature optima, one might expect these legumes to complement a C_4 crop such as corn that has a high light saturation point and high temperature optimum (Ofori and Stern 1987). Instead of competing for N, legumes may instead contribute N to the main crop (Fujita et al. 1992). Because of their different root systems (less fibrous and often having a taproot), competition for the immobile nutrients P and K can be expected to be limited (Ofori and Stern 1987; Vandermeer 1990). Legumes are therefore promising candidates for living mulches in agroecosystems.

Generally, the competitive ability is an obvious characteristic determining the suitability of a plant species as a living mulch. For example, tall and vigorously growing legumes with relatively large leaves and rapid leaf expansion might be detrimental to the associated crop, whereas poorly competing species will be out-competed and will therefore contribute little to improving soil fertility (Akanvou et al. 2001).

9.2 Application of appropriate practices

Appropriate management is essential to avoid or decrease the interspecific competition between intercropped species. Several approaches have been used to reduce competition between the living mulch and main crop species without eliminating the desirable attributes and benefits of the living mulch.

The classical attempts to reduce competition in living mulch systems have focused on chemical or mechanical suppression of mulch growth or screening for less competitive living mulches. Reducing interference between a white clover living mulch and sweet corn (*Zea mays* L. var. *saccharata*) by chemical suppression or mechanical suppression has been reported by Vrabel (1983) and by Grubinger and Minotti (1990), respectively. Reduced interference by mechanical suppression of white clover and subterranean clover living mulches in white cabbage (*Brassica oleracea* var. *capitata* L.) is also reported by Brandsæter et al. (1998).

Timely mowing of a clover (*Trifolium spp.*) living mulch prevented the competition in transplanted broccoli (*Brassica oleracea* L.) (Costello and Altieri 1994). Ilnicki and Enache

(1992) also found that mowing of a subterranean clover mulch was necessary to reduce early competition when sweet corn, tomato (*Lycopersicon esculentum* Mill.) and cabbage crops were planted into it. Mulongoy and Akobundo (1990) proposed the use of growth retardants to reduce growth of the associated legumes in maize. Werner (1988) investigated the influence of different living mulch species on weed density and diversity. Weed numbers were reduced and maize yield was not affected where growth of the living mulch was reduced by cutting or flaming treatments.

Another way to avoid or decrease the competition in such systems is to intercrop a main crop and a living mulch with a synchronized onset of maximum vegetative growth. This synchronization of living mulch and main crop could be achieved in different ways (Brandsæter and Netland 1999). Muller-Scharer and Potter (1991) concluded that living mulches should be seeded to emerge in the middle of the vegetation period of the main crop. De Haan et al. (1994) have studied the opposite way to avoid interference problems in living mulch systems in the north central region of the U.S.. They tried to develop a spring-seeded living mulch that had been selected for its ability to suppress weeds without affecting crop yield. This living mulch flowered 3 weeks after emergence and began senescence 5 weeks after emergence.

Shifting the relative sowing dates of the various intercropped components in a crop-living mulch system is an important means to ensure a better use of available resources and to minimize yield loss of the main crop (Midmore 1993). Usually, delaying the sowing time of living mulches might reduce the interaction effects. For example, velvetbean planted as living mulch 20 days after corn reduced weed biomass by 68% with no negative effects on corn yield (Caamal-Maldonado et al. 2001). Corn grain yield was not reduced when living mulch seeding was delayed until the corn was 15 to 30 cm in height (Scott et al. 1987), suggesting that yield can be maintained by delaying the seeding date of the living mulch. In another study, annual medics interseeded several weeks after corn planting did not affect corn yield (De Haan et al. 1997). Moreover, delaying the planting time of the main crop until senescing of living mulch might also decrease the interspecific competition. For example, competition was not a problem when dwarf beans (*Phaseolus vulgaris* L.) were planted into a clover mulch as it began senescing (Ilnicki and Enache 1992).

In general, the efficient management approaches to prevent or reduce the competition between living mulches and a main crop include:

1. Using low-growing living mulch that competes primarily for light. In this case, as long as the living mulch becomes established before the weeds, it would maintain weed suppression by excluding light but would not impact taller growing crops and would not compete with the crop excessively for soil resources such as water and nutrients.
2. Timely planting the living mulch so that the time of peak growth of the living mulch does not coincide with the critical period during which competition would have the greatest impact on main crop yield.
3. Reducing crop row spacing and/or increase crop population to enhance the competitiveness of the main crop relative to the living mulch.
4. Providing supplemental water and nitrogen to compensate for resources used by living mulch plants. Usually, soil moisture depletion by living mulches will become the primary management consideration in those areas of the world where soil moisture is

the limiting factor in crop production. Therefore, preparation of sufficient water for these cropping systems is very essential.

5. Suppressing the living mulch so as to reduce its competitiveness with the crop using the following methods:

a. A broadcast application of an herbicide at a rate that is suppressive but not lethal.

b. A banded application of a herbicide to kill the living mulch in the crop row so as to reduce competition within the row area but permit weed suppression by the living mulch between rows.

c. Strip tillage to provide suitable planting conditions without competition within the crop row but to permit weed suppression by the living mulch between rows.

d. Timely mowing to reduce the height and vigour of the living mulch (Teasdale 2003).

It can be concluded that although living mulches are efficient tools to suppress weeds in cropping systems, but an appropriate management program is very essential to reduce the competition with the main crop for environmental resources and enhance the potential benefits of living mulch such as weed suppressing ability.

10. References

Akanvou, R., L. Bastiaans, M. J. Kropff, J. Goudriaan and M. Becker. 2001. Characterization of Growth, Nitrogen Accumulation and Competitive Ability of Six Tropical Legumes for Potential Use in Intercropping Systems. Journal of Agronomy and Crop Science 187, 111-120.

Akemo, M.C., Regnier, E.E. and Bennett, M.A. 2000. Weed suppression in spring-sown rye-pea cover crop mixes. Weed Technology 14, 545-549.

Akobundo, I. O., 1993. Integrated weed management techniques to reduce soil degradation. In, Australian Crop Protection Council (ACPC) (eds), Proc. 1st Weed Control Congress, 1992, Melbourne, Australia, pp. 278-284. International Weed Science Society, Oregon State University, Corvallis.

Akobundu, I.O., Udensi, U.E. and Chikoye, D. 2000. Velvetbean suppresses speargrass and increases maize yield. International Journal of Pest Management 46, 103-108.

Ballare, C.L., R.A. Sanchez, A.L. Scopel, J.J. Casal and C.M. Ghersa. 1987. Early detection of neighbour plants by phytochrome perception of spectral changes in reflected sunlight. Plant Cell Environ. 10, 551–557.

Batool S., Hamid R. 2006. Effect of cover crops mulch on weed control in orchards. Pakistan J. Weed Sci. Res., 12, 347-352.

Beard, J. 1973. Turfgrass, Science and Culture Prentice-Hall Inc. Englewood Cliffs, NJ.

Begonia, G.B., R.J. Aldrich and C.J. Nelson. 1988. Effects of simulated weed shade on soybean photosynthesis, biomass partitioning and axillary bud development. Photosynthetica, 22, 309–319.

Bellinder, R.R., G. Gummeson and C. Karlson. 1994. Percentage-driven government mandates for pesticide reduction, The Swedish model. Weed Technol. 8, 350-359.

Blackshaw, R.E., Moyer, J .R., Doram, R.C. and Boswell, A.L. 2001. Yellow sweetclover, green manure, and its residues effectively suppress weeds during fallow. Weed Science 49, 406-413.

Blanchart E., Villenave C., Viallatoux A., Barthes B., Girardin C., Azontonde A., Feller C. 2006. Long-term effect of a legume cover crop (Mucuna pruriens var. utilis) on the communities of soil macrofauna and nematofauna, under maize cultivation, in southern Benin. Eur. J. Soil Biol., 42, 136-144.

Bollero G. A. and Bullock D. G. 1994. Cover cropping systems for the central corn belt. J. Product. Agric. 7, 55–58.

Bond W. and Grundy A. C. 2001. Non-chemical weed management in organic farming systems. Weed Res., 41, 383-405.

Bottenberg H, Masiunas J, Eastman, Eastburn D. 1997. Yield and quality constraints of cabbage planted in rye mulch. Biological Agriculture and Horticulture 14, 323-342.

Brainard, D.C. and Bellinder, R.R. 2004. Weed suppression in a broccoli-winter rye intercropping system. Weed Science 52, 281-290.

Brandsæter L. O. and J. Netland. 1999. Winter Annual Legumes for Use as Cover Crops in Row Crops in Northern Regions, I. Field Experiments. Crop Science 39,1369–1379.

Brandsæter, L.O., J. Netland and R. Meadow. 1998. Yield, weeds, pests and soil nitrogen in a white cabbage-living mulch system. Biol. Agric. Hortic. 16,291–309.

Brennan, E. B. and Smith, R. E 2005. Winter cover crop growth and weed suppression on the central coast of California. Weed Technology 19, 1017-1024.

Bridges, D. C. 1994. Impact of weed on human endeavors. Weed Technology 8, 392-395.

Brown R. W., Varvel G. E. amd Shapiro C. A. 1993. Residual effects of interseeded hairy vetch on soil nitrate–nitrogen levels. Soil Sci. Soc. Am. J. 57, 121–124.

Bruce R.R., Langdale G.W., West L.T. and Miller W.P. 1992. Soil surface modification by biomass inputs affecting rainfall infiltration. Soil Sci. Soc. Am. J. 56, 1614-1620.

Buhler, D. D., Kohler, K. A., and Foster, M. S. 2001. Corn, soybean, and weed responses to spring-seeded smother plants. J. Sustain. Agric. 18, 63-79.

Caamal-Maldonado, J. A., J. J. Jime´nez-Osornio, A. Torres-Barraga´n, and A. L. Anaya. 2001. The use of allelopathic legume cover and mulch species for weed control in cropping systems. Agron. J. 93,27–36.

Calkins, J. B. and B. T. Swanson. 1995. Comparison of conventional and alternative nursery weed management strategies. Weed Technol. 9,761-767.

Choi, B. S. and Daimon, H. 2008. Effect of hairy vetch incorporated as green manure on growth and N uptake of sorghum crop. Plant Prod. Sci. 11, 211-216.

Corak, S. J., W. W. Frye and M. S. Smith. 1991. Legume mulch and nitrogen fertilizer effects on soil water and corn production. Soil Sci. Am. J. 55,1395-1400.

Costello M. J. and Altieri M. A. 1994. Living mulches suppress aphids in broccoli. California Agriculture 48 (4), 24-28.

Costello, M. J. 1994. Broccoli growth, yield and level of aphid infestation in leguminous living mulches. Biol. Agric. Hortic. 10,207–222.

Creamer, N. G. and Baldwin, K. R. 2000. An evaluation of summer cover crops for use in vegetable production systems in north Carolina. Hort. Science 35, 600-603.

Cripps, R. W. and H. K. Bates. 1993. Effects of cover crops on soil erosion in nursery aisles. J. Environ. Hortic. 11,5-8.

Cromar, H. E., Murphy, S. D., and Swanton, C. J. 1999. Influence of tillage and crop residue on post dispersal predation of weed seeds. Weed Science, 47, 184-194.

Dabney S. M., Murphree C. E. and Meyer L. D. 1993. Tillage, row spacing, and cultivation affect erosion from soybean cropland. Trans. Am. Soc. Agric. Eng. 36, 87–94.

Davis, A. S. and Liebman, M. 2003. Cropping system effects on giant foxtail demography. I. Green manure and tillage timing. Weed Science 51, 919-929.

Davis, J. H. C., and S. Garcia, 1983. Competitive ability and growth habit of intermediate beans and maize for intercropping. Field Crops Res. 6, 59-75.

Davis, M. H. and S. R. Simmons. 1994. Tillering response of barley to shifts in light quality caused by neighboring plants. Crop Sci. 34, 1604–1610.

De Gregorio R. E. and R.A. Ashley. 1986. Screening living mulches/ cover crops for no-till snap beans. Proc. Northeast. Weed Sci. Soc. 40,87-91.

De Haan R. L., D. L. Wyse, N. J. Ehlke, B. D. Maxwell, and D. H. Putnam. 1994. Simulation of spring-seeded smoother plants for weed control in corn (Zea mays). Weed Sci. 42,35–43.

De Haan, R. L., C. C. Sheaffer and D. K. Barnes. 1997. Effect of annual medic smother plants on weed control and yield in corn. Agron. J. 89, 813-821.

Decker A. M., Clark A. J., Meisinger J. J., Mulford F. R. and McIntosh M. S. 1994. Legume cover crop contribution to no till corn. Agron. J. 86, 126–135.

Duiker, S. W. and N. L. Hartwig. 2004. Living Mulches of Legumes in Imidazolinone-Resistant Corn. Agronomy Journal 96, 1021–1028.

Duke, S. O. 1999. Weed management, Implications of herbicide resistant crops. Paper presented at the "Workshop on Ecological effects of Pest Resistance Genes in Managed Ecosystems" in Bethesda, MD, January 31-February 3, 1999.

Eberlein, C. V., C. C. Sheaffer, and V. F. Oliveira. 1992. Corn growth and yield in an alfalfa living mulch system. J. Prod. Agric. 5,332–339.

Echtenkamp, G. W., and R. S. Moomaw. 1989. No-till corn production in a living mulch system. Weed Technol. 3,261–266.

Elkins, D., D. Frederking, R. Marashi, and B. McVay. 1983. Living mulch for no-till corn and soybeans. J. soil Water Conserv, 38, 431-433.

Enache, A. J., and R. D. Ilnicki. 1990. Weed control by subterranean clover (Trifolium subterraneum) used as a living mulch. Weed Technol. 4,534–538.

Exner D. N. and Cruse R. M. 1993. Interseeded forage legume potential as a winter ground cover, nitrogen source, and competitor. J. Prod. Agric. 6, 226–231.

Facelli, J. M., and S. T. A. Pickett. 1991. Plant litter, Its dynamics and effects on plant community structure. Bot. Rev. 57,1–32.

Favero, C., Jucksch, I. and Alvarenga, R. C. 2001. Modifications in the population of spontaneous plants in the presence of green manure. Pesquisa Agropecuaria Brasileira 36, 1355-1362.

Fischer, A., and L. Burrill. 1993. Managing interference in a sweet corn–white clover living mulch system. Am. J. Alternative Agric. 8(2),51–56.

Freyman S. 1989. Living mulch ground covers for weed control between raspberry rows. Acta Horticulturae 262, 349-356.

Fujii Y. 2001. Screening and future exploitation of allelopathic plants' alternative herbicides with special reference to hairy vetch. J. Crop Prod. 4, 257–275.

Fujii, Y. 1999. Allelopathy of hairy vetch and Macuna; their application for sustainable agriculture. pp.289-300. In C.H. Chou et al. Biodiversity and Allelopathy from Organisms to Ecosystems in the Pacific. Academia Sinica, Taipei.

Fujita, K., K. G. Ofosu-Budu, and S. Ogata. 1992. Biological nitrogen fixation in mixed legume–cereal cropping systems. Plant Soil 141, 155–175.

Fuller L. G., Gon T. B. and Oscarson D. W. 1995. Cultivation effects on dispersible clay of soil aggregates. Can. J. Soil Sci 75, 101–107.

Gallagher, R. S., E. C. M. Fernandes, and E. L. McCallie. 1999. Weed management through short-term improved fallows in tropical agroecosystems. Agrofor. Syst. 47,197–221.

Gallandt, E. R., Molloy, T., Lynch, R. P. and Drummond, F. A. 2005. Effect of cover-cropping systems on invertebrate seed predation. Weed Science 53, 69-76.

Ghersa, C. M., M. A. Martinez-Ghersa, J. J. Casal, M. Kaufman, M. L. Roush and V. A. Deregibus. 1994. Effect of light on winter wheat (*Triticum aestivum*) and Italian ryegrass (*Lolium multiflorum*) competition. Weed Technology 8, 37–45.

Ghosheh, H. Z., E. Y. Bsoul, A. Y. Abdullah. 2004. Utilization of alfalfa (*Medicago sativa* L.) as a smother crop in field corn (*Zea mays* L.). Journal of Sustainable Agriculture, 25,5-17.

Grafton-Cardwell E. E., Ouyang Y., Bugg R. L. 1999. Leguminous cover crops to enhance population development of *Euseius tularensis* (acari, phytoseiidae) in citrus. Biol. Control, 16(1), 73-80.

Grimmer, O. P. and Masiunas, J. B. 2004. Evaluation of winter-killed cover crops preceding snap pea. Hort. Technology 14, 349-355.

Grubinger, V. P., and P. L. Minotti. 1990. Managing white clover living mulch for sweet corn production with partial rototilling. Am. J. Alternative Agric. 5, 4–12.

Hallberg G. R. 1989. Pesticide pollution of ground water in the humid United States. Agric. Ecosyst. Environ. 26, 299–367.

Hanson, J. C., E. Lichtenberg, A. M. Decker and A. J. Clark. 1993. Profitability of no-tillage corn following a hairy vetch cover crop. J. Prod. Agric. 6, 432-437.

Haramoto, E. R. and Gallandt, E. R. 2005. Brassica cover cropping. II. Effects on growth and interference of green bean and redroot pigweed. Weed Science 53, 702-708.

Hardin, G. 1960. The competitive exclusion principle. Science (Washington, DC) 131, 1292-1297.

Harris R. F., Chesters G. and Allen O. N. 1966. Dynamics of soil aggregation. Adv. Agron. 18, 107-169.

Hartmann K. M. and Nezadal W. 1990. Photocontrol of weeds without herbicides. Naturwissenschaften 77, 158-163.

Hartwig, N. L. 1989. Influence of a crownvetch living mulch on dandelion invasion in corn. Proc. Northeast Weed Sci. Soc. 43, 25.

Hoffman, M. L., E. E. Regnier and J. Cardina. 1993. Weed and corn (*Zea mays*) responses to a hairy vetch (*Vicia villosa*) cover crop. Weed Technol. 7, 594-599.

Holderbaum J. F., Decker A. M., Meisinger J. J., Mulford F. R. and Vough L. R. 1990. Fall-seeded legume cover crops for no-tillage corn in the humid east. Agron. J. 82, 117–124.

Holt J. S. and LeBaron H. N. 1990. Significance and distribution of herbicide resistance. Weed Technol. 4, 141-149.

Hoyt G. D., Hargrove W. L. 1986. Legume cover crops for improving crop and soil management in the southern united states. Hort. Sci., 21, 397-402.

Huang, J. Z., A. Shrestha, M. Tollenaar, W. Deen, H. Rahimian and C. J. Swanton. 2000. Effects of photoperiod on the phenological development of redroot pigweed (*Amaranthus retroflexus* L.). Can. J. Plant Sci. 80, 929–938.

Ilnicki, R. D., and A. J. Enache. 1992. Subterranean clover living mulch, An alternative method of weed control. Agric., Ecosys. Environ. 40, 249–264.

Ingels C., Van Horn M., Bugg R. L. and Miller P. R. 1994. Selecting the right cover crop gives multiple benefits. California Agriculture 48, 43-48.

Jeranyama, P., O. B. Hesterman and C. C. Sheaffer. 1998. Medic planting date effect on dry matter and nitrogen accumulation when clear-seed or intercropped with corn. Agron. J. 90, 601-606.

Johnson G. A., Defelice M. S. and Helsel Z. R. 1993. Cover crop management and weed control in corn (Zea mays).Weed Technol. 7, 425–430.

Kasperbauer, M. J. 1987. Far-red light reflection from green leaves and effects on phytochrome-mediated assimilate partitioning under field conditions. Plant Physiol. 85, 350-354.

Kitis Y. E., O. Koloren and F. N. Uygur. 2011. Evaluation of common vetch (Vicia sativa L.) as living mulch for ecological weed control in citrus orchards. African Journal of Agricultural Research 6, 1257-1264.

Kropff, J. M. and H. Van Laar. 1993. Modelling Crop-Weed Interactions. Cab International Publishers, Wallingford, UK.

Kruidhof, H. M., Bastiaans L. and Kropff M. J. 2008. Ecological weed management by cover cropping, effects on weed growth in autumn and weed establishment in spring. Weed Res., 48, 492-502.

Kurtz, T., S. W. Melsted, and R. H. Bray. 1952. Importance of nitrogen and water in reducing competition between intercrops and corn. Agron. J. 44, 13–17.

Latif M. A., Mehuys G. R., Mackenzie A. F., Alli I. and Faris M. A. 1992. Effect of legumes on soil physical quality in a maize crop. Plant and Soil 140, 15–23.

Leather, G. R. 1983. Sunflowers (Helianthus annus) are allelopathic to weeds. Weed Science 31, 37-42.

Lehman, M. E. and Blum, U. 1997. Cover crop debris effects on weed emergence as modified by environmental factors. Allelopathy J. 4, 69–88.

Liebel, R. A. and Worsham, A. D. 1983. Inhibition of pitted morning glory (Ipomoea lacunosa L.) and certain weed species by phytotoxic components of wheat straw. Journal of Chemical Ecology 9, 1027-1043.

Liebman M. and Davis A. S. 2000. Integration of soil, crop, and weed management in low-external-input farming systems. Weed Research 40, 27–47.

Liebman, M. and Staver, C. P. 2001. Crop diversification for weed management. pp. 322-374.In M. Liebman et al. Ecological Management of Agricultural Weeds. New York. Cambridge University Press.

Linares, J. C., J. M. S. Scholberg, C. A. Chase, R. M. McSorley, K. J. Boote, and J. J. Ferguson. 2008. Cover crop management and cover crop weed index as an indicator of weed suppression in organic citrus orchards. Hort. Science 43, 27–34.

Lybecker D. W., Schweizer E. E. and King R. P. 1988. Economic analysis of four weed management systems. Weed Sci. 36, 846–849.

Malik, R. K., Green T. H., Brown G. F. and Mays D. 2000. Use of cover crops in short rotation hardwood plantations to control erosion. Biomass and Bioenergy 18, 479-487.

Marks, M. J. 1993. Preliminary results of an evaluation of alternatives to the use of herbicides in orchards. Proceedings Brighton Crop Protection Conference - Weeds, Brighton, UK, 461-466.

Martin, R. C., P. R. Greyson, and R. Gordon. 1999. Competition between corn and a living mulch. Can. J. Plant Sci. 79, 579–586.

Masiunas, J. B., D. M. Eastburn, V. N. Mwaja, and C. E. Eastman. 1997. The impact of living and cover crop mulch systems on pests and yields of snap beans and cabbage. J. Sustainable Agric. 9, 61–89.

Mayer J. B. and Hartwig N. L. 1986. Corn yield in crown vetch relative to dead mulch, Proceedings of the Annual Meeting of the Northeastern. Weed Sci. Soc. 40, 34–35.

McLachlan, S. M., M. Tollenaar, C. J. Swanton and S. F. Weise. 1993. Effect of corn-induced shading on dry matter accumulation, distribution, and architecture of redroot pigweed (*Amaranthus retroflexus*). Weed Science 41, 568–573.

McVay K. A., Radcliffe D .E. and Hargrove W. L. 1989. Winter legume effects on soil properties and nitrogen fertilizer requirements. Soil Sci. Soc. Am. J. 53, 1856–1862.

Meschede, D. K., Ferreira, A. B. and Ribeiro Junior, C. C. 2007. Evaluation of weed suppression using different crop covers under Brazilian cerrado soil conditions. Planta Daninha, 25, 465-471.

Midmore, D. J. 1993. Agronomic modification of resource use and intercrop productivity. Field Crop Res. 34, 357-380.

Minnesota Agricultural Statistics, 2001, 2000 Soybean Chemical Usage. Minnesota Agricultural Statistical Service, St. Paul, MN.

Mohammadi, G. R. 2009. The effects of legumes as living mulches on weed control and plant traits of corn (*Zea mays* L.). Korean Journal of Weed Science 29, 222-228.

Mohammadi, G. R. 2010. Weed control in corn (*Zea mays* L.) by hairy vetch (*Vicia villosa* L.) interseeded at different rates and times. Weed Biology and Management 10, 25-32.

Moonen, A. C. and Barberi P. 2002. A system-oriented approach to the study of weed suppression by cover crops and their residues. In, Proceeding of the 5th EWRS Workshop on Physical Weed Control, pp. 184-191.

Mortensen, D. A., Bastiaans, L. and Sattin, M. 2000. The role of ecology in the development of weed management systems, an outlook. Weed Research 40, 49-62.

Moynihan, J. M., S. R. Simmons, and C. C. Sheaffer. 1996. Intercropping annual medic with conventional height and semidwarf barley grown for grain. Agron. J. 88, 823–828.

Muller-Scharer, H., and C. A. Potter. 1991. Cover plants in field grown vegetables, Prospects and limitations. p. 599–604. *In* Proceedings 1991 British Crop Protection Conference, Weeds. The British Crop Protection Council, Brighton, England.

Mulongoy, K., and I. O. Akobundo. 1990. Agronomic and economic benefits of nitrogen contributed by legumes in live-mulch and alley cropping systems. In, P. M. Gressho., L. R. Roths, G. Stacey, and W. E. Newton (eds), Nitrogen Fixation, Achievements and Objective, pp. 625-632. Chapman & Hall, New York.

Nakatsubo, A., Kato, W., Tanaka, K. and Sugiura, T. 2008. Effect of living mulch cultivation with spring sowing at the forage maize (*Zea mays* L.) field. Japanese J. Grassland Sci., 54, 31-39.

National Research Council 1989. *Alternative Agriculture.* National Academy Press,Washington, DC.

Newenhouse A. C. and Dana M. N. 1989. Grass living mulch for strawberries. Journal of the American Society for Horticultural Science 114, 859-862.

Ngouajio, N. and Mennan, H. 2005. Weed populations and pickling cucumber (*Cucumis sativus*) yield under summer and winter cover crop systems. Crop Prot. 24, 521-526.

Nissanka, S. P., M. A. Dixon and M. Tollenaar. 1997. Canopy gas exchange response to moisture stress in old and new maize hybrids. Crop Science 37, 172-181.

Nordquist, P. T. and G. A. Wicks. 1974. Establishment methods for alfalfa in irrigated corn. Agron. J. 66, 377-380.

Ofori, F., and W. R. Stern. 1987. Cereal–legume intercropping systems. Adv. Agron. 41, 41-90.

Oliver, L. R., T. E. Klingaman, and I. L. Eldridge. 1992. Influence of hairy vetch on weed control and soybean yield. Arkansas Farm Res. Arkansas Exp. Stn 41, 8-9.

Overland, L. 1966. Role of allelopathic substances in smother crop barley. Am. J. Bot. 53, 423-432.

Palada, M. C., S. Ganser, R. Hofstetter, B. Volak and M. Culik. 1982. Association of interseeded cover crops and annual row crops in year-round cropping systems. In, (ed. By Lockeretz W.). The Fourth IFOAM Conference, Cambridge, MA, USA, 193-213.

Peachey, R. E., William, R. D. and Mallory-Smith, C. 2004. Effect of no-till or conventional planting and cover crop residues on weed emergence in vegetable row crops. Weed Technology 18, 1023-1030.

Pendleton, J. W., J. A. Jackobs, F. W. Slife, and H. P. Bateman. 1957. Establishing legumes in corn. Agron. J. 49, 44-48.

Phatak, S. C. 1992. An integrated sustainable vegetable production system. Hort. Sci. 27, 738-741.

Putnam, A.R. and DeFrank, J. 1983. Use of phytotoxic plant residues for selective weed control. Crop Protection 2, 173-181.

Rajalahti, R. M. and Bellinder, R. R. 1996. Potential of interseeded legume and cereal cover crops to control weeds in potatoes. Xe Colloque International sur la Biologie des Mauvaises Herbes, Dijon, France, 349-354.

Rajalahti, R. M., Bellinder, R. R. and Hoffman, M. P. 1999. Time of hilling and interseeding affects weed control and potato yield. Weed Science 47, 215-225.

Rajcan, I. and Swanton C. J. 2001. Understanding maize–weed competition, resource competition, light quality and the whole plant. Field Crops Research. 71, 139-150.

Reddy, K. N. and Koger, C. H. 2004. Live and killed hairy vetch cover crop effects on weeds and yield in glyphosate-resistant corn. Weed Technology 18, 835-840.

Redfearn, D. D., D. R. Buxton and T. E. Devine. 1999. Sorghum intercropping effects on yield, morphology and quality of forage soybean. Crop Science 39, 1380-1384.

Regnier, E. E. and Janke, R. R. 1990. Evolving strategies in managing weeds. pp. 174-202. In C. A. Edwards, R. Lal, P. Madden, R. H. Miller and G. House (eds.) *Sustainable Agricultural Systems*. Soil and Water Conservation Society, Ames, Iowa, USA.

Richie, S. W., J. J. Hanway, and G. O. Benson. 1993. How a corn plant develops. Spec. Rep. 48. Iowa State Univ. Coop. Ext. Ser., Ames.

Ross, S. M., King, J. R., Izaurralde, R. C. and O'Donovan, J. T. 2001. Weed suppression by seven clover species. Agronomy Journal 93, 820-827.

Sainju, U. M., Singh, B. P. and Whitehead, W. F. 2001. Long-term effects of tillage, cover crops, and nitrogen fertilization on organic carbon and nitrogen concentrations in sandy loam soils in Georgia, USA. Soil Tillage Res., 63, 167-179.

SAN, Sustainable Agriculture Network. 1998. Managing cover crops profitably. SAN Handb. Ser. Book 3. SAN, Beltsville, MD.

Schaller F. W. and Larson W. E. 1955. Effect of wide spaced corn rows on corn grain yields and forage establishment. Agron. J. 47, 271-276.

Scott, T. W., J. Mt. Pleasant, R. F. Burt, and D. J. Otis. 1987. Contribution of ground cover, dry matter, and nitrogen from intercrops and cover crops in a corn polyculture system. Agron. J. 79,792-798.

Severino, F. J. and P. J. Christoffoleti. 2004. Weed suppression by smother crops and selective herbicides. Sci. Agric. 61, 21-26.

Sheaffer, C. C., Gunsolus, J. L., Grimsbo Jewett, J. and Lee, S. H. 2002. Annual Medicago as a smother crop in soybean. Journal of Agronomy and Crop Science 188, 408-416.

Skora Neto, E. 1993. Controle de plantas daninhas através de coberturas verdes consorciadas com milho. Pesquisa Agropecuaria Brasileira 28, 1165-1171.

Smith, H., J. J. Casal and G. M. Jackson. 1990. Reflection signals and the perception by phytochrome of the proximity of neighbouring vegetation. Plant Cell Environ. 13, 73-78.

Steenwerth, K. and Belina, K. M. 2008. Cover crops enhance soil organic matter, carbon dynamics and microbiological function in a vineyard agroecosystem. Appl. Soil Ecol., 40, 359-369.

Steinmaus, S., Elmore, C. L., Smith, R. J., Donaldson, D., Weber, E. A., Roncoroni, J. A. and Miller, P. R. M. 2008. Mulched cover crops as an alternative to conventional weed management systems in vineyards. Weed Res. 48, 273-281.

Stivers-Young, L. 1998. Growth, nitrogen accumulation, and weed suppression by fall cover crops following early harvest of vegetables. Hort. Science 33, 60-63.

Stoller, E. W., Wax, L. M. and Alm D. M. 1993. Survey results on environmental issues and weed science research priorities within the corn belt. Weed Technol. 7, 763-770.

Stringfield, G. H., and L. E. Thatcher. 1951. Corn row spaces and crop sequences. Agron. J. 43,276-281.

Sung, J. K., J. A. Jung, B. M. Lee, S. M. Lee, Y. H. Lee, D. H. Choi, T. W. Kim and B. H. Song. 2010. Effect of incorporation of hairy vetch and rye grown as cover crops on weed suppression related with phenolics and nitrogen contents of soil. Plant Production Science 13, 80-84.

Teasdale, J. R. 1993. Reduced-herbicide weed management systems for no-tillage corn (*Zea mays*) in a hairy vetch (*Vicia villosa*) cover crop. Weed Technol. 7, 879-883.

Teasdale, J. R. and Daughtry, C. S. T. 1993. Weed suppression by live and desiccated hairy vetch. Weed Sci. 41, 207-212.

Teasdale, J. R. 1996. Contribution of cover crops to weed management in sustainable agricultural systems. J. Prod. Agric. 9, 475-479.

Teasdale, J. R. 1998. Cover crops, smother plants, and weed management. pp. 247-270. *In* J. L. Hatfield *et al. Integrated Weed and Soil Management.* Ann Arbor Press, Chelsea, MI, USA.

Teasdale, J. R. and C. L. Mohler. 1993. Light transmittance, soil temperature, and soil moisture under residue of hairy vetch and rye. Agron. J. 85, 673-680.

Teasdale, J. R. 2003. Principles and practices of using cover crops in weed management systems. In, R. Labrada (ed), Weed Management for Developing Countries. Food and Agriculture Organization of the United Nations (FAO), Rome, Italy. pp. 169-178.

Teasdale, J. R., C. E. Beste and W. E. Potts. 1991. Response of weeds to tillage and cover crop residue. Weed Sci. 39, 195-199.

Teasdale, J. R. L. O. Brandsaterfi A. Calegari and F. Skora Neto. 2007. Cover crops and weed management. In, M. K. Upadhyaya and R. E. Blackshaw (eds.), Non-chemical Weed Management, principles, concepts and technology. CAB International. Biddles Ltd, King's Lynn. UK. pp. 49-64.

Teasdale, J. R., R. C. Rosecrance, C. B. Coffman, J. L. Starr, I. C. Paltineanu, Y. C. Lu, and B. K. Watkins. 2000. Performance of reduced-tillage cropping systems for sustainable grain production in Maryland. Am. J. Alternative Agric. 15, 79–87.

Thiessen-Martens, J. R., J. W. Hoeppner and M. H. Entz. 2001. Legume cover crops with winter cereals in southern Manitoba, Establishment, productivity, and microclimate effects. Agron. J. 93, 1086-1096.

Triomphe, B. L. 1996. Seasonal nitrogen dynamics and long-term changes in soil properties under the *Mucuna pruriens*/maize cropping system on the hillsides of Northern Honduras. PhD Thesis, Cornell University, Ithaca.

Triplett, G. B., Jr. 1962. Intercrops in corn and soybean cropping systems. Agron. J. 54,106–109.

Vandermeer, J. H. 1990. Intercropping. p. 481–516. *In* C.R. Carroll et al. (ed.) Agroecology. McGraw-Hill Publ. Co., New York.

Vesteeg, M. N., and V. Koudopon. 1990. *Mucuna pruriens* helps control *Imperata* in southern Benin.West Africa Farming Syst. Network Bull. 7, 7-8.

Vrabel, T. E. 1983. Effects of suppressed white clover on sweet corn yield and nitrogen availability in a living mulch cropping system. Ph.D Thesis (Diss. abstr. DA 8321911), Cornell Univ., Ithaca, NY.

Wall, G. J., Pringle, E. A. and Sheard, R. W. 1991. Intercropping red clover with silage corn for soil erosion control. Can. J. Soil Sci. 71, 137–145.

Weber, E. and Gut, D. 2005. A survey of weeds that are increasingly spreading in Europe. Agronomy for Sustainable Development 25, 109-121.

Werner, A. 1988. Biological control of weeds in maize. Proceedings 6th International Conference I.F.O.A.M. Global perspectives on agroecology and sustainable agricultural systems, 487-502.

Weston, L.A. 2005. History and current trends in the use of allelopathy for weed management. Hort. Technology 15, 529–534.

Weston, L. A., R. Harmon, and S. Mueller. 1989. Allelopathic potential of sorghum-sudangrass hybrid (Sudex). J. Chem. Ecol. 15, 1855–1865.

Weston, L. A. 1996. Utilization of allelopathy for weed management in agroecosystems. Agronomy Journal 88, 860-866.

Westra, E. P. 2010. Can Allelopathy be incorporated into agriculture for weed suppression? http,//www.colostate.edu/Depts/Entomology/courses/en570/papers_2010/westra.pdf.

White R. H., Worsham A. D. and Blum U. 1989. Allelopathic potential of legume debris and aqeous extracts. Weed Sci. 37, 674–679.

Yenish, J. P., Worsham, A. D. and Chilton, W. S. 1995. Disappearance of DIBOA-glucoside, DIBOA, and BOA from rye cover crop residue. Weed Science 43, 18-20.

Zemenchik, R. A., K. A. Albrecht, C. M. Boerboom, and J. G. Lauer. 2000. Corn production with kura clover as a living mulch. Agron. J. 92, 698–705.

4

Ridge Planted Pigeonpea and Furrow Planted Rice in an Intercropping System as Affected by Nitrogen and Weed Management

Manoj Kumar Yadav et al.[*]

Department of Geophysics, Banaras Hindu University, Varanasi, India

1. Introduction

Producing more food to feed the burgeoning population from shrinking agricultural land and water resources will be a challenge. Recently, intercropping has received more attention as a means to increase productivity of crops in per unit area and per unit time.

Intercropping is a crop management system involving the growing of two or more dissimilar crops in distinct row combinations simultaneously on the same land area. In intercropping, the component crop species are usually sown in parallel lines enabling mechanical crop production, maintenance, and harvest. Intercropping involves crop intensification in respect to both time and space dimensions (Ahlawat and Sharma, 2002). Conceptually, an intercropping system helps for risk avoidance from epidemic of insect-pest and diseases and overcome adverse environmental conditions in agro-climatologically unstable regions along with increasing solar radiation utilization and inputs including fertilizer and water utilization compared to monoculture crops. Intercropping not only reduces the risk associated with input costs but also increases profit potential (Rathi and Verma, 1979). Moreover, it provides several major advantages namely, diversification reduces risk associated with crop failure, increased productivity per unit area and time, offers greater yield stability and utilizes the available growth resources more efficiently and sustainably. Furthering rationales of this practice, it caters to the multiple needs of the farmer, is a self-provisioning device, is a mechanism to spread labour peaks, and keeps weeds under check (Singh and Jha, 1984). A number of researchers (Enyi, 1973; Sengupta et al., 1985) reported greater land use efficiency utilizing intercropping and reductions of weed growth through competition. The yield advantage obtained through intercropping has been reported mainly due to efficient utilization and optimization of available natural growth

[*] R.S. Singh[2], Gaurav Mahajan[3], Subhash Babu[4], Sanjay Kumar Yadav[5],
Rakesh Kumar[6], Mahesh Kumar Singh[7], Amitesh Kumar Singh[8] and Amalesh Yadav[9]
[1]*Department of Geophysics, Banaras Hindu University, Varanasi, India*
[2,6,7,8] *Department of Agronomy, Banaras Hindu University, Varanasi, India*
[3]*Department of Agronomy, Jawaharlal Nehru Krishi Vishwavidyalaya, Rewa, India*
[4]*Division of Agronomy, Indian Agricultural Research Institute, New Delhi, India*
[5]*Central Potato Research Station (ICAR), Shillong, India*
[9]*Department of Botany, University of Lucknow, Lucknow, India*

resources including water (Donald, 1963; Singh and Gupta, 1994), nutrients (Donald, 1963; Dalal, 1974); light (Donald, 1963; Nelliet et al., 1974; Singh and Gupta, 1994) as well as air and space (Singh and Gupta, 1994). In addition, intercropped species can be selected that produce allelopathic effect (Risser, 1969; Rice, 1974). Similarly, Willey (1979) made critical analysis of the yield advantages accrued from the intercropping. He explained that yield advantage occurs because the component crops differ in their use of growth resources in such a way that when they are grown in combination they are able to complement each other to make better overall use of resources than when grown separately. Annidation is the complementary use of resources by exploiting the environment in different ways by the components of a community. Maximizing intercropping advantage is a matter of maximizing the degree of complementarity between the components and minimizing intercrop competition.

Pigeonpea (*Cajanus cajan* L. Millsp.) is seldom or never grown in monoculture except on a very small scale, and mixed cropping is standard on field scale (Aiyer, 1949). Pigeonpea is commonly intercropped with cereals such as sorghum (*Sorghum bicolor* L. Moench), maize (*Zea mays* L.), pearl millet (*Pennisetum typhoides* L.), finger millet (*Eleusine coracana* Gaertn) and rice (*Oryza sativa* L.); grain legumes like black gram (*Vigna mungo* L. Hepper), green gram (*Vigna radiata* L. Wilczek), soybean (*Glycine max* L. Merrill) and oilseed such as sesamum (*Sesamum indicum* L.), groundnut (*Arachis hypogaea* L.) (Jena and Misra, 1988; Parida et al., 1988; Gouranga Kar, 2005; Behera et al., 2009; Ashok et al., 2010).

Practice of intercropping of pigeonpea with different short duration companion crops in India is very common. Being deep rooted, pigeonpea is very well suited for intercropping with the shallow rooted ones. Intercropping besides offering an insurance against failure of the crop due to disease, pests and frost, enables the farmers to obtain a variety of crops of their needs from the same piece of land. Pigeonpea is generally grown with wide row spacing of about 75-80 cm. However, the initial growth is quite slow and the grand growth period starts after 60-70 DAS. A lot of inter-row spaces, therefore, remain vacant during the early stages and get infested by weeds. The space between the rows could be profitably utilised by growing short duration crops such as black gram, green gram, cowpea, rice etc. The row arrangement that utilises a high proportion of the early crop to maximise its yield and allows the late maturing component to fully cover the ground should normally give the highest productivity. Based on the per cent of plant population used for each crop in intercropping system, it is divided into two types *viz.* additive and replacement series. In additive series, one crop is sown with 100% of its recommended population in pure stand which is known as the base crop. Another crop known as intercrop is introduced into base crop by adjusting or changing crop geometry. The population of intercrop is less than its recommended population in pure stand. In replacement series, both the crops called component crops. By scarifying certain proportion of population of one component, another component is introduced. Soybean+pigeonpea (4:2) is one of the example of intercropping in replacement series (Kasbe et al., 2010) and pigeonpea+greengram (1:2) is in additive series (Arjun Sharma et al., 2010).

A new concept of pigeonpea +rice intercropping system under ridge-furrow method of planting has been developed for rice ecosystem of Varanasi in India in additive series (Singh, 2006a). Since both upland rice and pigeonpea are sensitive to moisture regime (rice to drought and pigeonpea to excess soil moisture); however in this system, pigeonpea and rice both receive their favourable micro-climate at field level. The major advantage of rice intercropping in furrow with ridge planted pigeonpea is that it can give greater yield stability compared to other intercropping choices because they are either adversely affected due to higher soil moisture or waterlogging at initial growth stage, thus the risk of a total crop failure is halved.

Weed infestation reduces grain yield directly and indirectly. Many crop and weeds have evolved with similar requirements for growth and development (Pujari et al., 1989; Yadav and Singh, 2009). Competition occurs when one of the resources (nutrients, light, moisture and space) fall short of total requirement of the crop and/or weeds. Weeds, by virtue of their high adoptability and faster growth, usually dominate the crop habitat and reduce the yield potential. Due to slow initial growth, wide crop row spacing is inefficient in fully utilizing light and moisture resources at initial growth stages and subsequently yield is reduced through competition with weeds. The inclusion of additional intercrop species can overcome this limitation. The presence of weeds is one of the major constraints to increase the seed yield in grain crops. Weeds are also an important factor responsible for low fertilizer use efficiency. Effective weed control measures are one of the several ways of increasing fertilizer use efficiency in crops in monoculture as well intercropping systems.

The nature and magnitude of crop-weed competition differs considerably between monoculture and intercropping systems. The crop species, population density, sowing geometry, duration and growth rhythm of the component crops, the moisture and fertility status of soil and tillage practices all influence weed flora in intercropping system (Moody and Shetty, 1981).

Since weeds are the main concern in many cultivated crops they should be controlled at the proper time. The most critical stage of crop-weed competition was observed between 15 to 45 days after sowing for the pigeonpea based intercropping system (Singh and Singh, 1995). Hand weeding, which is common practice, is very effective if repeated, though it is tedious, time consuming and costly. Moreover, present labour availability for such operations has decreased due to rapid industrialization, increased literacy and migration of labour to urban areas. Further, manual weed control methods are usually initiated after weeds have attained size and thus already competed for some time with the crop. Continuous rains in the rainy season make weed control by hand more difficult due to improper field conditions. In such situations, herbicide use likely will control weeds from the beginning of crop growth and can increase the crop yields. Herbicides not only control weeds and reduce labour cost, but also allow coverage of more area in a relatively shorter time period thus protecting yield potential (Ampong-Nyarko and De Datta, 1993). Many herbicides are crop specific; a herbicide that does not harm both the component crops, usually does not control a broad spectrum of weed species. The herbicides used in intercropping are selective in action for both component crops, but likely have narrow spectrum of weed control, leaving the other weeds to develop and compete with the crop. In addition, herbicidal soil activity expires before the critical period of crop-weed competition. A long duration crop of pigeonpea responded positively to two manual weeding and a pre-emergence herbicide likely substitute first one out of these two (Maheswarappa and Nanjappa, 1994). Higher yield attributes and yield of pigeonpea were also observed in different intercropping system under two sequential hand weedings or by integrated use of herbicides and hand weeding (Dwivedi et al., 1991; Rafey and Prasad, 1995; Rana et al., 1999).

The objectives of our present investigation entitled "Ridge Planted Pigeonpea and Furrow Planted Rice in an Intercropping System as Affected by Nitrogen and Weed Management" were to study the growth pattern and yields and nutrient uptake as affected by nitrogen and weed management in pigeonpea+ rice intercropping system.

2. Materials and methods

2.1 Physiographic situation

The Agricultural Research Farm is situated in the South eastern part of Varanasi city, India at an altitude of 125.93 meter above the MSL, 20^0 18' north latitude and 80^0 36' eastern longitude. The experiment was established at the Agricultural Research Farm, Institute of Agricultural Sciences, Banaras Hindu University. The area accurately reflects the agro-climatic conditions of North Gangetic Alluvial Plains with annual rainfall of about 1100 mm.

2.2 Climatic condition

Varanasi's climate is sub-tropical and is subjected to extremes of weather conditions i.e. heat of summer (33.4-41.4 ^{0}C) and cold in winter (9.3-11.8 ^{0}C). The temperature increases from mid-February and reaches its maximum by May/June but has a tendency to decrease from July onwards reaching the minimum in December/January. The normal period for the onset of monsoon in the region is third week of June which lasts up to the end of September or sometimes into the first week of October. The area occasionally experiences some winter cyclonic rains during December/ February. The period between March and May is generally dry. The normal annual rainfall of the region is about 1081.4 mm. In terms of percentage of total rainfall, 88 per cent is received from June to September, 5.7 per cent from October to December, 3.3 per cent from January to February and 3 per cent from March to May as per monsoon rains. The mean relative humidity is 62 per cent which rises up to 82 per cent during July to September and fall down to 28 per cent during the end of April and early June.

2.2.1 Rainfall

The cumulative rainfall received during the period of investigation was 683.0 mm and 783.3 mm in the year 2004-05 and 2005-06, respectively. The distribution of rainfall was more uniform during second year as compared to first year during crop production. The month-wise distribution of the rainfall indicated that July and August of second year received more rain than the corresponding period of the first year.

2.2.2 Temperature

The weekly mean maximum temperature ranged from 20.0 to 38.6 ^{0}C with an average of 30.0 ^{0}C during 2004-05 and 18.8 to 44.1 ^{0}C with an average of 30.8 ^{0}C during 2005-06. The weekly mean minimum temperature ranged from 8.3 ^{0}C to 27.4 ^{0}C with an average of 19.1 ^{0}C during 2004-05 and 7.4 to 30.4 ^{0}C with an average of 18.7 ^{0}C during 2005-06. The mean fluctuation in maximum and minimum temperature was almost normal during both the years.

2.2.3 Relative humidity

The weekly mean maximum relative humidity varied from 62 to 95% with an average of 84% during 2004-05 and it varied from 37 to 92% with an average of 82% during 2005-06. The weekly mean minimum relative humidity varied from 18 to 81%-with an average of 55% during 2004-05 and it varied from 18 to 83% with an average of 52% during 2005-06. The relative humidity indicated considerable variation throughout the growing season during both the years. Data also indicated that the first year was comparatively more humid as compared to second year.

Soil Physical and Chemical Properties	Value	Analysis Method Employed
Soil separates 0 – 15 cm (%) Sand Silt Clay	43.68 30.66 25.66	Hydrometer method (Bouyoucos, 1962)
Textural class	Sandy clay loam	Textural triangle (Black, 1967)
pH (1:2.5 soil water ratio)	7.3	Glass electrode pH meter (Jackson, 1973)
Electrical conductivity (d S/m at 25⁰C)	0.29	Systronics electrical conductivity meter (Jackson, 1973)
Organic carbon (%)	0.35	Chromic acid rapid titration method (Walkley and Black, 1934)
Available nitrogen (kg N/ha)	208.5	Alkaline permanganate method (Subbiah and Asija, 1973)
Available phosphorus (kg P/ha)	18.21	0.5 M $NaHCO_3$ extractable Olsen's colorimetric method. (Olsen et al., 1954)
Available potassium (kg K /ha)	185.02	Flame photometric method (Ammonium acetate extract) (Jackson, 1973)

Table 1. Physio-chemical properties of the experimental field

2.2.4 Sunshine duration

The average duration of bright sunshine day was 6.9 and 7.2 hours in first and second year, respectively. The range of maximum and minimum mean weekly bright sunshine duration was ranged from 2.9 to 10.2 hours during 2004-05 and it ranged from 1.9 to 10.0 hours during 2005-06.

2.2.5 Evaporation

The evaporation data recorded from a United States Weather Bureau class A pan evaporimeter revealed that the weekly average evaporation per day varied from 6.4 to 1.4 mm/day in 2004-05 and 9.5 to 1.5 mm/day in 2005-06. The total evaporation during crop growing period was 942.2 mm in 2004-05 and 1061.9 mm in 2005-06.

2.3 Soil and soil analysis

In order to know the initial fertility status of the experimental plot, soil sample from 0-15 cm were collected and analysed for mechanical composition and chemical constituents. Data obtained are reported in Table 1. The experimental plot area soil was classified as sandy clay loam in texture, low in nitrogen, and medium in available phosphorus and potassium.

2.4 Technical programme

Considering the nature of factors evaluated and the convenience of agricultural operation, the experiment was laid out in split plot design with three replications. Six main plot treatments (consisting of all possible combinations of two nitrogen levels in pigeonpea including one control and another 25 kg N/ha as starter application with three nitrogen levels in rice i.e. 50, 75 and 100 kg N/ha) and four sub plot treatments(weed management) were established. Weed management treatments included: 1) a weedy check, 2) pendimethalin at 1.0 kg/ha, 3) pendimethalin at 1.0 kg/ha followed by one hand weeding at 45 days after seeding (DAS) or 4) two sequential hand weeding at 15 and 45 DAS. The whole field was divided into three blocks, each representing a replication. Each block was further divided in six main plots where main plots treatments were randomly allocated within them. Then each main plot was again divided into four equal sub plots and the sub plot treatments were again allocated randomly.

2.4.1 Field preparation

Proper field preparation is essential for a healthy pigeonpea and rice crop in intercropping. The experimental area was ploughed with tractor drawn mould board plough followed by two passes with a disc. Finally the field was levelled.

2.4.2 Ridge and furrow establishment

Ridges and furrows were established manually by spade.

2.4.3 Fertilizer application

The recommended doses of P_2O_5 and K_2O for pigeonpea were 40 and 30, and for rice were 40 and 40 kg/ha, respectively. Quantity of P_2O_5 and K_2O/ha were applied on row basis to each crop separately in the form of single super phosphate and muriate of potash, respectively. Full doses of phosphorus and potassium were applied to pigeonpea and rice as basal applications. Nitrogen was applied as per treatment through urea. Full nitrogen dose of pigeonpea and 75% nitrogen dose of rice were applied as basal and remaining nitrogen dose of rice was top dressed at it's tillering growth stage.

2.4.4 Seed and sowing

Seed rate for pigeonpea and rice were 20 and 60 kg/ha, respectively. Pigeonpea seeds were sown on top of the ridges and rice seeds were sown in two rows in each furrow at the same date. The crops were sown on 8th July in 2004 and 12th July in 2005 using full season pigeonpea variety 'Bahar' and early rice variety 'NDR 97'. Row to row spacing of pigeonpea was 75 cm and plant to plant spacing of pigeonpea and row to row spacing of rice were 20 cm.

2.4.5 Herbicide application

The required quantity (1.0 kg/ha) of pendimethalin was mixed in water and sprayed with a backpack sprayer using the spray volume of 600 litres of water/ha as per treatment. Pendimethalin was applied pre-emergence (1 DAS).

2.4.6 Thinning

The extra plants were thinned out at 30 days after sowing to maintain the plant to plant spacing of 20 cm for pigeonpea.

2.4.7 Hand weeding

Hand weeding was accomplished as per the treatment in the experiment. The weeds were removed from hand weeded plots twice at 15 and 45 DAS or once at 45 DAS in integration with herbicidal treatment as per treatment 3, respectively. Weedy check plots were kept weed infested condition until crop maturity.

2.4.8 Plant protection

There was no serious incidence of any major pest or disease during period of crop growth. However, as a preventive measure against leaf folder, pod borer attack, two applications of Endosulfan 35 EC at the rate of 2 litres/ha dissolved in 800 litres of water were applied at 65 DAS and at pod formation growth stage.

2.4.9 Harvesting

The crops were harvested at physiological maturity growth stage. Rice was harvested on 10th and 21st October in 2004 and 2005 respectively and pigeonpea on 20th and 28th March in 2005 and 2006 respectively. Firstly, the border rows were harvested and separated. Following border row harvest, crop from net plot was harvested and sun dried. The harvested material from each net plot was bundled, tagged and threshed separately.

2.4.10 Threshing, cleaning and weighing

The individual net plot's harvested crop bundles were weighed after drying prior to threshing. The grain yield was recorded separately after threshing, winnowing and cleaning. The straw/stalk yield was calculated by subtracting grain yield from the bundle weight and was converted to t/ha based on net plot size harvest.

2.5 Observation

The following observations were taken during the study periods which are described below:

2.5.1 Studies on pigeonpea and rice

2.5.1.1 Shoot dry matter/ plant or/meter row(g)

The five pigeonpea plants randomly selected from the sample row were cut carefully at the ground surface and then sun dried. After sun drying, plant samples were collected in paper bags after being cut into smaller pieces and placed in an electric oven at 70 ⁰C for drying to obtain a constant dry weight. The dry weight of the samples then obtained was expressed in g/plant. For dry matter production by rice, all plant samples from 0.50 meter running row length were selected from the sampling rows (leaving aside one border row from the each side) at harvest. The plants were cut at the collar region. The collected sample tillers were oven dried at 60 ⁰C for 48 hours and weighted. The weight of sample tillers thus obtained was converted into g/ running meter by multiplying with conversion factor.

2.5.1.2 Grain yield (t/ha)

The harvested crop from each net plot was threshed separately. After proper cleaning and drying, grain yield was recorded in kg/plot and finally converted into t/ha by multiplying with conversion factor.

2.5.1.3 Stalk/straw yield (t/ha)

Stalk/straw yield for each net plot was calculated by subtracting the grain yield from total biological yield and finally expressed in terms of t/ha.

2.5.2 Weed assessment

Weeds were collected from each individual plot during each year of the investigation for identification. Weed samples were collected by placing a quadrate (0.50 m x 0.50 m) randomly at two places in each plot at 60 DAS.

2.5.2.1 Weed population

Species wise weed counts were recorded at 60 DAS of crops from the two randomly place quadrates of 0.50 m x 0.50 m (0.25 m^2) in each net plot. Thus, weed population/m^2 was calculated from total number of all weed species of two quadrates multiplied with conversion factor.

2.5.2.2 Weed dry matter production/m^2 (g)

Weed enclosed in a quadrate of 0.25 m^2 (0.50 m x 0.50 m) were removed from the sampling rows at 60 DAS. After sun drying the samples were placed in an oven at 60 ⁰C for 48 hours. The dry weight was multiplied with conversion factor to express in g/m^2.

2.5.2.3 Weed control efficiency (%)

Weed control efficiency (WCE) was calculated at 60 DAS using the formula USDA/ICAR (AICRPWC, 1994).

$$WCE = \frac{DMC\text{-}DMT}{DMC} \times 100$$

Where,

DMC= Dry matter production of weeds/m^2 in weedy check.

DMT= Dry matter production of weeds/m^2 in the treatment to be compared.

WCE has been expressed in percentage.

2.6 Chemical analysis of crops and weeds

The plant samples from crops (pigeonpea and rice) and weed flora collected within each treatment at crop harvest and thus the maximum growth stage of weeds (60 DAS), respectively were washed with tap water followed by 0.1N HCl, distilled water and then with double distilled water. Plants were first dried under shade then in hot air oven at 60 ^{0}C for 48 hours. After recording oven dry weight, plant samples were individually grinded in Willey' Mill and stored in butter paper covers. The powder of plant samples was analysed for nitrogen, phosphorus and potassium as per the methods described in Table 2.

Element	Method employed
Total nitrogen	Modified Kjeldahl method (Jackson,1973)
Total phosphorus	Vanadomolybdo phosphoric yellow colorimetric method (Jackson,1973)
Total potassium	Flame photometric method (Jackson,1973)

Table 2. Methods used for determination of chemical composition of crops and weeds

2.6.1 Nutrient uptake (kg/ha)

Nutrient content (N,P and K) in grain and stalk/straw of each crop and in entire weed complex were analysed separately using procedure given in Table 2. Nutrient uptake by grain and straw of crops and that of by weeds were calculated in kg/ha by multiplying the corresponding dry matter and nutrient content (Black, 1967).

2.7 Pigeonpea grain equivalent yield (kg/ha)

Pigeonpea grain equivalent yield (PGEY) was calculated as follows:

$$PGEY = \sum_{i=1}^{n} (Yi.ei)$$

Where,

Yi= Grain yield i^{th} component

ei= equivalent price of ith component

PGEY has been expressed in tonne/hectare

2.8 Statistical analysis

The data pertaining to each of the treatments and interactions were analyzed statistically by applying the procedure as described by Gomez and Gomez (1984).

3. Results and discussion

3.1 Weeds

The weed flora of the experimental field included: Jungle rice [*Echinochloa colona* (L.) Link.], barnyard grass [*Echinochloa crusgalli* (L.) Beauv.], bermuda grass [*Cynodon dactylon* (L.) Pers], goose grass [*Eleusine indica* (L.) Gaerth.], crab grass [*Digitaria sanguinalis*(L.) Scop.], crowfoot grass [*Dactyloctenium aegypticum* (L.) P. Beauv.], purple nutsedge *(Cyperus rotundus* Linn.), variable flatsedge (*Cyperus difformis* Linn.), ricefield flatsedge (*Cyperus iria* Linn.), grass-like fimbry [*Fimbristylis miliacea* (L.) Vahl.], goat weed (*Ageratum conyzoides* L.), dayflower (*Commelina benghalensis* Linn.), climbing dayflower (*Commelina diffusa* L.), hairy spurge *(Euphorbia hirta* Linn.), asian spiderflower (*Cleome viscosa* L.), wild carrot weed (*Parthenium hysterophorus* L.), pink node flower (*Caesulia axillaris* Roxb.), silver cock's comb (*Celosia argentea* L.), gale of the wind (*Phyllanthus niruri* Linn.), false daisy [*Eclipta alba*(L.) Hassk.] and wild jute (*Corchorus acutengulus* lamk).

Application of 25 kg N/ha in pigeonpea increased weed population and their dry weight/m² as compared to control (Table 3). Analysis further reveals that weed density and dry weight/m² increased with increasing levels of nitrogen to rice up to 100 kg N/ha (Table 3). This likely due to weeds utilizing a greater quantity of applied and available nutrients. Thus, higher dose of nitrogen accelerated weed emergence and growth. Weed control efficiency increased with increasing nitrogen levels for pigeonpea and rice. Similar findings were also reported by Pujari et al. (1989) and Yadav and Singh (2009).

Data presented in Table 3 also indicates that two hand weeding at 15 and 45 DAS resulted in the lowest density and dry weight of weeds/m² followed by pendimethalin + one hand weeding at 45 DAS; both treatments were superior over other weed management treatments. This result was likely owing to better indiscriminate control of all types of weeds by hand weeding. These findings were in close agreement with those of Shetty and Krantz (1976), Ampong-Nyarko and De Datta(1993) and Reddy et al. (2007). Reflecting minimum density and dry weight results, maximum weed control efficiency was obtained with two hand weeding at 15 and 45 DAS. This finding is in agreement with finding of Sinha et al. (1989a and b), Goyal et al. (1991), Parthi et al. (1991), Mahapatra (1991), Prasad and Srivastava (1991), Maheswarappa and Nanjappa (1994), Rafey and Prasad (1995), Patil and Pandey (1996), Mishra et al. (1998), Singh et al. (1998c), Singh et al. (1999), Rana and Pal (1999), Rana et al. (1999), Manickam et al. (2000), Reddy et al. (2007) and Singh (2007).

3.2 Growth, yields and pigeonpea grain equivalent yield

Dry matter production, grain and straw yield of pigeonpea and rice, and pigeonpea grain equivalent yield were increased with application of 25 kg N/ha to pigeonpea over control (Table 3 and 4). Similar findings have been reported earlier by Singh et al. (1978), Bhandhari et al. (1989), Chittapur et al. (1994), Patel and Patel (1994), Singh et al.(1998a and b), Mandal et al. (1999) and Singh (2006b). Dry matter production, grain and straw yield of rice were increased significantly up to 75 kg N/ha applied to rice (Table 3 and 4). The improvement in the dry matter production and yields of rice might be attributed to the adequate supply of photosynthate to sink under sufficient supply of nitrogen. These results were supported by the findings of Samui et al. (1979), Reddy et al. (1986), Abdulsalam and Subramaniam (1988), Purushotham et al. (1988), Raju et al. (1990), Dubey et al. (1991), Bhattacharya and Singh

(1992), Mazid et al. (1998), Panda et al. (1999) and Bindra et al. (2000). Many researchers also reported that cereal component in legumes based intercropping yielded more at higher levels of nitrogen application (Reddy et al. 1980; Ramesh and Surve 1984; Ofori and Stern 1986; Ezumah et al. 1987; Rao et al. 1987; Kaushik and Gautam 1987; Chowdhury and Rosario 1992; Rafey and Prasad 1992; Bhagat and Dhar 1995; Kushwaha and Chandel 1997; Mandal et al. 2000; Sarwagi and Tripathi 1999; Shivay et al. 1999; Shivay and Singh 2000; Singh, 2006b). Whereas, same were failed to show its effect on pigeonpea (Table 3 and 4). This might be due to the fact that localized placement of nitrogen made was first available to that crop for which it was applied. Forage area of pigeonpea at initial growth stage (50 DAS) was slow due their slow growth habit. Contrary to this, forage area of short duration rice was higher due to faster initial growth rate and planting in furrow between ridges, likely taking most of applied nitrogen easily by themselves in comparison to pigeonpea. Mahapatra et al. (1990) and Singh (2006b) were also find the similar result.

Treatment	Pigeonpea dry matter/ plant at harvest (g)	Rice dry matter /m row at harvest (g)	Weed number /m²	Weed dry weight /m²(g)	Weed control efficiency (%)
N level in Pigeonpea(kg/ha)					
0	151.4	187.3	256.9	193.9	76.4
25	168.0	209.9	286.7	225.0	77.9
CD(P=0.05)	7.6	11.9	13.9	10.9	0.5
N level in Rice(kg/ha)					
50	156.1	175.2	256.3	194.7	76.6
75	160.5	204.0	274.4	211.7	77.5
100	162.6	216.7	284.7	222.1	77.5
CD(P=0.05)	NS	14.6	17.0	13.4	0.6
Weed Management					
Weedy check	136.1	60.4	545.0	498.8	-
Pendimethalin@ 1kg / ha	159.0	178.6	266.1	205.2	58.7
Pendimethalin@ 1kg / ha +one hand weeding at 45 DAS	171.0	255.1	161.7	79.7	83.9
Two hand weedings at 15 and 45 DAS	172.9	300.4	114.3	54.3	89.0
CD(P=0.05)	3.2	6.5	8.0	7.7	0.3

Table 3. Effect of nitrogen levels and weed management practices on crop growth and weed and weed control efficiency under pigeon pea + rice intercropping system (mean of two years)

Treatment	Pigeonpea		Rice		Pigeonpea grain equivalent yield (t/ha)
	Grain yield (t/ha)	Stalk yield (t/ha)	Grain yield (t/ha)	Straw yield (t/ha)	
N level in Pigeonpea(kg/ha)					
0	1.9	6.3	0.7	1.2	2.1
25	2.3	7.0	0.8	1.4	2.5
CD(P=0.05)	0.2	0.5	0.1	0.1	0.17
N level in Rice(kg/ha)					
50	2.0	6.4	0.7	1.2	2.2
75	2.1	6.7	0.8	1.3	2.3
100	2.1	6.9	0.8	1.4	2.4
CD(P=0.05)	NS	NS	0.1	0.1	0.21
Weed Management					
Weedy check	1.5	6.0	0.2	0.4	1.6
Pendimethalin@ 1kg / ha	2.0	6.6	0.7	1.2	2.2
Pendimethalin@ 1kg / ha +one hand weeding at 45 DAS	2.3	7.0	1.0	1.6	2.6
Two hand weeding at 15 and 45 DAS	2.4	7.1	1.2	1.9	2.7
CD(P=0.05)	0.1	0.2	0.1	0.1	0.07

Table 4. Effect of nitrogen levels and weed management practices on yields under pigeon pea +rice intercropping system (mean of two years). DAS: days after sowing

Among weed management practices, two sequential hand weeding recorded maximum dry matter accumulation and yields of pigeonpea and rice and minimum weed density and dry weight which was followed by pendimethalin + one hand weeding at 45 DAS (Table 3 and 4). This was likely owing to minimum weed competition for water, nutrient and space etc. (Fig. 1). Similar observations were seen by Dwivedi et al. (1991), Mahapatra (1991), Parthi et al. (1991), Dahama et al. (1992), Varshney (1993), Rafey and Prasad (1995), Mahalle (1996), Patil and Pandey (1996), Mishra et al. (1998), Rana and Pal (1999), Rana et al. (1999) and Reddy et al. (2007). The minimum yields were attained in the weedy check. This was again likely owing to higher weed competition for water, nutrient and space etc. (Fig. 1). Similar results were also reported by Ghobrial (1981), Dwivedi et al. (1991), Mahapatra (1991), Rafey and Prasad (1995) and Chandra Pal et al. (2000).

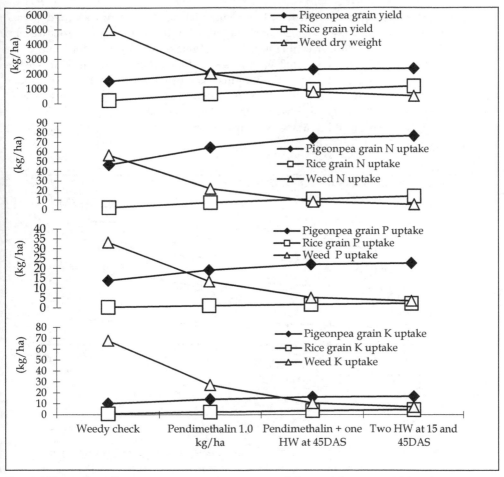

Fig. 1. Effect of weed management practices on weed dry weight, crop yield and nutrient
uptake under pigeon pea + rice intercropping system (mean of two years).

3.3 Nutrient uptake

Application of 25 kg N/ha to pigeonpea increased NPK uptake by grain as well by
stalk/straw of pigeonpea and rice over the control. Weed NPK uptake were also higher with
25 kg N/ha applied to pigeonpea. NPK uptake by rice grain and straw increased
significantly with each successive increase in nitrogen level applied to them up to 75 kg
N/ha (Table 5). Nitrogen levels applied to rice also increased weed NPK uptake up to 100
kg/ha. This is likely due to the optimum nitrogen application and ultimately resulted in
subsequent uptake of other nutrients (phosphorus and potassium) due to increased growth.
The maximum nutrient uptake under higher nitrogen dose might be due to better root
establishment and thus enhanced translocation of absorbed nutrients from soil to plant
ultimately resulting in higher growth and yield. Singh (2006b) and Yadav and Singh (2009)
also observed similar results.

Among the weed management practices, two sequential hand weeding recorded higher NPK uptake by grain and straw of pigeonpea and rice and this treatment was followed by with pendimethalin + one hand weeding at 45 DAS (Table 5). Contrary to this, minimum NPK removals by weed were associated with these treatment and maximum with weedy check (Table 5). This might be due to applied inputs assimilated efficiently by weeds under weedy condition and by crops under weed free condition (Table 5 and Fig. 1). These results are in agreement with findings of Singh et al. (1980), Singh and Singh (1985), Sinha et al.(1989a and b), Goyal et al. (1991), Maheswarappa and Nanjappa (1994), Singh et al. (1998c) and Singh (2007).

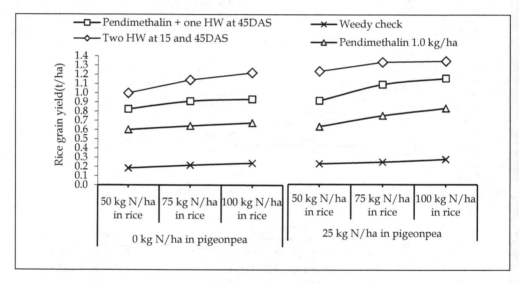

Fig. 2. Interaction effect of nitrogen levels and weed management practices on rice grain yield underpigeon pea + rice intercropping system (mean of two years).

3.4 Interaction effect

Interaction effect of nitrogen levels in pigeonpea and rice and weed management practices was significant in respect to grain yield of rice (Fig. 2). Grain yield of rice increased with increasing level of nitrogen applied to rice up to 100 kg N/ha with or without 25 kg N/ha applied to pigeonpea in combination with all weed management treatments except weedy check where all nitrogen level failed to show any significant increase in grain yield of rice. This might be due to fact that rice grew better even with lower nitrogen addition where weeds were controlled than weedy check. In case of weedy check, weeds were dominant competitor to applied nitrogen over crop. Soundara and Mahapatra (1978) found that the maximum grain yield of direct seeded rice with application of 100 kg N/ha along with two sequential hand weedings where weeds effectively controlled. Sharma (1997) observed that grain yield of rice increased significantly with N application up to 60 kg/ha when weeds were controlled. He also observed that grain yield of rice remained unaffected with N application under weedy conditions due to severe competition.

Treatment	Nitrogen uptake(kg/ha)					Phosphorus uptake(kg/ha)					Potassium uptake(kg/ha)				
	Pigeonpea		Rice		W	Pigeonpea		Rice		W	Pigeonpea		Rice		W
	G	S	G	S		G	S	G	S		G	S	G	S	
N level in Pigeonpea(kg/ha)															
0	58.4	43.3	8.1	5.2	21.1	17.4	7.0	1.3	1.4	12.2	12.7	57.6	2.6	18.1	25.6
25	73.1	49.8	9.5	5.8	25.2	21.5	8.0	1.5	1.6	15.4	15.8	65.2	3.1	20.4	30.7
C.D.(P=0.05)	6.2	5.0	0.5	0.3	2.0	1.7	1.0	0.1	0.2	1.1	1.5	6.0	0.3	1.9	2.3
N level in Rice(kg/ha)															
50	63.0	44.1	7.8	5.0	21.2	18.7	7.0	1.2	1.4	12.5	13.7	58.1	2.5	17.6	25.8
75	66.3	46.9	9.0	5.6	23.3	19.6	7.6	1.4	1.6	13.9	14.4	62.0	2.9	19.6	28.5
100	67.9	48.7	9.7	5.9	24.9	20.0	7.9	1.5	1.7	15.1	14.7	64.0	3.1	20.7	30.1
C.D.(P=0.05)	NS	NS	0.6	0.4	2.4	NS	NS	0.2	0.2	1.4	NS	NS	0.4	2.4	2.8
Weed Management															
Weedy check	46.7	39.6	2.3	1.7	56.7	13.9	5.8	0.4	0.4	33.2	10.1	53.1	0.8	6.1	67.8
Pendimethalin@ 1kg / ha	64.8	46.1	7.6	5.0	22.0	19.2	7.4	1.1	1.4	13.4	14.0	60.8	2.5	18.0	27.2
Pendimethalin@ 1kg / ha +one hand weeding at 45 DAS	74.6	49.8	11.2	6.9	8.3	22.1	8.3	1.7	1.9	5.2	16.2	65.2	3.6	23.9	10.5
Two hand weeding at 15 and 45 DAS	76.9	50.8	14.2	8.5	5.5	22.7	8.4	2.2	2.4	3.5	16.7	66.4	4.6	29.1	7.0
C.D.(P=0.05)	2.6	2.1	0.3	0.2	1.5	0.7	0.4	0.1	0.1	0.8	0.6	2.6	0.2	1.0	1.7

Table 5. Effect of nitrogen levels and weed management practices on nutrient uptake by component crops and weed under pigeon pea +rice intercropping system (mean of two years).G: Grain, S: Straw/Stalk, W: Weed, DAS: days after sowing

Interaction effect of nitrogen levels in pigeonpea and weed management practices was significant in respect to grain yield and grain nutrient uptake by pigeonpea (Fig. 3). Application of 25 kg N/ha to pigeonpea under two hand weeded plots resulted in maximum yield and nutrient uptake by pigeonpea and this treatment was similar to application of 25 kg N/ha to pigeonpea with pendimethalin + one hand weeding at 45 DAS. Further, there was minimum removal of NPK by weeds (Fig. 6) in this treatment which was ultimately utilized by the crop and promoted its growth and yield. All weed management practices along with no nitrogen application in pigeonpea gave higher pigeonpea grain yield and nutrient uptake over the weedy check along with application of 25 kg N/ha in pigeonpea. This might be due pigeonpea growing better even without nitrogen addition where weeds were controlled. In case of weedy check, weeds were dominant competitor to applied nitrogen over crop.

Interaction effect of nitrogen levels in pigeonpea and weed management practices was significant in respect to pigeonpea grain equivalent yield, rice grain yield and rice grain nutrient uptake (Fig. 4). Application of 25 kg N/ha to pigeonpea following two sequential hand weedings gave maximum pigeonpea grain equivalent yield, rice grain yield and rice grain nutrient uptake and minimum yield and nutrient uptake resulting from no nitrogen application under weedy condition. These results agree with findings of Soundara and Mahapatra (1978) and Sharma (1997). The significant increase in pigeonpea grain equivalent yield, rice grain yield and rice grain N, P and K uptake with N application were observed only in weed controlled plots (Fig. 4). This might be due to rice compete strongly with pigeonpea for nitrogen in absence of weeds when first at its log phase and second at its lag phase of growth.

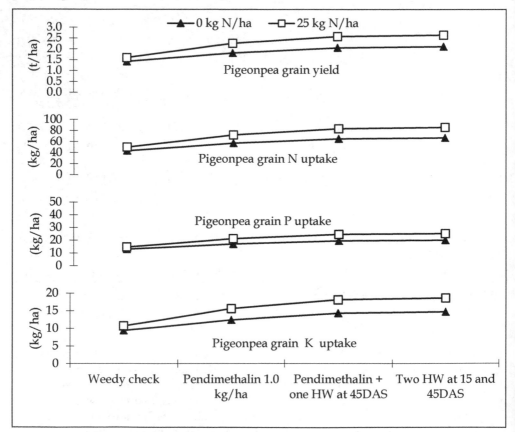

Fig. 3. Interaction effect of nitrogen levels in pigeonpea and weed management practices on pigeopea grain yield and pigeopea grain nutrient uptake under pigeon pea + rice intercropping system (mean of two years).

Interaction effect of nitrogen levels in rice and weed management practices was significant in respect to grain yield and grain nutrient uptake by rice (Fig. 5). Application of 100 kg

N/ha to rice under two hand weeded plots resulted in maximum grain yield and grain nutrient uptake by rice which was similar to the application of 75 kg N/ha under two sequential hand weeded plots and superior than rest of the other treatment combination (Fig. 5). All weed management practices in combination with 50 kg N/ha applied in rice produced higher grain yield and NPK uptake by rice over weedy check in combination with 100 kg N applied in rice. This might be due to crop plants utilizing nitrogen more efficiently even at lower level of nitrogen (50 kg/ha) in absence of weeds than higher level of nitrogen (100 kg/ha) in presence of weeds because most of which was utilized by weeds. Similar results were also reported by Sharma (1997).

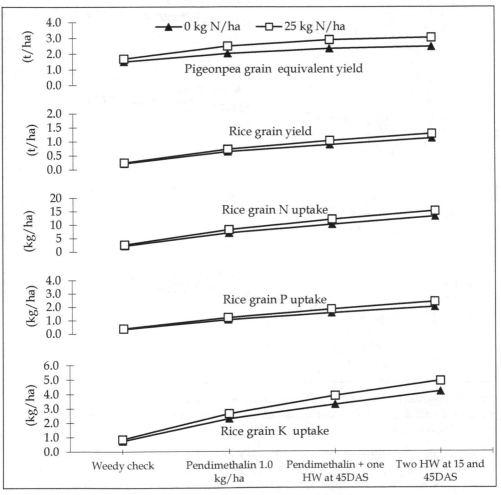

Fig. 4. Interaction effect of nitrogen levels in pigeonpea and weed management practices on pigeonpea grain equivalent yield, rice grain yield and rice grain nutrient uptake under pigeon pea + rice intercropping system (mean of two years).

Interaction effect of nitrogen levels in pigeonpea and weed management practices was significant in respect to weed dry weight and weed nutrient uptake (Fig. 6). The weed dry weight and weed nutrient uptake were recorded lower with or without application of 25 kg N/ha to pigeonpea under weed controlled plots than with or without application of 25 kg N/ha to pigeonpea under weedy check. Whereas, application of 25 kg N/ha to pigeonpea increased the weed dry weight and weed nutrient uptake only under weedy check. This due to one would exert severe competition on another under their dominance.

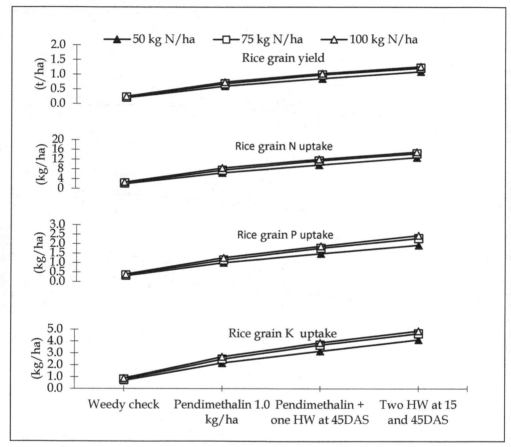

Fig. 5. Interaction effect of nitrogen levels in rice and weed management practices on rice grain yield and rice grain nutrient uptake under pigeon pea + rice intercropping system (mean of two years).

Interaction effect of nitrogen levels in rice and weed management practices was significant in respect to weed dry weight and weed nutrient uptake (Fig. 7). All weed management practices recorded lower nutrient removal by weeds irrespective of nitrogen levels applied in rice over weedy check which had maximum nutrient removal by weeds. This might be due to lower weed density and their dry weight in weed free condition ultimately resulting in lower NPK removal by weeds. Varying nitrogen levels applied in rice either along with

two hand weeding at 15 and 45 DAS or pendimethalin + one hand weeding at 45 DAS did
not cause any significant variation in NPK removal by weeds. Maximum weed dry weight
and weed nutrient uptake were recorded with application of 100 kg N/ha to rice under the
weedy check (Fig. 7). This might be due to weeds utilizing more inputs than crop plant
under severe competition.

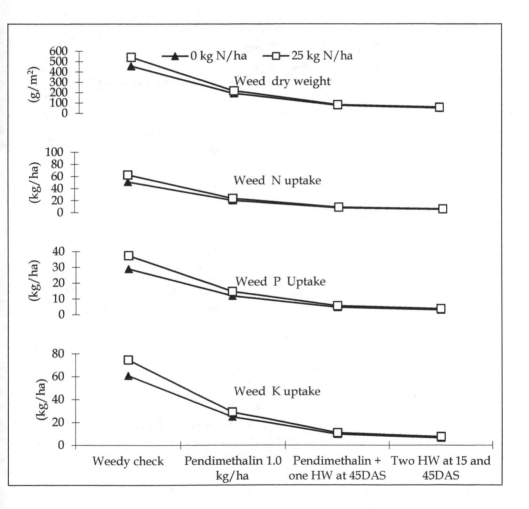

Fig. 6. Interaction effect of nitrogen levels in pigeonpea and weed management practices on
weed dry weight and weed nutrient uptake under pigeon pea + rice intercropping system
(mean of two years).

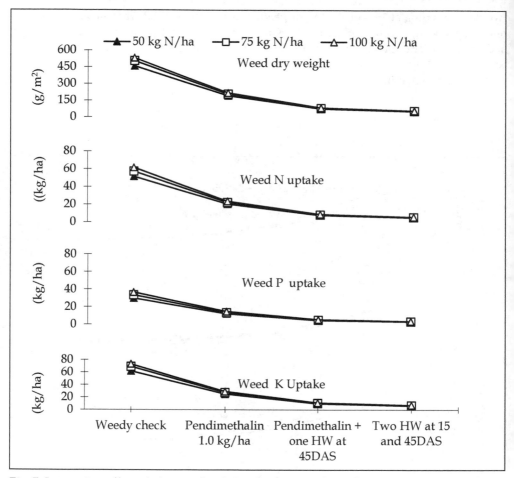

Fig. 7. Interaction effect of nitrogen levels in pigeonpea and weed management practices on weed nutrient uptake under pigeon pea + rice intercropping system (mean of two years)

4. Conclusion

Pigeonpea and rice could be fertilized with 25 kg N/ha and 75 kg N/ha, respectively, in an intercropping system integrated with two sequential hand weeding at 15 and 45 DAS for higher growth, yield and nutrient uptake by the crops. The next most effective treatment was application of 25 kg N/ha to pigeonpea and 75 kg N/ha to rice in the intercropping system integrated with a pre emergence application of pendimethalin at the rate of 1.0 kg/ha followed by one hand weeding at 45 DAS.

5. Future research

Studies are required to investigate the effect of rice cultivars, nitrogen levels under weeded and weedy condition in a pigeonpea+ rice intercropping system.

6. Acknowledgment

The senior author would like to express his gratitude to University Grants Commission,
New Delhi for providing research fellowship during his Ph.D. programme.

7. References

Abdulsalam, M. and Subramaniam, S. (1988). Influence of nitrogen, zinc and their
interaction on the yield and nutrient uptake of IR-20 rice (*Oryza sativa*) in different
seasons. *Indian Journal of Agricultural Sciences* 58: 190-193

Ahlawat, I.P.S. and Sharma, R.P. (2002). *Agronomic Terminology*. Indian Society of
Agronomy, Division of Agronomy, Indian Agricultural Research Institute, New
Delhi, India

AICRPWC (1994). Annual report. All India Co-ordinated Research Programme on Weed
Control. G.B.P.U.A. and T., Pantnagar. pp. 5-9

Aiyer, A.K.Y.N. (1949). Mixed cropping in India. *Indian Journal of Agricultural Sciences* 19(4):
439-443

Arjun Sharma, Rathod, P. S. and Mohan Chavan (2010). Integrated nutrient management
in pigeonpea (*Cajanus cajan*) based intercropping systems under rainfed conditions.
Karnataka Journal of Agricultural Sciences 23(4): 584-589

Ampong-Nyarko, K. and De Datta, S.K. (1993). Effect of nitrogen application on growth,
nitrogen use efficiency and rice-weed interaction. *Weed Research* 33(3): 269-276

Ashok, E. G.; Dhananjaya, B. N.; Kadalli, G. G.; Kiran Mathad, V. and Gowda, K. T. K.
(2010). Augmenting production and profitability of finger millet+pigeonpea
intercropping system. *Environment and Ecology* 28(1): 28-33

Behera, B.; Sankar, G. R. M.; Mohanty, S. K.; Mishra, A. and Chari, G. R. (2009). Sustainable
and effective fertilizer management for rice+pigeonpea intercropping under sub-
humid alfisols. *E-planet* 7(1): 20-25

Bhagat, R.K. and Dhar, V. (1995). Fertilizer management in upland rice and pigeonpea+rice
intercropping system. *Journal of Research, Birsa Agricultural University* 7(2): 159-160

Bhandhari, A. L.; Rana, D. S. and Sharma, K. N. (1989). Effect of fertilizer application on
rainfed blackgram (*Phaseolus mungo*), lentil (*Lens culinaris*) and pigeonpea (*Cajanus
cajan*) in farmar's fields. *Indian Journal of Agricultural Sciences* 59(11): 709-712

Bhattacharya, H.C. and Singh, K.N. (1992). Response of direct-seeded rice (*Oryza sativa*) to
levels and time of nitrogen application. *Indian Journal of Agronomy* 37 (4): 681-684

Bindra, A.D.; Kalia, B.D. and Kumar, S. (2000). Effect of N-levels and dates of transplanting
on growth, yield and yield attributes of scented rice. *Advances in Agricultural
research in India* 10 : 45-48

Black, C.A. (1967). Soil-Plant Relationship. 2nd edn. John. Willey and Sons. Pub. New York,
U.S.A. pp. 515-516

Bouyoucos, G.J. (1962). Hydrometer method for measuring particle size analysis of soil.
Agronomy Journal 54: 464-465

Chandra Pal; Kaushik, S.K. and Gautam, R.C. (2000). Weed control studies in pearlmillet
(*Pennisetum glaucum*)+pigeonpea (*Cajanus cajan*) intercropping system under
rainfed condition. *Indian Journal of Agronomy* 45(4): 662-668

Chittapur, B.M.; Kulkarni, B.S.; Hiremath, S.M. and Hosmani, M.M. (1994). Influence of nitrogen and phosphorus on growth and yield of short-duration pigeonpea (*Cajanus cajan*). *Indian Journal of Agronomy* 39(4): 657-659

Chowdhury, M.K. and Rosario, E.L. (1992). Utilization efficiency of applied nitrogen as related to yield advantage in maize+mungbean intercropping. *Field Crop Research* 30(1 and 2): 41-51

Dahama, A.K.; Bhagat, K.L. and Singh, H. (1992). Weed management in direct- seeded upland rice (*Oryza sativa*). *Indian Journal of Agronomy* 37(4): 705-709

Dalal, R.C. (1974). Effect of intercropping maize with pigeonpea on grain yield and nutrient uptake. *Experimental Agriculture* 10: 219-225

Donald, C.M. (1963). Competition among crop and pasture plants. *Advances in Agronomy* 15: 11-18

Dubey, O.P.; Upadhyaya, S.P.; Dixit, J.P. and Garg, D.C. (1991). Response of short duration paddy varieties to nitrogen under rainfed conditions. *Indian Journal of Agronomy* 36(2): 255-256

Dwivedi, V.D.; Pandey, R.P. and Namdev, K.N. (1991). Weed control in pigeonpea-sorghum intercropping system. *Indian Journal of Agronomy* 36(2): 234-238

Enyi, B.A.C.(1973). Effect of plant population on growth and yield of soybean. *Journal of Agricultural Sciences(Cambridge)* 81(1): 131-138

Ezumah, H.C.; Nam, N.K. and Walkar, P. (1987). Maize-cowpea intercropping as affected by nitrogen fertilization. *Agronomy Journal* 79(2): 275-280

Ghobrial ,G. I. (1981). Weed control in irrigated dry seeded rice. *Weed Research* 21: 201-204

Gomez, K.A. and Gomez, A.A. (1984). Statistical procedure for agricultural research, 2nd ed. John Willey and Sons, New York, U.S.A., pp. 241-271

Gouranga Kar (2005). Radiation interception and rain water utilization efficiency of legume based intercropping in rainfed upland rice area of eastern India. *Journal of Agrometeorology* 7(1): 84-89.

Goyal, S.N.; Tikka, S.B.S.; Patel, N.L.; Patel, N.M. and Ahlawat, R.P.S. (1991). Integrated weed management in pigeonpea. *Indian Journal of Agronomy* 36(1): 52-54

Jackson, M.L. (1973). Soil Chemical Analysis. Prentice Hall of India Pvt. Ltd., New Delhi, India, pp. 183-204

Jena, D. and Misra, C.(1988). Effect of crop geometry(row proportions) on the water balance of the root zone of a pigeonpea and rice intercropping system. *Experimental Agriculture* 24: 385-391

Kasbe, A. B.; Karanjikar, P. N. and Thete, N. M. (2010). Effect of planting pattern and intercropping of soybean-pigeonpea on growth and yield. *Journal of Maharashtra Agricultural Universities* 35(3): 381-384

Kaushik, S.K. and Gautam, R.C. (1987). Effect of nitrogen and phosphorus on the production potential of pearlmillet-cowpea or greengram intercropping systems under rainfed conditions. *Journal of Agricultural Sciences* 108(2): 361-364

Kushwaha, H.S. and Chandel, A.S. (1997). Effect of soybean(*Glycine max*) intercropping under different nitrogen levels on yield, yield attributes and quality of maize (*Zea mays*). *Indian Journal of Agricultural Sciences* 67(6): 249-252

Mahalle, S.S. (1996). Monetary analysis of chemical weed control in direct-seeded rice (*Oryza sativa*). *Indian Journal of Agronomy* 41(4): 591-594

Mahapatra, P.K. (1991). Weed control in intercropping systems based on pigeonpea (*Cajanus cajan*). *Indian Journal of Agricultural Sciences* 61(12): 885-888

Mahapatra, P.K.; Satpathy, D.; Hati, N. and Senapati, P.C. (1990). Effect of nitrogen on pigeonpea(*Cajanus cajan*) and rice(*Oryza sativa*) intercropping system. *Indian Journal of Agricultural Sciences* 60(8): 519-522

Maheswarappa, H.P. and Nanjappa, H.V. (1994). Relative efficacy of herbicides in controlling the weeds infesting pigeonpea(*Cajanus cajan*). *Indian Journal of Agronomy* 39(4): 662-664

Mandal, B.K.; Saha, S. and Jana, T.K. (2000). Yield performance and complementarity of rice (*Oryza sativa*) with greengram (*Phaseolus radiatus*), blackgram (*Phaseolus mungo*) and pigeonpea (*Cajanus cajan*) under different rice-legume associations. *Indian Journal of Agronomy* 45(1): 41- 47

Mandal, B.K.; Saha, Sanjay and Jana, T.K. (1999). Growth and resource use of rice, greengram, blackgram and pigeonpea in different inter and mixed cropping systems under rainfed upland situation. *Oryza* 36(4): 358-363

Manickam, G.; Durai, R. and Gnanamurthi, P. (2000). Weed characteristics, yield attributes and crop yield as influenced by integrated weed management in groundnut (*Arachis hypogaea*) based intercropping system. *Indian Journal of Agronomy* 45(1): 70-75

Mazid Miah, M.A.; Faiz, S.M.A. and Panaullah, G.M. (1998). Effect of tillage and nitrogen levels on dry matter yield, harvest index and nitrogen use efficiency of wetland rice. *Thai Journal of Agricultural Science* 31(3): 411-422

Mishra , R. K ; Bajpai , R. P.; Pandey , V. K. and Choudhary, S. K. (1998). Response of pigeonpea and soybean to planting pattern and weed control measures. *Indian Journal of Weed Science* 30(1/2): 1- 4

Moody, K and Shetty, S.V.R. (1981). Weed management in intercropping system. In: *Proceedings of the International Workshop on intercropping*, ICRISAT Center, India, January 1981

Nelliet, V.; Barappa, K.V. and Naur, P.K.R. (1974) Multiple cropping. A-new discussion in multiple cropping for coconut plantation. *World Crops Books*. pp. 262-266

Ofori, Francis and Stern, W.R. (1986). Maize + cowpea intercrops: Effect of nitrogen fertilizer on productivity and efficiency. *Field Crops Research* 14: 247-267

Olsen, S.R.; Cole, C.W.; Watanabe, F.S. and Dean, L. A. (1954). Estimation of available phosphorus in soils by extraction with sodium bicarbonate. *USDA Circular No. 939.* pp. 19-23

Panda, S.C.; Patro, H.; Panda, P.C. and Reddy, G.M.V. (1999). Effect of integrated nitrogen management on rice yield and physio-chemical properties of soil. *Crop Research (Hissar)* 18(1): 25-28

Parida, D.; Dikshit, U.N.; Satpathy, D. and Mahapatra, P.K. (1988). Pigeonpea genotypes and rice yield in an intercropping system. *International Rice Reseach Newsletter* 13: 26-27

Parthi, A.K.; Sahoo, B.K. and Das K.C. (1991). Effect of weed management on yield of rainfed direct-seeded upland rice (*Oryza sativa*). *Indian journal of Agricultural Sciences* 61(1): 27-30

Patel, J.R. and Patel, Z.G. (1994). Effect of irrigation, *Rhizobium* and nitrogen on yield, quality and economics of pigeonpea (*Cajanus cajan*). *Indian Journal of Agronomy* 39(4): 659-661

Patil, B.M. and Pandey, J. (1996). Chemical weed control in pigeonpea (*Cajanus cajan*) intercropped with short duration grain legumes. *Indian Journal of Agronomy* 41(4): 529-535

Prasad, K. and Srivastava, V.C. (1991). Weed management in pure and mixed crops of pigeonpea (*Cajanus cajan*) and soybean (*Glycine max*). *Indian Journal of Agricultural Sciences* 61(6): 374-378

Pujari, B.T.; Hosamani, M.M; Sharma, K.M.S.; Goudreddy, B.S. and Patel, V.C. (1989). Response of paddy to nutrients levels and weed management in Upghat region of Karnataka. *Journal of Maharashtra Agricultural Universities* 14(2): 189-192

Purushotham, S.; Kulakarmi, K.P. and Sharma, K.M.S. (1988).Comparative performance of paddy varieties under irrigation at different nitrogen levels during *kharif* season in Tumkur district. *Current Research. University of Agricultural Science, Bangalore* 17: 143-146

Rafey, A. and Prasad, N.K. (1992). Biological potential and economic feasibility of maize(*Zea mays*)+pigeonpea(*Cajanus cajan*) intercropping system in drylands. *Indian Journal of Agricultural Sciences* 62(2): 110-113

Rafey, A. and Prasad, N.K. (1995). Influence of weed management practices in pigeonpea (*Cajanus cajan*) intercropped with upland rice(*Oryza sativa*). *Indian Journal of Agricultural Sciences* 65(4): 281-282

Raju, U.R.; Jaganathan, A. and Rao, R.S. (1990). Performance of scented rice varieties under different levels of nitrogen in Andhra Pradesh. *Indian Journal of Agricultural Sciences* 60: 694-696

Ramesh, D.G. and Surve, D.N. (1984). Intercropping of legumes in sorghum with different levels of nitrogen. *Journal of Maharashtra Agricultural Universities* 9(3): 313-315

Rana, K.S. and Pal, M. (1999). Effect of intercropping systems and weed control on crop weed competition and grain yield of pigeonpea. *Crop Research (Hissar)* 17(2): 179-182

Rana, K.S.; Mahendra Pal and Rana, D.S. (1999). Nutrient depletion by pigeonpea (*Cajanus cajan*) and weeds as influenced by intercropping systems and weed management under rainfed conditions. *Indian Journal of Agronomy* 44(2): 267-270

Rao, M.R.; Rego, T.J. and Willey, R.W. (1987). Response of cereals to nitrogen in sole cropping and intercropping with different legumes. *Plant and Soil* 101(2): 167-177

Rathi, K.S. and Verma, V.S. (1979). Potato and mustard- A new companionship. *Indian Farming* 28(11): 13-14

Reddy, G.R.S.; Reddy, G.B.; Ramaiah, N.V. and Reddy, G.V. (1986). Effect of rates and source of nitrogen on transplanted lowland rice. *Indian Journal of Agronomy* 31(4): 416-418

Reddy, K.C.S.; Hussain, M.M. and Krantz, B.A. (1980). Effect of nitrogen levels and spacing on sorghum intercropped with pigeonpea and greengram in semi-arid land. *Indian Journal of Agricultural Sciences* 50(1): 17-22

Reddy, M.M; Madhavilatha, A. and Rao, L.J. (2007). Integrated weed management in pigeonpea(*Cajanus cajan*)+soybean(*Glycine max*) intercropping system in vertisols under rainfed conditions. *Indian Journal of Agricultural Sciences* 77(3): 177-178

Rice, E.L. (1974). Allelopathy. Academic Press, New York, U.S.A.

Risser, P.G. (1969). Competitive relationship among herbaceous grassland plants. *Biological Review* 35: 351

Samui, R.C.; Maiti, B.K. and Jana, P.K. (1979). Effect of nitrogen on pre- *kharif* direct-seeded rice. *Indian Journal of Agronomy* 24(1): 77-80

Sarwagi, S.K. and Tripathi, R.S. (1999). Planting geometry and nitrogen management in rice(*Oryza sativa*) and soybean (*Glycine max*) intercropping. *Indian Journal of Agronomy* 44(4): 681-687

Sengupta, K.; Bhattacharrya, K.K. and Chatterjee, B.N. (1985). Intercropping of upland rice with blackgram. *Journal of Agricultural Sciences (Cambridge)* 104: 217-221

Sharma, A.R. (1997). Effect of integrated weed management and nitrogen fertilization on the performance of rice under flood prone lowland conditions. *Journal of Agricultural Sciences (Cambridge)* 129(4): 409-414

Shetty, S.V.R. and Krantz, B.A. (1976). *Weed Research Annual Report*. 1976, ICRISAT, Hyderabad, India

Shivay, Y.S. and Singh, R.P. (2000). Growth, yield attributes, yields and nitrogen uptake of Maize (*Zea mays*) as influenced by cropping systems and nitrogen levels. *Annals of Agricultural Research* 21(4): 494-498

Shivay, Y.S.; Singh, R.P. and Pandey, C.S. (1999). Response of nitrogen in maize(*Zea mays*) based intercropping system. *Indian Journal of Agronomy* 44(2): 261-266

Singh, G.; Singh, R.K.; Singh, V.P.; Singh, B.B. and Nayak, R. (1999). Effect of crop-weed competition on yield and nutrient uptake by direct-seeded rice (*Oryza sativa*) in rainfed lowland situation. *Indian Journal of Agronomy* 44(4): 722-727.

Singh, G.V.; Rana, N.S. and Ahlawat, I.P.S. (1998a). Effect of nitrogen, *Rhizobium* inoculation and phosphorus on growth and yield of pigeonpea (*Cajanus cajan*). *Indian Journal of Agronomy* 43(2): 358-361

Singh, H.P.; Saxena, M.C.; Yadav, D.S. and Sharma, R.P. (1980). Chemical and mechanical weed control in pigeonpea under humid-subtropical conditions of Pantnagar. *Legume Research* 3: 22-26

Singh, K.; Singh, M. and Singh, R.S. (1978). Response of pigeonpea to fertilizers and *Rhizobium* inoculation under rainfed conditions. *Legume Research* 1(2):87-91

Singh, O.N. and Singh, R.S. (1985). Chemical weed control in pigeonpea. *International Pigeonpea Newsletter* 4: 26

Singh, O.N. and Singh, R.S. (1995). Effect of crop-weed competition on seed yield in pigeonpea+urdbean intercropping system. *Indian Journal of Pulses Research* 8(1): 29-32

Singh, R.S. (2006a). Studies on pigeonpea+rice intercropping under ridge-furrow method of planting. 8th Indian Agricultural Scientists and Farmer's Congress, February 21-22, 2006, B.H.U., Varanasi, Abst:147.

Singh, R.S. (2006b). Effect of row ratio and nitrogen levels on rice intercropped with ridge planted pigeonpea. *Indian Journal of Pulses Research* 19(2): 222-224

Singh, R.S. (2007). Integrated weed management in pigeonpea. *Environment and Ecology* 25(3): 780-782

Singh, R.V. and Gupta, P.C. (1994). Production potential of wheat and mustard cropping systems under adequate water supply conditions. *Indian Journal of Agricultural Research* 28 (4): 219-224

Singh, S.P. and Jha, D. (1984). Stability of sorghum-based intercropping system under rainfed conditions. *Indian Journal of Agronomy* 29(1): 101-106

Singh, S.P.; Singh, S.P. and Misra, P.K. (1998b). Response of short duration pigeonpea(*Cajanus cajan*) to nitrogen, *Rhizobium* inoculation and phosphorus. *Indian Journal of Agronomy* 43(4): 681-684

Singh, V.K.; Singh, N.P.; Sharma, B.B. and Sahu, J.P. (1998c). Effect of planting method and weed control practice on weed management and productivity of pigeonpea (*Cajanus cajan*) in foothills region. *Indian Journal of Agronomy* 43(4): 685-688

Sinha, A.C.; Mandal, B.B. and Jana, P.K. (1989a). Effect of time of sowing, row spacing and weed control treatments on weeds and grain yield of pigeonpea(*Cajanus cajan*). *Indian Journal of Agricultural Sciences* 59(6): 353-358

Sinha, A.C.; Mandal, B.B. and Jana, P.K. (1989b). Effect of time of sowing, row spacing and weed control practices on production of pigeonpea. *Indian Journal of Agronomy* 34(3): 283-285

Soundara, M.S.R. and Mahapatra, I.C. (1978). Relative efficiency of slow release and split application of nitrogenous fertilizers and weed control methods on yield of direct-seeded upland rice. *Oryza* 15(2): 117-123

Subbiah, B.V. and Asija,G.L. (1973). A rapid procedure for estimation of available nitrogen in soils. *Current Science* 28(8): 259-260

Varshney, J.G. (1993). Weed management in pigeonpea(*Cajanus cajan*) and greengram (*Phaseolus radiatus*) intercropping system. *Indian journal of Agricultural Sciences* 63(1): 4-7

Walkley, A. and Black, I.A. (1934). An examination of the Digtijareff method for determining soil organic matter and a proposed modification of the chromic acid titration method. *Soil Science* 37: 29-33

Willey, R.W. (1979). Intercropping - its importance and research needs - 1. Competition and yield advantage. *Field Crop Abstracts* 32(1): 1-10

Yadav, M.K. and Singh, R.S. (2009). Effect of nitrogen levels and weed management practices on pigeonpea(*Cajanus cajan*) and rice(*Oryza sativa*) intercropping system under ridge-furrow planting system. *Indian Journal of Agricultural Sciences* 79(4):268-272

5

Utilization of Sunn Hemp for Cover Crops and Weed Control in Temperate Climates

Andrew J. Price[1], Jessica Kelton[2] and Jorge Mosjidis[2]
[1]United States Department of Agriculture,
[2]Auburn University,
USA

1. Introduction

The use of smother crops or cover crop residue to suppress weed growth in agriculture is not a recent innovation; yet, only recently have smother, or cover crops, received considerable attention. The need to develop increasingly integrated pest management and sustainable food production systems has encouraged a greater interest to thoroughly evaluate effective utilization of cover crops in agricultural systems. In addition to providing a measure of weed control through physical obstruction and/or biochemical suppression, cover crops provide numerous environmental benefits that can promote long-term viability of farm lands (Jordan et al. 1999; Phatak et al. 2002; Yenish et al. 1996). Implementation of cover crops can reduce soil erosion, reduce runoff and improve water availability, improve soil structure, enhance soil organic matter, and increase diversity of soil biota (Bugg and Dutcher 1989; Reeves 1994; Wang et al. 2002a). These soil improvements, along with weed suppression capabilities, have made cover crops ideally suited for use in current and future sustainable agronomic systems.

Autumn-seeded cover crops include cereal grains, such as oat (*Avena sativa* L.) or rye (*Secale cereale* L.), Brassicas, like mustard (*Brassica* spp.) and radish (*Raphanus sativus* L.), or legumes, like clover (*Trifolium* spp.) or vetch (*Vicia* spp.) (SARE 2007). Each type of cover provides ecological benefits; however, leguminous cover crops are capable of providing biologically fixed nitrogen (N) which is available for uptake by the succeeding cash crop (Balkcom and Reeves 2005; Cherr et al. 2006; Karlen and Doran 1991; Wang et al. 2005). This source of nitrogen can greatly reduce N fertilizer applications necessary for the subsequent crop, and is of particular interest in low-input agriculture systems (Deberkow and Reichelderfer 1988). The drawback when utilizing legume cover crops, in comparison to grain covers, is acceleration of residue decomposition (Cherr et al. 2006; Somda et al. 1991). For weed control purposes, cover crops with plant portions containing relatively high C:N and high residue levels, such as cereal grains or sunn hemp (*Crotalaria juncea* L.), offer increased weed suppression for a relatively lengthier period of time during the growing season compared to cover crops with low C:N ratios (Cherr et al. 2006; Vigil and Kissel 1995). Legume cover crops have low C:N ratios, thus generally decompose more rapidly than cereal grains and require substantial biomass for extended ground cover. To resolve this issue, research has examined the use of tropical legume cover crops in temperate regions to facilitate N fixation while achieving suitable levels of biomass (Balkcom et al. 2011; Gallaher et al. 2001; Marshall et al. 2002; Mosjidis and Wehtje 2011).

Sunn hemp, a tropical legume that most likely originated from the Indo-Pakistani sub-continent, has been identified as a potential alternative to traditional legume cover crops employed in the southern portion of the United States (Cook and White 1996; Mansoer et al. 1997; Montgomery 1954; Mosjidis and Wetje 2011). As a tropical legume, sunn hemp can produce larger quantities of biomass in a shorter time period than winter legumes from temperate zones, while still providing an agronomically important amount of fixed N (Mansoer et al. 1997; Reeves et al. 1996; Wang et al. 2002b). The increased biomass production from sunn hemp would improve and extend weed control compared to other legume covers. Research continues worldwide to evaluate this species to determine its potential for widespread use in sustainable agricultural production, as well as to identify any limitations with the use of sunn hemp.

This chapter briefly explores the biological features of sunn hemp that make it a suitable cover crop and reviews previous research concerning weed suppression by sunn hemp. It also outlines current research projects targeting constraints on extensive adoption which include plant breeding studies and herbicide evaluations to improve sunn hemp production for seed availability. In order to improve weed managment options, it is necessary to continue investigating alternative methods to achieve effective, yet sustainable, weed control.

2. Cover crops

As stated previously, cover crops provide numerous environmental and weed suppression benefits. Implementation of cover crops into a production system is often in response to the need to reduce soil erosion and water runoff (Hartwig and Ammon 2002). However, with current advances toward sustainable growing practices as well as a need to reduce input costs, growers have begun to integrate cover crops for their weed control capabilities. The use of cover crops is typically found in conservation agriculture settings; cover crop residue left on the soil surface at planting provides a measure of weed control through shading of the soil and/or through allelopathy, chemical inhibition of plant germination, and as physical barrier for weed growth (Creamer et al. 1996; Price et al. 2007; Teasdale 1996).

To maximize weed suppression, high-residue cover crop systems that provide at least 4,500 kg ha^{-1} of biomass for ground cover are generally utilized (Balkcom et al. 2007). In these instances, winter cereal grain crops such as rye or oat are employed to attain the greatest amounts of residue prior to cash crop planting to maintain ground cover for an extended period into the growing season (Duiker and Curran 2005; Price et al. 2006, 2007; Ruffo and Bollero 2003). Although cereal crops can be established with relatively low costs and offer maximum biomass production for weed suppression, some systems would benefit more from the use of fall or winter legume cover crops.

3. Legume cover crops

Leguminous cover crops provide many benefits also achieved with other cover crop species including erosion control, improved water filtration, and improved soil organic matter.

However, a major constraint to the use of winter legumes covers is the lack of ample growing time between cash crops (Mansoer et al. 1997). Traditional planting windows for cover crops do not allow for maximum growth of cover crop species prior to the onset of cold temperatures; earlier planting of legumes would require a harvest of summer crops before maturity. In addition, planting cash crops often interferes with maturation of cover crops. Current limitations with legume biomass production have warranted research to resolve these issues in order to make use of the nitrogen fixation properties offered by legumes.

3.1 Nitrogen fixation in legumes

With the majority of atmospheric nitrogen present in a form unavailable for plant use (N_2), biological fixation of N_2 to NH_3 by bacteria is a critical process for contributing nitrogen to the soil environment (Meyer et al. 1978; Novoa and Loomis 1981). With legumes, bacteria fix atmospheric nitrogen within root nodules while the plant provides needed carbohydrates to facilitate the process (Figure 1). This symbiotic relationship allows many leguminous crops to be grown without the addition of synthetic fertilizers (Lindemann and Glover 2003; Phillips 1980).

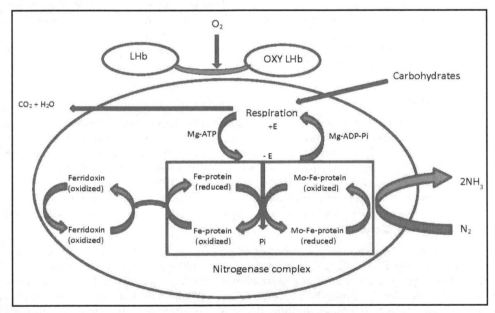

Fig. 1. Nitrogen fixation by bacteria, such as *Rhizobium*, occurs in root nodules of legumes.

As cover crops, legumes release nitrogen accumulations through vegetative decomposition, making NH_3 available for succeeding crops (Lindemann and Glover 2003). The rising cost of synthetic fertilizer, and the desire for viable alternatives to traditional high-input production practices, has boosted interest in the use of a number of legumes for cover crops.

The amount of N biologically fixed by a legume is dependent on a number of environmental conditions and management practices that affect biomass production (Holderbaum et al.

1990; Reeves 1994; Wagger 1989). Traditional fall-seeded legume cover crops have limited biomass production prior to cold temperatures, which can limit N accumulation and availability to subsequent crops. Utilization of tropical legumes, such as sunn hemp, may allow for greater biomass production and N accumulation during the available growing season between fall harvest and onset of winter in temperate climates (Mansoer et al. 1997).

3.2 Weed control with legume cover crops

Weed control obtained through legume cover crops has been well researched for a variety of cash crops (Caamal-Maldonado et al. 2001; DeGregoria and Ashley 1986; Teasdale 1988). Common legume covers have been shown to suppress growth of many species like pigweed (*Amaranthus* spp.), foxtail (*Setaria* spp.), and morningglory (*Ipomoeae* spp.) (Collins et al. 2007; Teasdale 1988; White et al. 1989). Either as a ground cover or through allelopathy, as noted in subterranean clover (*Trifolium subterraneum* L.), weed control achieved through legume cover crops has the potential to reduce early-season herbicide use in agricultural systems (Hartwig and Ammon 2002; Leather 1983; Mosjidis and Wehtje 2011). However, in most climates, the inability to produce high biomass with winter legumes, coupled with rapid decomposition, weed suppression is largely obtained during active cover growth and just after cover crop termination (Reddy 2001; Teasdale 1996). The use of sunn hemp immediately behind an early harvest summer cash crop like corn (*Zea mays* L.) may allow for extended post-harvest weed control through increased biomass production as well as slower decomposition rates in comparison to some other legume choices (Cherr et al. 2006; Cobo et al. 2002; Mansoer et al. 1997).

4. Sunn hemp

Sunn hemp, or Indian hemp, has become an important crop in regions such as India and Brazil that have climates well suited to the tropical, herbaceous annual (Bhardwaj et al. 2005; Duke 1981) (Figure 2). Typically utilized as a green manure due to its nitrogen accumulation, sunn hemp is also grown as a fiber crop; it can also be grown for forage since this *Crotalaria* species is nontoxic to animals (Rotar and Joy 1983). As a vigorously growing, relatively drought tolerant plant species, sunn hemp has been shown to thrive in a variety of soil types and with variable rainfall, but it is still most successful in tropical or subtropical environments (Wang et al. 2002b).

The utilization of this species in cooler, temperate climates, such as those found in the continental United States (US), began in the early 1930's in response to sunn hemp's potential as a green manure and for suppression of root-knot nematodes (Cook and White 1996; Cook et al. 1998; Dempsey 1975). At the onset of World War II, increased demands for rope fiber drew more attention to sunn hemp as an alternative for imported cordage material (Cook and White 1996; Wilson et al. 1965). During the 1950's and 1960's, US research placed particular emphasis on sunn hemp production as a quickly-renewed source of fiber for paper materials (Nelson et al. 1961). Although most attention for nonwood fiber sources has been concentrated on kenaf (*Hibiscus cannabinus* L.), some research continues to identify sunn hemp as a potential source that can be produced in the US, particularly in Hawaii, southern Texas, and south Florida (Cook and Scott 1998; Webber and Bledsoe 1993). In temperate regions of the US, however, more recent research has evaluated the cover crop potential of sunn hemp as a frost-terminated, late-summer alternative to winter legume covers such as clover and vetch or between crop harvest and cereal cover crop planting (Balkcom and Reeves 2005; Mansoer et al. 1997).

4.1 Sunn hemp cover crops

Sunn hemp is readily used as a rotational crop in tropical regions with cash crops such as rice, cotton, and corn (Purseglove 1974). Its value as a cover crop is due to its biomass production, N accumulation, reduced pests and pathogen infestation, and weed suppression achieved when planted (Wang et al. 2002b). While sunn hemp is not winter hardy, its adaptability to various soil types and precipitation amounts has allowed sunn hemp to be grown in temperate regions as a green manure (Dempsey 1975). In the US, sunn hemp was used in this manner during the 1930's until reduced seed availability caused interest in the crop to diminish (Cook and White 1996). Recently, sunn hemp has again received attention as a potential alternative to winter annual legume cover crops in these temperate climates (Balkcom and Reeves 2005; Mansoer et al. 1997). Sunn hemp's vigorous growth and nitrogen production can provide needed ground cover to control erosion and an N source to succeeding cash crops. To fully utilize sunn hemp nitrogen release, investigations have also been conducted to determine the suitability of sunn hemp grown as a late summer cover crop between harvest and winter planting of cash crop or cereal cover crop (Balkcom et al. 2011; Creamer and Baldwin 2000; Schomberg et al. 2007).

The rapid growth of sunn hemp in a relatively short period of time allows for a relatively high amount of biomass production prior to the onset of cool temperatures in mild climates across the southeastern US. Previous research has reported sunn hemp biomass to average between 1 and 9 Mg ha^{-1} in 45 to 90 days after planting, respectively (Mansoer et al. 1997; Morris et al. 1986; Schomberg et al. 2007; Reeves et al. 1996; Yadvinder et al. 1992). Although environmental conditions affect potential biomass production, substantial amounts of biomass can be achieved under typical late-summer and autumn conditions to aid in erosion control, nematode and weed suppression, and N accumulation before frost occurs.

As a legume cover crop, sunn hemp can fix atmospheric nitrogen that is available over time to succeeding crops as it decomposes. With high fertilizer prices and sustainability concerns with synthetic soil amendments, the potential use of sunn hemp to provide nitrogen to crops such as cotton, corn, and rice has prompted research to determine N availability from a sunn hemp cover crop (Balkcom and Reeves 2005; Chung et al. 2000; Sangakkara et al. 2004; Schomberg et al. 2007). Nitrogen production with sunn hemp varies depending on many factors; however, reported N values in sunn hemp range between 110 and 160 kg ha^{-1} (Balkcom and Reeves 2005; Mansoer et al. 1997; Marshall et al. 2001). In most investigations, sunn hemp nitrogen content equals or exceeds N content of traditional winter legume cover crops (Reeves 1994).

With nitrogen release during winter months reported at approximately 75 kg ha^{-1}, remaining N may still be utilized by spring planted crops (Reeves et al. 1996). Due to winter nitrogen loss from leaching and utilization, however, an alternative scenario mentioned previously for sunn hemp use is to employ the crop as a late summer cover prior to grain cover crop or winter cash crop planting in the fall (Balkcom et al. 2011; Schomberg et al. 2007). Released nitrogen from sunn hemp residue would be available to the subsequent crop while minimizing N losses during winter months.

Fig. 2. The vigorous growth of sunn hemp cultivars developed at Auburn University, Alabama, can be seen here, 70 days after planting (Photo by J.A. Mosjidis).

4.2 Weed suppression by sunn hemp

It has been noted by many researchers that *C. juncea* can suppress populations of pests such as nematodes and vigorous weed species (Collins et al. 2007; Fassuliotis and Skucas 1969; McSorley et al. 1994; Taylor 1985; Wang et al. 2001). Much of the pest control potential of sunn hemp has focused primarily on nematode control of such species as *Meloidogyne* spp., *Rotylenchulus reniformis*, *Radopholus similis*, and *Heterodera glycines* (Birchfield and Bristline 1956; Desaeger and Rao 2000; Good et al. 1965; Marla et al. 2008; Robinson et al. 1998; Wang et al. 2004). Although some questions remain concerning suppression by sunn hemp for specific nematode species, extensive research has provided considerable knowledge as to how *C. juncea* reduces certain nematode populations (Halbrendt 1996; Kloepper et al. 1991; LaMondia 1996; Rodriguez-Kabana 1994; Wang et al. 2002b).

In contrast, weed suppression specifically by sunn hemp cover crops has been minimally investigated and only recently has it received more attention. General comments concerning the potential of sunn hemp to suppress weed species have been reported in several studies (Reeves et al. 1996; SARE 2007). Weed control by sunn hemp has been mostly attributed to vigorous plant growth and rapid shading of the ground (Duke 1981). In fact, Mosjidis and Wehtje (2011) demonstrated that there was a progressive reduction in weed biomass as a sunn hemp stand increased up to 100 plants/m^2. Furthermore, recent research has suggested allelopathic compounds released from sunn hemp also cause weed suppression (Adler and Chase 2007; Collins et al. 2007; Leather and Forrence 1990; Price et al. 2008). More research is necessary to determine the extent of allelochemical functions in sunn hemp.

Significant weed control can be achieved under moderate to high levels of sunn hemp (Mosjidis and Wehtje 2011; SARE 2007; Severino and Christoffoleti 2004). However, several weed species, such as nutsedge (*Cyperus* spp.), morningglory (*Ipomoea* spp.), and bermudagrass (*Cynodon dactylon*), are capable of thriving in a sunn hemp stand (Chaudhury et al. 2007; Collins et al. 2007; McKee et al. 1946). It is expected that, as sunn hemp utilization for cover crops and weed control grows, research efforts to fully understand weed suppression by *C. juncea* will continue.

4.3 Weed control in sunn hemp production

Little research has been conducted to determine weed control strategies in sunn hemp production. *C. juncea* grown as a green manure or cover crop typically does not require extensive weed management but production of sunn hemp for seed production may benefit from additional weed control practices. Due to sunn hemp's rapid growth and possible allelopathic effects, it can be easily established and can out-compete neighboring weed species. In fact, areas that rely on hand removal of weed species in agricultural productions generally do not employ this practice in sunn hemp grown for fiber, since weed competition is minimal (Chaudhury et al. 2007). However, in regions that utilize herbicides for weed control, early season weed suppression with herbicide applications may increase seed production in sunn hemp stands.

At present, no herbicides are labelled for use in sunn hemp production, but research by Mosjidis and Wehtje (2011) identified a preemergent herbicide, pendimethalin, as a potential treatment that would provide effective weed control during establishment. This research also found that sunn hemp could also tolerate 2,4-DB as a postemergent application (Mosjidis and Wehtje 2011). Imazethapyr was also determined to be safe for use in sunn hemp and effective against yellow nutsedge (*Cyperus esculentus* L.), which is not well suppressed by sunn hemp (Collins et al. 2007; Mosjidis and Wehtje 2011). With advancements in sunn hemp breeding and increased utilization of this species in a broader geographical range, continued research efforts to determine additional herbicide programs will be critical for successful production in the future.

4.4 Breeding and seed availability

As a native tropical species, sunn hemp use in temperate regions faces some challenges. Although sunn hemp experiences rapid vegetative growth in a short time frame, viable seed production typically requires a longer season than can be achieved before winter conditions in temperate regions (Li et al. 2009)(sunn hemp seed pictured in Figure 3). The lack of seed production outside of tropical and subtropical climates severely limits seed availability to producers in cooler climates. Moreover, with seed costs ranging from $90 to $130 (US) per hectare, implementation of C. *juncea* as a cover crop can be an expensive task for growers (Li et al. 2009; Petcher 2009).

Fig. 3. Sunn hemp seed production can yield 450 to 1000 kg of seed per hectare (Photo by J.A. Mosjidis).

A good deal of breeding research has been conducted in countries throughout the world that grow sunn hemp for fiber production (Kundu 1964; Ram and Singh 2011). In India, particularly, cultivars are developed for high fiber yield and resistance to wilt diseases (Chaudhury et al. 2007). Most commonly used in this region is 'Kharif sunn' or 'K-12' which produce high yield of good quality fiber (Chaudhury et al. 2007). Other important cultivars in India include 'SS-11' and 'T-6' which is a day-neutral variety, while most varieties are short day plants (Chaudhury et al. 2007). In other regions of sunn hemp production, varieties such as 'Somerset' in South Africa and 'KRC-1' in Brazil are commonly used.

Due to elevated seed costs and limited seed production in the US, sunn hemp breeding research has focused on developing a C. *juncea* cultivar that can produce seeds in climatic conditions prevalent in the southeastern US. For successful seed production in traditional sunn hemp regions, characteristics of seed production locations should typically be below 24° N latitude, not fall below 10° C, have ample sunlight and not receive rainfall during fruit set (Chaudhury et al. 2007). In cooler climates above this latitude, temperatures below optimal usually occur before sunn hemp seed can mature. The recent development of cultivar, 'AU Golden' and 'AU Durbin' has been shown to produce viable seed in these temperate regions; research is on-going to determine best management practices for implementing these sunn hemp cultivars (Balkcom et al. 2011; Mosjidis 2007, 2010).

5. Conclusions

The progression of agricultural systems towards more sustainable, yet high yielding production has required researchers to identify numerous alternative weed management practices that can be employed along with traditional weed control tactics. The use of sunn hemp as a cover crop, either as a substitute for winter annual legumes or as a late summer cover between harvest and winter crops, delivers effective weed control while providing ground cover and a nitrogen source for subsequent crops. Continued research with sunn hemp and crops similar to this species may provide even more benefits to weed control efforts in the future.

6. References

Adler, M.J., and C.A. Chase. 2007. A comparative analysis of the allelopathic potential of leguminous summer cover crops: cowpea, sunn hemp and velvetbean. HortScience 42: 289-293.

Balkcom, K.S. and D.W. Reeves. 2005. Sunn hemp utilized as a legume cover crop for corn production. Agronomy Journal 97: 26-31.

Balkcom, K.S., H. Schomberg, D.W. Reeves, and A. Clark. 2007. Managing cover crops in conservation tillage systems. *In* A. Clark (ed.). *Managing Cover Crops Profitably*. SARE. College Park, MD: 44-61.

Balkcom, K.S., J.M. Massey, J.A. Mosjidis, A.J. Price, and S.F. Enloe. 2011. Planting date and seeding rate effects on sunn hemp biomass and nitrogen production for a winter cover crop. International Journal of Agronomy 2011: 1-8.

Bhardwaj, H.L., C.L. Webber, and G.S. Sakamoto. 2005. Cultivation of kenaf and sunn hemp in the mid-Atlantic United States. Industrial Crops and Products 22: 151-155.

Birchfield, W. and F. Bristline. 1956. Cover crop in relation to the burrowing nematode, *Radopholus similis*. Plant Disease Reporter 40: 398-399.

Bugg, R.L. and J.D. Dutcher. 1989. Warm-season cover crops for pecan orchards: horticultural and entomological implications. Biology, Agriculture, and Horticulture 6: 123-148.

Caamal-Maldonado, J.A., J.J. Jimenez-Osornio, A. Torres-Barragan, and A.L. Anaya. 2001. The use of allelopathic legume cover and mulch species for weed control in cropping systems. Agronomy Journal 93: 27-36.

Chaudhury, J., D.P. Singh, and S.K. Hazra. 2007. Sunn Hemp (*Crotalaria juncea* L.). Central Research Institute for Jute and Allied Fibres (ICAR). available at: http://assamagribusiness.nic.in/Sunnhemp.pdf

Cherr, C.M., J.M.S. Scholberg, and R. McSorley. 2006. Green manure as nitrogen source for sweet corn in a warm-temperate environment. Agronomy Journal 98: 1173-1180.

Chung, R., C. Wang, C.W. Wang, and Y.P. Wang. 2000. Influence of organic matter and inorganic fertilizer on the growth and nitrogen accumulation of corn plants. Journal of Plant Nutrition 23: 297-311.

Collins, A.S., C.A. Chase, W.M. Stall, and C.M. Hutchinson. 2007. Competitiveness of three leguminous cover crops with yellow nutsedge (Cyperus esculentus) and smooth pigweed (Amaranthus hybridus). Weed Science 55: 613-618.

Cobo, J.G., E. Barrios, D.C.L, Kass and R.J. Thomas. 2002. Decomposition and nutrient release by green manures in a tropical hillside agroecosystem. Plant and Soil 240: 331-342.

Cook, C.G. and G.A. White. 1996. Crotalaria juncea: A potential multi-purpose fiber crop. In J. Janick (ed.) Progress in New Crops. ASHS Press. Arlington, VA: 389-394.

Cook, C.G. and A.W. Scott. 1998. Plant population effects on stalk growth, yield, and bark fiber content of sunn hemp. Industrial Crops and Products 8: 97-103.

Cook, C.G., A.W. Scott, P. Chow. 1998. Planting date and cultivar effects on growth and stalk yield of sunn hemp. Industrial Crops and Products 8: 89-95.

Creamer, N.G., M.A. Bennett, B.R. Stinner, J. Cardina, and E.E. Regnier. 1996. Mechanisms of weed suppression in cover crop-based production systems. HortScience 31: 410-413.

Creamer, N.G. and K.R. Baldwin. 2000. An evaluation of summer cover crops for use in vegetable production systems in North Carolina. HortScience 35: 600-603.

Deberkow, S.G. and K.H. Reichelderfer. 1988. Low-input agriculture: trends, goals, and prospects for input use. American Journal of Agricultural Economics 70: 1159-1166.

Dempsey, J.M. 1975. Fiber Crops. University Presses of Florida: Gainesville, FL.

DeGregorio, R.E. and R.A. Ashley. 1986. Screening living mulches/cover crops for no-till snap beans. Proceedings of the Northeast Weed Science Society 40: 87- 91.

Desaeger, J., and M.R. Rao. 2000. Parasitic nematode populations in natural fallows and improved cover crops and their effects on subsequent crops in Kenya. Field Crops Research 65: 41-56.

Duiker, S.W. and W.S. Curran. 2005. Rye cover crop managemenr for corn production in the northern Mid-Atlantic region. Agronomy Journal 97: 1413-1418.

Duke, J.A. 1981. Handbook of Legumes of World Importance. Plenum Press: New York, NY.

Fassuliotis, G., and G.P. Skucas. 1969. The effect of pyrrolizidine alkaloid ester and plants containing pyrrolizidine on Meloidogyne incognita acrita. Journal of Nematology 1: 287-288.

Gallaher, R.N., A. Higuera, and A.J. Marshall. 2001. Dry matter and nitrogen accumulation in fall planted sunn hemp. Proceedings of annual American Society of Agronomy, Crop Science Society of America, and Soil Science Society of America meeting. Charlotte, N.C. 21-25 October, 2001.

Good, J.M., N.A. Minton, and C.A. Jaworski. 1965. Relative susceptibility of selected cover crops and Coastal Bermudagrass to plant nematodes. Phytopathology 55: 1026-1030.

Halbrendt, J.M. 1996. Allelopathy in the management of plant-parasitic nematodes. Journal of Nematology 28: 8-14.

Hartwig, N.L. and H.U. Ammon. 2002. Cover crops and living mulches. Weed Science 50: 688-699.

Holderbaum, J.F., A.M. Decker, J.J. Meisinger, F.R. Mulford, and L.R. Vough. 1990. Fall-seeded legume cover crops for no-tillage corn in the humid East. Agronomy Journal 82: 117-124.

Jordan, D.L., P.K. Bollich, M.P. Braverman, and D.E. Sanders. 1999. Influence of tillage and *Triticum aestivum* cover crop on herbicide efficacy in *Oryza sativa*. Weed Science 47: 332-337.

Karlen, D.L. and J.W. Doran. 1991. Cover crop management effects on soybean and corn growth and nitrogen dynamics in an on-farm study. American Journal of Alternative Agriculture 6: 71-82.

Kloepper, J.W., R. Rodriguez-Kabana, J.A. McInroy, and D.J. Collins. 1991. Analysis of populations and physiological characterization of microorganisms in rhizospheres of plants with antagonistic properties to phytopathogenic nematodes. Plant and Soil 136: 95-102.

Kundu, B.C. 1964. Sunn-hemp in India. Proceedings of the Soil and Crop Society of Florida. 24: 396-404.

LaMondia, J.A. 1996. Trap crops and population management of *Globoder tabacum tabacum*. Journal of Nematology 28: 238-243.

Leather, G.R. 1983. Weed control using allelopathic crop plants. Journal of Chemical Ecology 9: 983-989.

Leather, G.R. and L.E. Forrence. 1990. Sunn hemp is allelopathic to leafy spurge. Proceedings and Progress Reports of the Leafy Spurge Symposium. Gillette, WY. 10-12 July, 1990.

Li, Y., Q. Wang, W. Klassen, and E.A. Hanlon. 2009. Sunn hemp-A cover crop in Florida. University of Florida Extension bulletin SL 306. Available at: http://edis.ifas.ufl.edu/pdffiles/TR/TR00300.pdf

Lindemann, W.C. and C.R. Glover. 2003. Nitrogen fixation by legumes. New Mexico State University Cooperative Extension Guide A-129. available at: http://aces.nmsu.edu/pubs/_a/a-129.pdf.

Marla, S.R., R.N. Huettel, and J. Mosjidis. 2008. Evaluation of *Crotalaria juncea* populations as hosts and antagonistic crops to manage *Meloidogyne incognita* and *Rotylenchulus reniformis*. Nematropica 38: 155-161.

Mansoer, Z., D.W. Reeves, and C.W. Wood. 1997. Sustainability of sunn hemp as an alternative late-summer legume cover crop. Soil Science Society of America Journal 61: 246-253.

Marshall, A.J., R.N. Gallaher, A. Higuera, and R.S. Tubbs. 2001. Seeding rate for fall planted sunn hemp. Proceedings Annual American Society of Agronomy, Crop Science Society of America, and Soil Science Society of America meeting. Charlotte, N.C. 21-25 October, 2001.

Marshall, A.J., R.N. Gallaher, K.H. Wang, and R. McSorley. 2002. Partitioning of dry matter and minerals in sunn hemp. *In* E. van Santen (ed.) *Making Conservation Tillage Conventional: Building a Future on 25 Years of Research*. Proceedings Annual Southern Conservation Tillage Conference for Sustainable Agriculture 25: 310-313.

McKee, R., G.E. Ritchey, J.L. Stephens, and H.W. Johnson. 1946. Crotalaria culture and utilization. United States Department of Agriculture Farmer's Bulletin Number 1980.

McSorley, R., D.W. Dickson, and J.A. Brito. 1994. Host status of selected tropical rotation crops to four populations of root-knot nematodes. Nematropica 24: 45-53.

Meyer, J., B.C. Kelly, P.M. Vignais. 1978. Nitrogen fixation and hydrogen metabolism in photosynthetic bacteria. Biochimie 60: 245-260.

Montgomery, B. 1954. Sunn fiber. In H.R. Mauersberger (ed.) Mathew's Textile Fibers. Wiley Publishing. New York, NY: 323-327.

Morris, R.A., R.E. Furoc, and M.A. Dizon. 1986. Rice responses to a short-duratin green manure. I. Grain yield. Agronomy Journal 78: 409-412.

Mosjidis, J.A. 2007. Breeding of annual and perennial legumes and their utilization as forage and crops. Field and Vegetable Crops Research 44: 7-11.

Mosjidis, J.A. 2010. Performance of Sunn Hemp cultivars developed at Auburn University. Annual Meeting Abstracts. Laguna Beach, CA: ASA, CSSA, SSSA. CD-ROM.

Mosjidis, J.A. and G. Wehtje. 2011. Weed control in sunn hemp and its ability to suppress weed growth. Crop Protection 30: 70-73.

Nelson, G.H., H.J. Nieschlag, M.E. Daxenbichler, I.A. Wolf, and R.E. Perdue. 1961. A search for new fiber crops. III. Laboratory scale pulping studies. Technical Association of the Pulp and Paper Industry 44: 319-325.

Novoa, R. and R.S. Loomis. 1981. Nitrogen and plant production. Plant and Soil 58: 177-204.

Petcher, R. 2009. Sunn hemp- a new cover crop for Alabama. Auburn University Extension Report. Available at: http://www.aces.edu/counties/Baldwin/sunhemp.php.

Phatak, S.C., J.R. Dozier, A.G. Bateman, K.E. Brunson, and N.L. Martini. 2002. Cover crops and conservation tillage in sustainable vegetable production. In E. van Santen (ed.) Making Conservation Tillage Conventional: Building a Future on 25 Years of Research. Proceedings Annual Southern Conservation Tillage Conference for Sustainable Agriculture 25: 401-403.

Phillips, D.A. 1980. Efficiency of symbiotic nitrogen fixation in legumes. Annual Review of Plant Physiology 31: 29-49.

Price, A.J., D.W. Reeves, M.G. Patterson, B.E. Gamble, K.S. Balkcom, F.J. Arriaga, and C.D. Monks. 2007. Weed control in peanut grown in a high-residue conservation-tillage system. Peanut Science 34: 59-64.

Price, A. J., D. W. Reeves, and M. G. Patterson. 2006. Evaluation of weed control provided by three winter cereals in conservation-tillage soybean. Renewable Agric. Food Sys. 21:159-164.

Price, A.J., M.E. Stoll, J.S. Bergtold, F.J. Arriaga, K.S. Balkcom, T.S. Kornecki, and R.L. Raper. 2008. Effect of cover crop extracts on cotton and radish radicle elongation. Communications in Biometry and Crop Science. 3: 60-66.

Purseglove, J.W. 1974. Tropical crops: Dicotylendons. Longman Group Limited, London.

Ram, H. and G. Singh. 2011. Growth and seed yield of sunnhemp genotypes as influenced by different sowing methods and seed rates. World Journal of Agricultural Sciences 7: 109-112.

Reeves, D.W. 1994. Cover crop and rotations. In J.L. Hatfield and B.A. Stewart (ed.) Crops Residue Management. Lewis Publishing. Boca Raton, FL: 125-172.

Reeves, D.W., Z. Mansoer, and C.W. Wood. 1996. Suitability of sunn hemp as an alternative legume cover crop. Proceedings of the New Technology and Conservation Tillage 96: 125-130.

Reddy, K.N. 2001. Effects of cereal and legume cover crop residues on weeds, yield, and net return in soybean (*Glycine max*). Weed Technology 15: 660-668.

Robinson, A.F., C.G. Cook, and A.C. Bridges. 1998. Comparative reproduction of *Rotylenchulus reniformis* and *Meloidogyne incognita* race 3 on kenaf and sunn hemp grown in rotation with cotton. Nematropica 28: 144.

Rodriguez-Kabana, R., N. Kokalis-Burelle, D.G. Roberson, P.S. King and L.W. Wells. 1994. Rotations with coastal bermudagrass, cotton, and bahiagrass for management of *Meloidogyne arenaria* and southern blight in peanut. Supplement to Journal of Nematology 26: 665-668.

Rotar, P.P. and R.J. Joy. 1983. 'Tropic Sun' sunn hemp. *Crotalaria juncea* L. University of Hawaii, College of Tropical Agricultural and Human Resources, Series 36. available at: http://scholarspace.manoa.hawaii.edu/bitstream/handle/10125/15089/RES-36.pdf.txt?sequence=3.

Ruffo, M.L. and G.A. Bollero. 2003. Modeling rye and hairy vetch residue decomposition as a function of degree-days and decomposition-days. Agronomy Journal 95: 900-907.

Sangakkara, U.R., M. Liedgens, A. Soldati, and P. Stamp. 2004. Root and shoot growth of maize (*Zea mays*) as affected by incorporation of *Crotalaria juncea* and *Tithonia diversifolia* as green manures. Journal of Agronomy and Crop Science 190: 339-346.

SARE (Sustainable Agriculture Research and Education). 2007. A. Clark (ed.). *Managing Cover Crops Profitably*. SARE: College Park, MD.

Schomberg, H.H., N.L. Martini, J.C. Diaz-Perez, S.C. Phatak, K.S. Balkcom, and H.L. Bhardwaj. 2007. Potential for using sunn hemp as a source of biomass and nitrogen for the Piedmont and Coastal Plain regions of the southeastern USA. Agronomy Journal 99: 1448-1457.

Severino, F.J. and P.J. Christoffoleti. 2004. Weed suppression by smother crops and selective herbicides. Scientia Agricola 61: 21-26.

Somda, Z.C., P.B. Ford, and W.L. Hargrove. 1991. Decomposition and nitrogen recycling of cover crops and crop residues. *In* W.L. Hargrove (ed.) *Cover Crops for Clean Water*. Soil and Water Conservation Society: Ankeny, IA.

Taylor, S.G. 1985. Interactions between six warm-season legumes and three species of root-knot nematodes. Journal of Nematology 17: 367-370.

Teasdale, J.R. 1988. Weed suppression by hairy vetch residue. Proceedings of the Northeast Weed Science Society 42: 73.

Teasdale, J.R. 1996. Contribution of cover crops to weed management in sustainable agriculture systems. Journal of Production Agriculture 9: 475-479.

Vigil, M.F. and D.E. Kissel. 1995. Rate of nitrogen mineralized from incorporated crop residues as influenced by temperature. Soil Science Society of America Journal 59: 1636-1644.

Wagger, M.G. 1989. Time of desiccation effects on plant composition and subsequent nitrogen release from several winter annual cover crops. Agronomy Journal 81: 236-241.

Wang, K.-H., B.S. Sipes, and D.P. Schmitt. 2001. Suppression of *Rotylenchulus reniformis* by *Crotalaria juncea, Brassica napus,* and *Tagetes erecta*. Nematropica 31: 235-249.

Wang, Q., W. Klassen, A.A. Abdul-Baki, H.H. Bryan, and Y. Li. 2002a. Influence of summer cover crops on soil nematodes in a tomato field. Proceedings of the Florida Soil Crop Science Society 62: 86-91.

Wang, K., B.S. Sipes, and D.P. Schmitt. 2002b. *Crotalaria* as a cover crop for nematode management: a review. Nematropica 32: 35-57.

Wang, K.-H., R. McSorley, A.J. Marshall, and R.N. Gallaher. 2004. Nematode community changes associated with decomposition of *Crotalaria juncea* amendment in litterbags. Applied Soil Ecology 27: 31-45.

Wang, Q., Y. Li, and W. Klassen. 2005. Influence of summer cover crops on conservation of soil water and nutrients in a subtropical area. Journal of Soil and Water Conservation 60: 58-63.

Webber, C.L. and V.K. Bledsoe. 1993. Kenaf production, harvesting and products. *In* J. Janick and J.E. Simon (eds.). New Crops. Wiley, New York, NY: 416-421.

White, R.H., A.D. Worsham, and U. Blum. 1989. Allelopathic potential of legume debris and aqueous extracts. Weed Science 37: 674-679.

Wilson, F.D., T.E. Summers, J.F. Joyner, D.W. Fishler, and C.C. Seale. 1965. 'Everglades 41' and 'Everglades 71', two new cultivars of kenaf (*Hibiscus cannabinus* L.) for the fiber and seed. Florida Agricultural Experiment Station Circular S168.

Yadvinder, S., S. Bihay, and C.S. Khind. 1992. Nutrient transformations in soil amended with green manures. Advances in Soil Science 20: 237-309.

Yenish, J.P., A.D. Worsham, and A.C. York. 1996. Cover crops for herbicide replacement on no-tillage corn (*Zea mays*). Weed Technology 10: 815-821.

6

Allelopathic Weed Suppression Through the Use of Cover Crops

Jessica Kelton[1], Andrew J. Price[2] and Jorge Mosjidis[1]
[1]Auburn University
[2]United States Department of Agriculture
USA

1. Introduction

There has long been observed an inhibitive response by plant species to certain neighboring plants. The Greek philosopher and botanist, Theophrastus, noted this effect from cabbage as early as 300 BC (Willis 1985). Since that time, others have documented similar plant interactions. In 1937, Austrian botanist, Hans Molisch, described this phenomenon as allelopathy, which he determined to be the result of biochemical interactions between plants (Molisch 1937; Putnam and Duke 1978). When first described, allelopathy referred to both deleterious and beneficial interactions between species; since that time, however, allelopathy has been applied to only adverse plant interactions, rather than to both. First described by a Roman scholar during the first century, black walnut (*Juglans nigra* L.) has long served as the common example of allelopathic effects with its ability to inhibit growth of surrounding plants either through decaying leaves or nuts or from the tree itself (Weir et al. 2004). Researchers have continued to examine allelopathy and the mechanism for biochemical inhibition, which was initially scrutinized by many since differentiation between this effect and plant competition remained uncertain (Weir et al. 2004). Subsequent bioassays involving specific chemical compounds extracted from plants have confirmed that certain species do, in fact, produce biochemicals that can inhibit plant germination and growth in the absence of resource competition (Einhellig 1994a).

With confirmation of allelopathy, many investigations have been conducted in order to determine how best to utilize this effect for possible weed control in agricultural settings (Khanh et al. 2005; Olofsdotter 2001; Weston 1996). The ability to inhibit weed growth through the implementation of cover crops into a crop rotation has been a focal point for this research for several reasons. In addition to weed suppression and control through allelopathy, as well as a mulching effect, cover crops provide substantial environmental benefits such as reduced erosion and water runoff (Price et al. 2006; Truman et al. 2003). Moreover, cover crops are readily available and easily adapted to many agricultural situations. Because of these many benefits, including natural weed suppression through allelopathy, the use of cover crops has become a vital component of sustainable agriculture systems, as well as organic production.

Ensuring sufficient food and fiber production for future generations can be hampered by limited options for weed control, particularly in developing countries where yields are reduced by up to 25% by weed competition. Identifying and describing sustainable weed control measures that can be implemented to reduce weed pressure in a number of settings can help safeguard the productivity of agriculture. Therefore, the objectives of this chapter are to describe the fundamentals of allelopathy and how to utilize allelopathic compounds for weed control through cover crop use. The chapter also highlights many of the identified biochemicals, their structures, and the respective cover crops in which they are found. Lastly, we describe the degree of allelopathic potential for a number of cover crops, as determined by laboratory testing.

2. Production and release of allelopathic compounds

Allelochemicals enter the environment from plants in a number of ways, such as plant degradation, volatilization, leaching from plant leaves, and from root exudation (Bertin et al. 2003; Weir et al. 2004). During active plant growth, particularly in early growth stages or during periods of stress, root exudation, either through diffusion, ion channels, or vesicle transport, is the primary method for release of many organic and inorganic compounds into the rhizosphere (Battey and Blackbourn 1993; Uren 2000). These compounds serve a multitude of functions such as improving nutrient uptake, root lubrication, plant growth regulation, microorganism defense, and waste removal (Bertin et al. 2003; Fan et al. 1997; Uren 2000).

A large proportion of identified allelochemicals are noted to be secondary compounds formed during photosynthetic processes (Einhellig 1994b; Swain 1977). Since many allelopathic chemicals appear to perform no primary metabolic functions, although some compounds such as cinnamic acid and salicylic acid do serve other functions within a plant, it is unclear at this point as to what regulates the release of these compounds (Einhellig 1994a). Many environmental plant stressors have been observed to increase allelochemical release but not necessarily chemical production (Bertin et al. 2003; Inderjit and Weston 2003; Sterling et al. 1987). Plant stressors such as elevated temperature, reduced water availability, and herbivory may cause increased allelochemical release; however, a definitive correlation between environmental factors and allelopathic compounds has yet to be made (Bertin et al. 2003; Pramanik et al. 2000). Continued research directed at isolating and identifying individual root exudates while manipulating environmental stress factors may help to increase our understanding of allelochemical release into the rhizosphere.

3. Allelopathic compounds

Many allelochemicals have been identified since experiments began to isolate and determine allelopathic potentials of plant compounds. Compounds that have been identified thus far include a variety of chemical classes such as phenolic acids, coumarins, benzoquinones, terpenoids, glucosinolates, and tannins (Chung et al. 2002; Putnam and Duke 1978; Seigler 1996; Swain 1977; Vyvyan 2002). These and other allelochemicals are found in many plant species from woody to herbaceous plants, grasses and broadleaves, weeds and crops. There are many details left to be determined such as regulation and production stimuli and mode of action for inhibition. It is also not readily understood to what extent allelopathic compounds interact with each other and other chemical compounds within the rhizosphere to inhibit surrounding plants. The following sections present several of the structural classes of recognized allelochemicals as well as specific compounds within each group.

3.1 Phenolic acids

Like most allelochemicals, phenolic acids are secondary plant compounds typified by a hydroxylated aromatic ring structure. To date, a number of phenolic acids have been determined to have allelopathic properties and have been measured in extracts from a variety of plant species (Figure 1). Species which have been noted to produce phenolic acids include: rice (*Oryza sativa* L.), wheat (*Triticum aestivum* L.), mango (*Mangifera indica* L.), and spotted knapweed (*Centaurea stoebe* L.) (Bais et al. 2003; Chung et al. 2002; El-Rokiek et al. 2010; Fitter 2003). Many species, such as rice, contain multiple phenolic compounds along with other allelopathic compounds. In two studies, researchers isolated nine individual phenolic acids from rice hull extracts and 14 different phenolic acids from buffalograss [*Buchloe dactyloides* (Nutt.) Engelm] (Chung et al. 2002; Wu et al. 1998). At this time, however, it is not clear to what degree individual allelochemicals interact to produce plant inhibition. Some reports show a synergistic effect when allelochemicals are in a mixture, while other studies indicate decreased plant inhibition in the presence of a mixture when compared to individual chemical inhibition (Chung et al. 2002; Einhellig 1996).

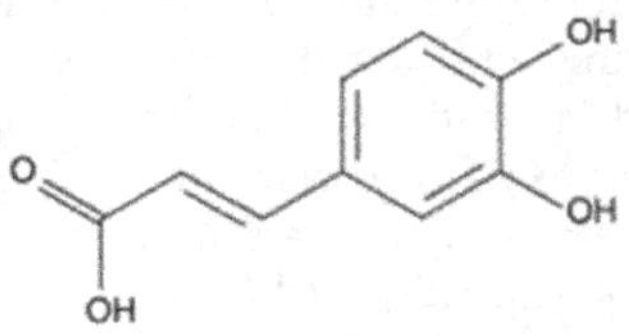

| Caffeic acid | Coumaric acid | Vanillic acid |

| Syringic acid | *p*-hydrobenzoic | Ferulic acid |

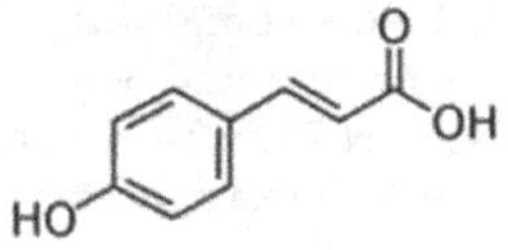

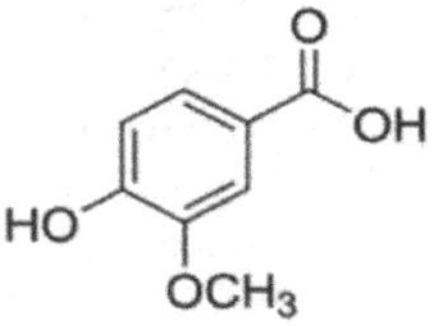

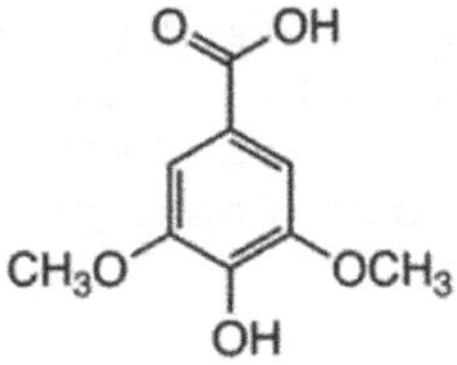

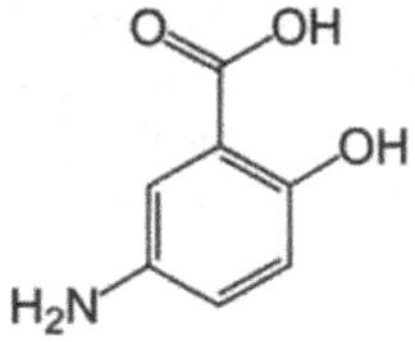

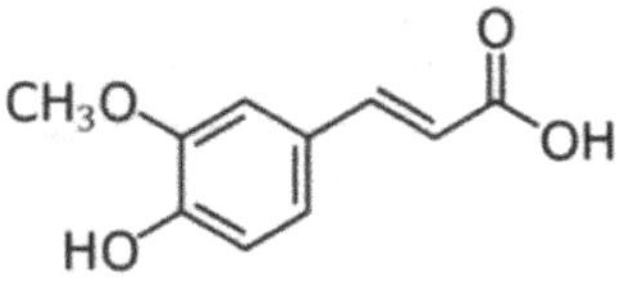

Fig. 1. Phenolic acids identified in many plant species, such as oat (*Avena sativa* L.) and rice (*Oryza sativa* L.), have been found to have allelopathic properties.

Although modes of action for allelopathic chemicals are not readily understood for each identified allelochemical, phenolic acids have been the focus of many studies designed to establish the basis of their allelopathy (Putnam 1985). Early research with phenolic acids indicated that some phenolic acids could function though increasing cell membrane permeability, thus affecting ion transport and metabolism (Glass and Dunlop 1974). More recent studies report disruption of cell division and malformed cellular structures in plants exposed to phenolic acids (Li et al. 2010). Reduced respiration and reduced photosynthetic rates, due to decreased photosynthetic products such as chlorophyll, have also been reported in the presence of phenolic acids (Patterson 1981; Yu et al. 2003). Other studies have cited altered plant enzymatic functions, inhibited protein synthesis, and inactivated plant hormones as inhibitory mechanisms from these allelochemicals (Batish et al. 2008; Li et al. 2010). Each mechanism of plant inhibition can lead to the reduced growth and/or death of an exposed plant; however, it is likely multiple functions within a plant are being affected simultaneously due to the mixture of allelochemicals released from a plant species. Despite the extensive research with phenolic acids, target sites for allelochemical activity within affected plant species remain to be determined for many phenolic compounds.

3.2 Glucosinolates

Glucosinolates occur in many plant species, but are widely known to be produced by species within the Brassicaceae family (Figure 2) (Haramoto and Gallandt 2005; Malik et al. 2008; Mithen 2001). Members of this family include: wild radish (*Raphanus raphanistrum* L.), white mustard (*Sinapis alba* L.), turnip (*Brassica campestris* L.), and rapeseed (*Brassica napus* L.). Glucosinolates, secondary metabolites containing sulfur and nitrogen, are enzymatically hydrolyzed by myrosinase in the presence of water to form isothiocynates, the active allelochemicals (Haramoto and Gallandt 2005; Norsworthy and Meehan 2005; Petersen et al. 2001; Price et al. 2005). Previous research examining extracts from glucosinolate-producing plant species have shown inhibition of other species through reduced germination, reduced seedling emergence and reduced size, as well as delayed seed germination (Al-Khatib et al. 1997; Brown and Morra 1996; Malik et al. 2008; Norsworthy et al. 2007; Wolf et al. 1984). Although specific modes of action have not been thoroughly investigated for each compound, it is evident that some plant species are able to tolerate these allelochemicals more readily than other species (Norsworthy and Meehan 2005). Some suggest that seed size variability plays a role in determining inhibitory effects of these allelochemicals; however, this may not be the only determinant for tolerance to these compounds (Haramoto and Gallandt 2005; Westoby et al. 1996). Future research with these allelopathic compounds will likely seek to answer this question, along with identifying the mode of action for plant inhibition, in order to utilize these compounds more effectively in agricultural production.

3.3 Coumarins

Coumarin compounds (Figure 3) are found in a range of plant species, particularly from the Apiaceae, Asteraceae and Fabaceae families (Razavi 2011). Coumarins and their derivatives have been identified in plants such as lettuce (*Lactuca sativa* L.), wild oat (*Avena sativa* L.), sweet vernalgrass (*Anthoxanthum odoratum* L.), and a number of other species (Abenavoli et al. 2004; Razavi 2011). Like many other allelochemicals, coumarins have been found to inhibit plant growth by reduced seedling germination and reduced root and shoot growth, likely with interference in photosynthesis, respiration, nutrient uptake and metabolism (Abenavoli et al. 2001; Abenavoli et al. 2004; Razavi et al 2010; Yamamoto 2008).

In addition to plant inhibition, biological activity of coumarins includes antibacterial, nematicidal, antifungal, and insecticidal activity; moreover, pharmacological activity of coumarins has been commonly noted in a number of instances with specific compounds functioning to reduce edema and inflammation (Casley-Smith and Casley-Smith 1992; Hoult and Paya 1996; Maddi et al. 1992; Razavi 2011). The broad activity of these compounds has made pharmaceutical use difficult due to the potential for non-target activity. Although allelopathic research has yet to indicate that the broad spectrum activity of coumarins could limit future use of these compounds for weed control, this may require further investigation as research moves forward.

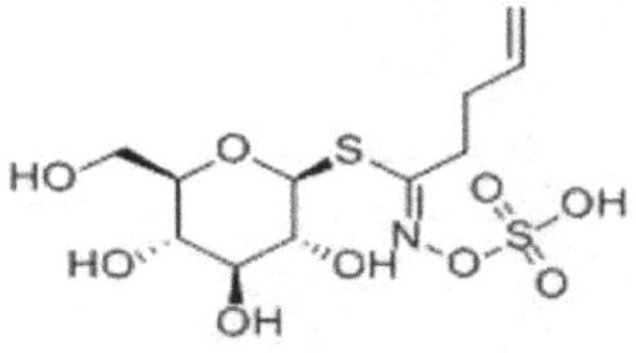

| Gluconapin | Glucobrassicin | Progoitrin |

| Sinigrin | Neoglucobrassicin | Gluconasturtin |

| Glucoiberin | Glucotropaeolin | Glucoraphenin |

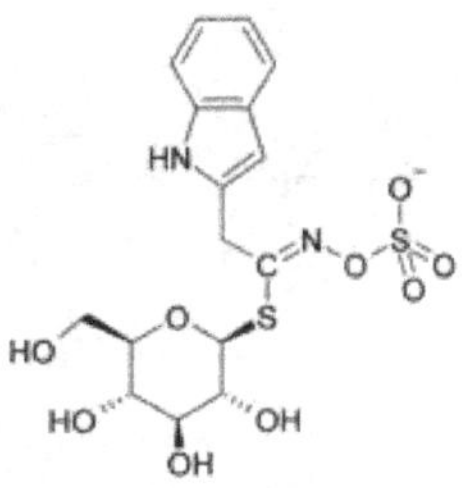

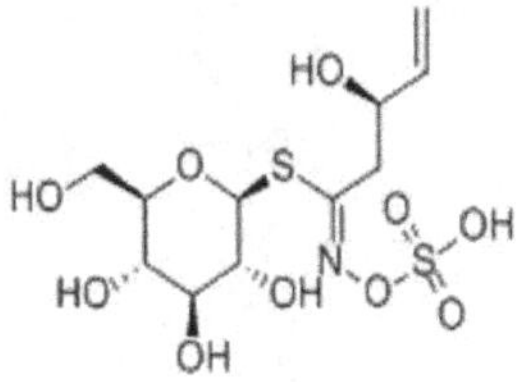

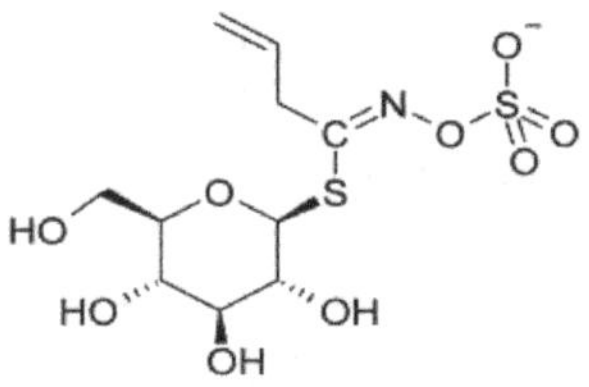

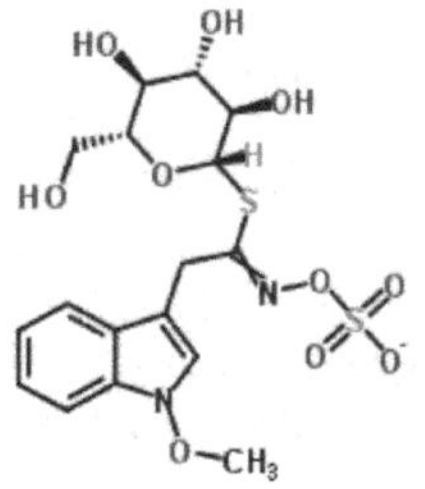

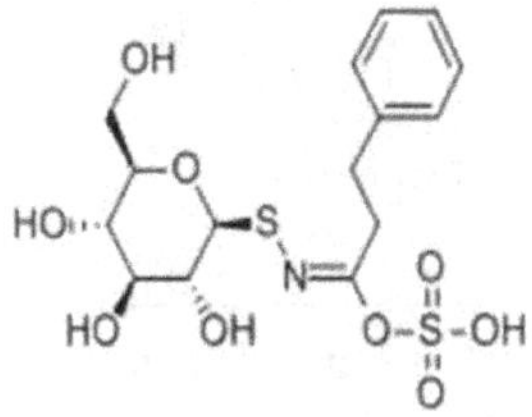

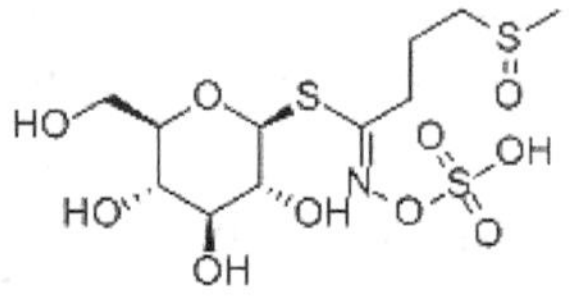

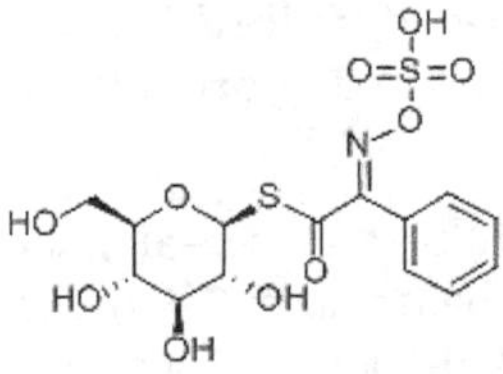

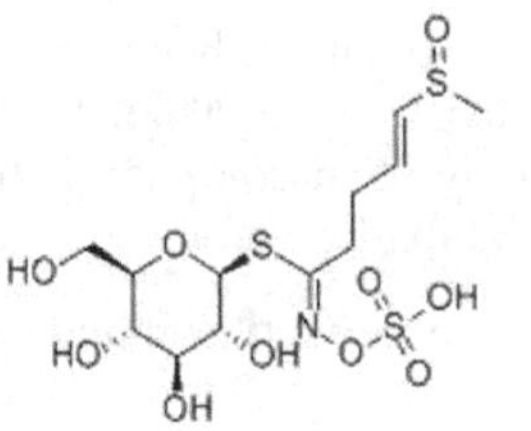

Fig. 2. Glucosinolates, allelopathic compounds known to be produced by plants in the Brassicaceae family as well as other families, are produced in both the root and shoot regions of plants.

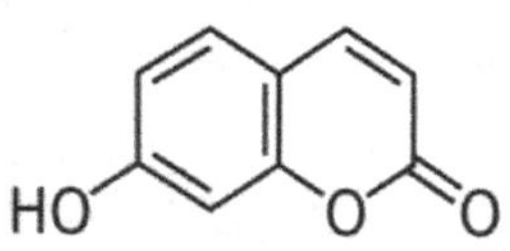

Umbelliferone 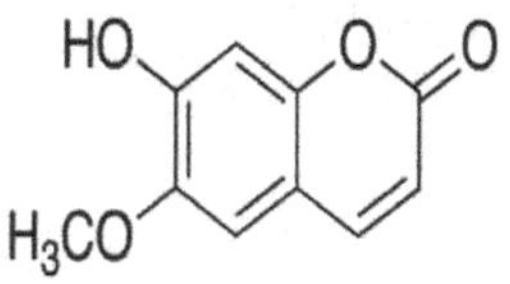   Scopoletin 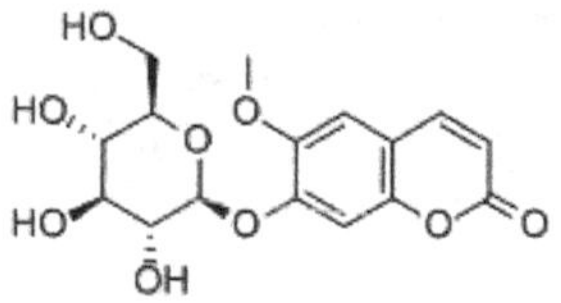   Scopolin

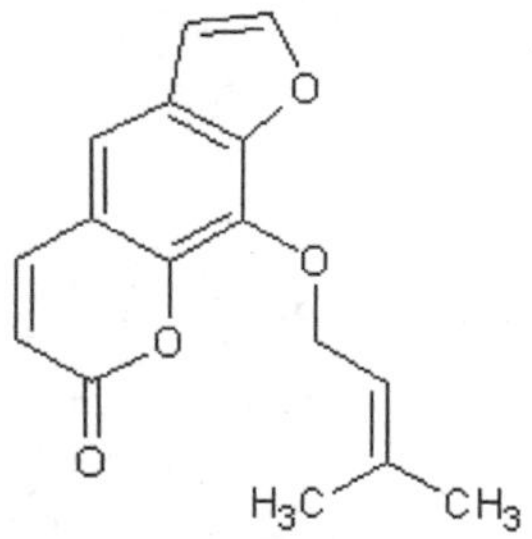

Imperatorin 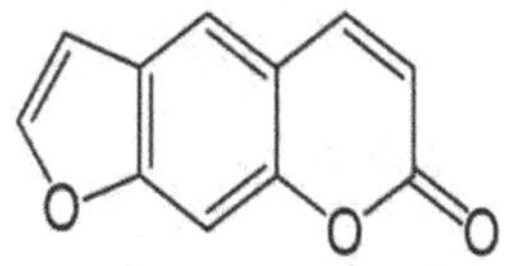   Psoralen 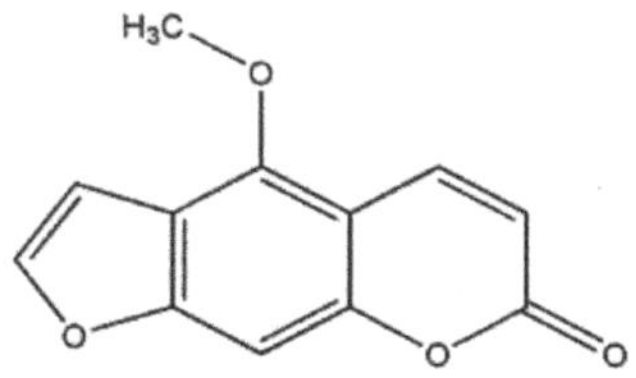   Bergapten

Fig. 3. Coumarins and their subgroups have been identified as allelopathic compounds in several plant families including Apiaceae and Fabaceae.

3.4 Other allelopathic compounds

Many other allelochemicals have been detected in a wide range of species; however, a few compounds have been more widely researched. Classes of allelochemicals under thorough investigation, such as the benzoxazinoids, heliannuols, and benzoquinones, offer potential benefits for weed control in agricultural systems (Figure 4) (Macias et al. 2005; Vyvyan 2002). These classes, described briefly below, represent only a few of the many other compounds that may one day provide substantial weed suppression through allelopathy.

Benzoxazinoid compounds, identified in cereal grains such as wheat and rye, include DIBOA [2,4-dihydroxy-(2H)-1,4-benzoxazin-3(4H)-one] and DIMBOA [2,3-dihydroxy-7-methoxy-(2H)-1,4-benzoxazin-3(4H)-one] (Burgos and Talbert 2000; Macias et al. 2005). These compounds are easily degraded into other allelopathic forms, BOA (2-benzoxazolinone) and MBOA (7-methoxy-2-benzoxazolinone), within the soil and can diminish plant germination and growth (Barnes et al. 1987; Burgos and Talbert 2000). In light of the allelopathic properties of BOA and MBOA, it is now recognized that continued research efforts are needed to understand the role of breakdown products of initial allelochemicals in inhibiting plant growth (Macias et al. 2005).

From the sunflower plant (*Helianthus annuus* L.), several compounds have been identified as being allelopathic (Leather 1983; Vyvyan 2002). The heliannuols are classified as phenolic sesquiterpenes and are noted for allelopathic as well as pharmacological activity (Vyvyan 2002). In addition to having been isolated from the sunflower, similarly structured compounds have been detected in animal species as well (Harrison and Crews 1997). Most notable about heliannuolic compounds is their ability to suppress plant growth at relatively low concentrations. Although they have been shown to inhibit growth of many broadleaf weed species, heliannuols appear to have a stimulating effect upon monocotyleden species (Weidenhamer 1996; Vyvyan 2002). This aspect of heliannuol activity may prove difficult when developing weed control applications of these compounds.

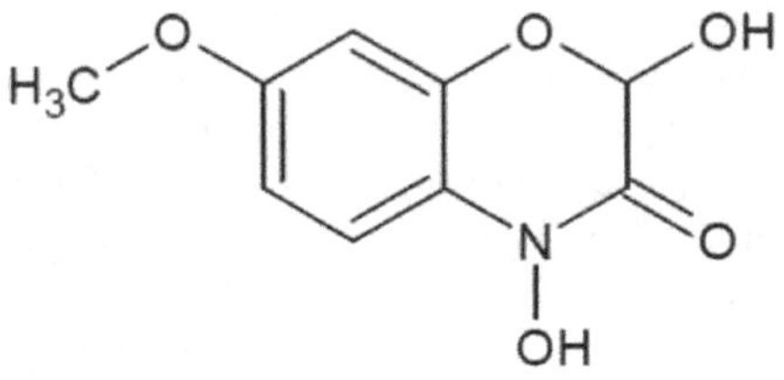

DIMBOA 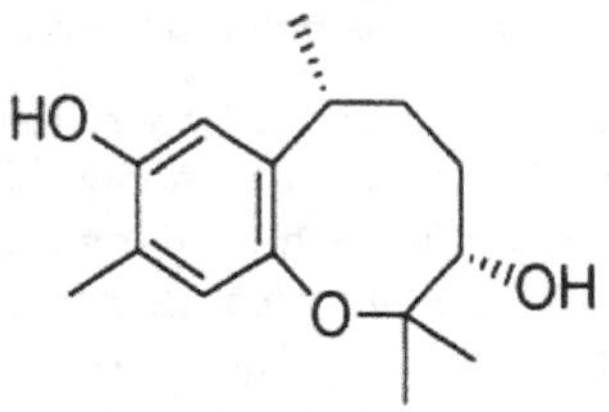   Heliannuol A

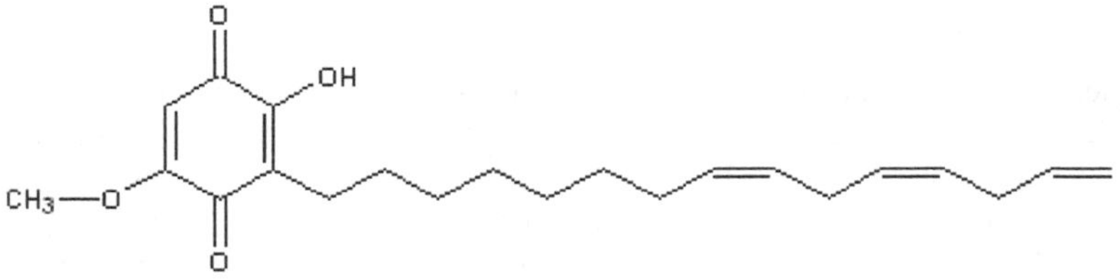

Sorgoleone

Fig. 4. Compounds, such as DIMBOA, heliannuol A, and sorgoleone, continue to be studied for their allelopathic properties.

Benzoquinone compounds, primarily sorgoleone, isolated from sorghum [*Sorghum bicolor* (L.) Moench], have also been determined to be highly allelopathic (Netzly et al. 1998). Research with this compound indicates plant growth inhibition is achieved through disruption of photosynthesis as well as reduced chlorophyll development (Einhellig and Souza 1992). Like some other compounds, sorgoleone exhibits selective activity with inhibition of many germinating seedlings but little activity against certain species such as morningglory (*Ipomoea spp.*) (Nimbal et al. 1996). Research conducted with sorghum root exudates compares sorgoleone activity to that of the herbicide, diuron, but has many target sites (Nimbal et al. 1996; Rimando et al. 1998). Thus far, characteristics of sorgoleone show that it is a promising compound for development into a natural herbicide as an alternative to synthetic herbicides.

4. Weed control through allelopathy

Ongoing research into allelopathy seeks to better understand the mechanisms of allelopathy in order to make use of these naturally occurring weed suppressants within agricultural areas. Benefits offered by employing allelopathy as some form of weed control could aid in developing more sustainable agricultural systems for future generations (Einhellig 1994a). Current efforts focus primarily on natural herbicide production and cover crops. Although these concepts are being utilized to some degree, there remains a great deal of research to fully utilize the potential of allelopathy.

The role of naturally derived compounds, or synthetically produced mimics, for use as pesticides has been widely adopted, particularly for insect control. Several plant derived compounds, such as pyrethrum, neem, and nicotine, are important chemicals for insect control in many areas (Isman 2006). Herbicide potentials of isolated plant extracts have been indicated by a number of researchers but to date, few have been marketed. Synthetic compounds, such as cinmethylin, and mesotrione, were developed based upon plant-derived allelochemicals, but release of subsequent plant-based herbicides has lagged (Lee et al. 1997; Macias et al. 2004; Secor 1994; Vyvyan 2002). Slow production and release of herbicides developed in this manner are most likely attributed to limited understanding of the modes of action for many identified allelochemicals. To date, a number of allelochemicals have been isolated and investigated to develop natural herbicides with these compounds. Understanding the mode of action for plant inhibition may aid in the development of new products for the market.

A great deal of research has been devoted to the use of cover crops for weed control. Until recently, however, the allelopathic potential of cover crops has received less attention due, in part, to the lack of knowledge about allelopathy in general. As the functions of allelopathic compounds are beginning to be understood, more focus is being given to the allelochemicals within cover crops. In agricultural settings, cover crops have been in use for a number of years as a ground cover to slow erosion and water runoff as well as to impede germination of weed seed by providing a physical barrier (Kaspar et al. 2001; Price et al. 2008; Sarrantonio and Gallandt 2003). The growing need for sustainable agricultural systems has necessitated increased cover crop research to better utilize these covers for effective weed control. As a result, recent investigations have sought to understand the role of allelopathy for weed suppression within various cover crops (Burgos and Talbert 2000; Khanh et al. 2005; Price et al. 2008; Walters and Young 2008).

5. Allelopathic potential of cover crops

Determining allelopathic potential of exudates of plant species can be difficult and time consuming to complete. Bioassays are generally conducted to identify allelopathic properties of compounds in order to differentiate between allelopathy and mulching effects. Our research has focused on determining the extent of allelopathic effects of available cover crops on weed species as well as crop species. Extract-agar bioassays conducted with radish

(*Raphanus sativus* L.), an indicator species, and cotton (*Gossypium hirsutum* L.) established levels of inhibition for radicle elongation by extracts from cover crops, primarily legumes and cereal grains.

Legume cover crops have the ability to fix atmospheric nitrogen that potentially provides a nitrogen source to the subsequent crop without the need for additional fertilizer applications (Balkcom et al. 2007; Hartwig and Ammon 2002). Legume species such as vetch (*Vicia villosa* Roth), clover (*Trifolium spp.*), black medic (*Medicago lupulina* L.), and winter pea (*Pisum sativum* L.) are typically used as cover crops in agricultural production in the United States (Figure 5) (SARE 2007). Other legume crops beginning to be researched as possible choices for cover crops are sunn hemp (*Crotalaria juncea* L.) and white lupin (*Lupinus albus* L.); however, their availability and use are not as widespread as the previously mentioned legumes. In addition to being a nitrogen source for primary crops, legume covers provide a weed control potential. Due to the rapid degradation of legume residue on the soil surface in comparison to cereal grain residue, weed control through a physical barrier may not last as long into the season as other cover crops.

Fig. 5. Legume cover crops, such as white lupin (in mixture with black oats), provide weed suppression and nitrogen benefits to the subsequent cash crop.

Determining allelopathic effects of legume cover crop extracts concluded that legume covers did inhibit radish and cotton radicle elongation; however, cotton root exhibited less inhibition than that of radish for all included crops (Price et al. 2008) (Figure 6). In our research, hairy vetch had the greatest inhibition while winter pea had the least effect on germinating seedlings. It is important to note that different varieties of cover crops are available for use in agricultural systems and the varieties of one species may differ in level of allelopathy. Although under field conditions, allelopathic performance of these species may fluctuate, it is apparent that these cover crops can provide additional weed control measures over systems that do not include a cover crop.

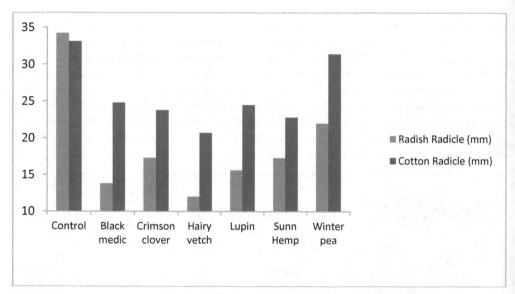

Fig. 6. Legume cover crops affect radicle elongation of different plant species to varying degrees.

Cereal grain crops such as black oat (*Avena strigosa* Schreb), rye, triticale (*X Triticosecale* Wittmack), and wheat, are utilized frequently in conservation systems as cover crops with effective ground cover and weed suppression (Figure 7). Rye is a commonly used cereal cover crop due to its ability to be sown later in the season while maintaining successful growth and its biomass production capability. With increased biomass on the soil surface, weed suppression will be increased as well. Cereal crops will also decay more slowly than more herbaceous plant species and provide some ground cover, and allelochemical release, further into the growing season. Additionally, rye has been noted to be less affected by plant diseases than other cover crops, and aids in reducing insect pests within a system (Wingard 1996).

Like legumes, cereal grain crop exudates in our study were able to significantly inhibit radicle elongation compared to the control (Figure 8). The disparity between radish and cotton radicle inhibition for each cover crop studied suggests that minimized interference with primary crops and increased weed suppression potential could be achieved with the use of cereal grain crops. These allelopathic effects, however, may be amplified or diminished depending on the field environment, plant stress levels, cover crop variety, and a number of other factors involved in determining allelochemical levels. Nevertheless, this research provides a base of allelopathic concentrations and impacts from various cover crops and may be an initial consideration when choosing a cover crop for inclusion in a system.

Fig. 7. Cotton growing in rolled black oat residue. Cereal grain cover crops, like black oat and rye, can be utilized to achieve a large quantity of plant residue on the soil surface.

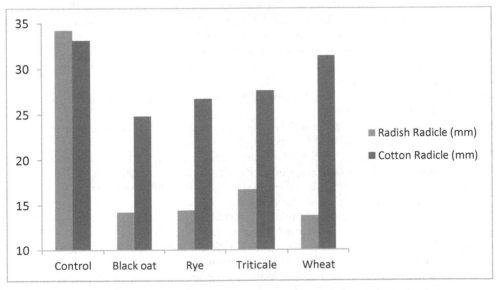

Fig. 8. Radish and cotton radicle elongation is reduced by cereal grain cover crops.

6. Conclusions

The growing demand for sustainable agricultural systems requires that researchers reevaluate current production methods and inputs. To ensure continued productivity and potentially reduce synthetic herbicide requirements, allelopathy has become a focal point for research in the agricultural community. Although, many questions have yet to be resolved, the utilization of allelochemicals for weed suppression remains a promising avenue for reducing herbicide usage. Whether through the development of natural herbicides from isolated allelochemicals or through the application of cover crops with allelopathic properties, allelopathy will most likely be a factor in providing sustainable systems in the future.

7. References

Abenavoli, M.R., C. De Santis, M. Sidari, A. Sorgona, M. Badiani, nad G. Cacco. 2001. Influence of coumarin on the net nitrate uptake in durum wheat (*Triticum durum* Desf. Cv. Simeto). *New Phytologist* 150: 619-627.

Abenavoli, M.R., A. Sorgona, S. Albano, and G. Cacco. 2004. Coumarin differentially affects the morphology of different root types of maize seedlings. *Journal of Chemical Ecology* 30: 1871-1883.

Al-Khatib, K., C. Libbey, and R. Boydston. 1997. Weed suppression with *Brassica* green manure crops in green pea. *Weed Science* 45: 439-445.

Bais, H.P., R. Vepachedu, S. Gilroy, R.M. Callaway, and J.M. Vivanco. 2003. Allelopathy and exotic plant invasion: From molecules and genes to species interactions. *Science* 301: 1377-1380.

Balkcom, K.S., H. Schomberg, W. Reeves, and A. Clark. 2007. Managing cover crops in conservation tillage systems. In *Managing Cover Crops Profitably*, 3rd Edition, ed. A. Clark, 44-65. College Park, MD: Sustainable Agriculture Research and Education.

Barnes, J.P., A.R. Putnam, B.A. Burke, and A.J. Aasen. 1987. Isolation and characterization of allelochemicals in rye herbage. *Phytochemistry* 26: 1385-1390.

Batish, D.R., H.P. Singh, S. Kaur, R.K. Kohli, and S.S. Yadav. 2008. Caffeic acid affects early growth and morphogenetic response of hypocotyl cuttings of mung bean (*Phaseolus aureus*). *Journal of Plant Physiology* 165: 297-305.

Battey, N.H., and H.D. Blackbourn. 1993. The control of exocitosis in plant cells. *New Phytology* 125: 307-308.

Bertin, C., X. Yang, and L.A. Weston. 2003. The role of root exudates and allelochemicals in the rhizosphere. *Plant and Soil* 256: 67-83.

Brown, P.D. and M.J. Morra. 1996. Hydrolysis products of glucosinolates in *Brassica napus* tissues as inhibitors of seed germination. *Plant and Soil* 181: 307-316.

Burgos, N.R. and R.E. Talbert. 2000. Differential activity of allelochemicals from *Secale cereale* in seedling bioassays. *Weed Science* 48: 302-310.

Casley-Smith, J.R. and J.R. Casley-Smith. 1992. Modern treatment of lymphedema II. The benzopyrones. *Australian Journal of Dermatology* 33: 69-74.

Chung, I.M., K.H. Kim, J.K. Ahn, S.C. Chun, C.S. Kim, J.T. Kim and S.H. Kim. 2002. Screening of allelochemicals on barnyardgrass (*Echinochloa crus-galli*) and identification of potentially allelopathic compounds from rice (*Oryza sativa*) variety hull extracts. *Crop Protection* 21: 913-920.

Einhellig, F.A. and I.F. Souza. 1992. Phytotoxicity of sorgoleone found in grain sorghum root exudates. *Journal of Chemical Ecology* 18: 1-11.

Einhellig, F.A. 1994a. Allelopathy: Current status and future goals. In *Allelopathy*, eds. D. Inderjit, K.M.M. Dakshini and F.A. Einhellig, 1-24. Washington, D.C.: American Chemistry Society.

Einhellig, F.A. 1994b. Mechanisms of action of allelochemicals in allelopathy. In *Allelopathy: Organisms, Processes, and Applications*, eds D. Inderjit, K.M.M. Dakshini and F.A. Einhellig, 96. Washington, D.C.: American Chemistry Society.

Einhellig, F.A. 1996. Interactions involving allelopathy in cropping systems. *Agronomy Journal* 88: 886-893.

El-Rokiek, K.G., R. Rafat, N.K. El-Masry, and S.A. Ahmed. 2010. The allelopathic effect of mango leaves on the growth and propagative capacity of purple nutsedge (*Cyperus rotundus* L.). *Journal of American Science* 6: 151-159.

Fan, T.W.M., A.M. Lane, D. Crowley, and R.M. Higashi. 1997.Comprehensive analysis of organic ligands in whole root exudate using nuclear magnetic resonance and gas chromatography-mass spectrometry. *Analytical Biochemistry* 257: 57.

Fitter, A. 2003. Making allelopathy respectable. *Science* 301: 1337-1338.

Glass, A.D.M. and J. Dunlop. 1974. Influence of phenolic acids on ion uptake. *Plant Physiology* 54: 855-858.

Haramoto, E.R. and E.R. Gallandt. 2005. Brassica cover cropping: 1. Effects on weed and crop establishment. *Weed Science* 53: 695-701.

Hartwig, N.L. and H.U. Ammon. 2002. Cover crops and living mulches. *Weed Science* 50:688-699.

Harrison, B. and P. Crews. 1997. The structure and probable biogenesis of helianane, a hetercyclic sesquiterpene, from the Indo-Pacific sponge *Haliclona fascigera*. *Journal of Organic Chemistry* 62: 2646-2648.

Hoult, J.R.S. and M. Paya. 1996. Pharmacological and biochemical actions of simple coumarins: Natural products with therapeutic potential. *General Pharmacology* 27: 713-722.

Inderjit, D. and L.A. Weston. 2003. Root exudation: an overview. In *Root Ecology*, eds. DeKroon and E.J.W. Visser. Heidelberg, Germany: Springer-Verlag.

Isman, M.B. 2006. Botanical insecticides, deterrents, and repellants in modern agriculture and an increasingly regulated world. *Annual Review of Entomology* 51: 45-66.

Kaspar, T.C., J.K. Radke, and J.M. Laflen. 2001. Small grain cover crops and wheel traffic effects on infiltration, runoff, and erosion. *Journal of Soil and Water Conservation* 56: 160-164.

Khanh, T.D., M.I. Chung, T.D. Xuan, and S. Tawata. 2005. The exploitation of crop allelopathy in sustainable agricultural productions. *Journal of Agronomy and Crop Science* 191: 172-184.

Leather, G.R. 1983. Sunflowers (*Helianthus annuus*) are allelopathic to weeds. *Weed Science* 31: 37-42.

Lee, D.L., M.P. Prisbylla, T.H. Cromartie, D.P. Dagartin, S.W. Howard, W.M. Provan, M.K. Ellis, T. Fraser, and L.C. Mutter. 1997. The discovery and structural requirements of inhibitors of p-hydroxyphenylpyrvate dioxygenase. *Weed Science* 45: 601-609.

Li, Z., Q. Wang, X. Ruan, C. Pan, and D. Jiang. 2010. Phenolics and plant allelopathy. *Molecules* 15: 8933-8952.

Macias, F.A., J.M.G. Molinillo, A. Oliveros-Bastidas, D. Marin, and D. Chinchilla. 2004. Allelopathy. A natural strategy for weed control. *Communications in Applied Biological Science* 69: 13-23.

Macias, F.A., A. Oliveros-Bastidas, D. Marin, D. Castellano, A.M. Simonet, and J.M.G. Molinillo. 2005. Degradation studies on benzoxazinoids. Soil degradation dynamics of (2R)-2-O-β-D-glucopyranosyl-4-hydroxy-(2H)-1,4-benzoxazin-3(4H)-one (DIBOA-Glc) and its degradation products, phytotoxic allelochemicals from Gramineae. *Journal of Agricultural and Food Chemistry* 53: 554-561.

Maddi, V., K.S. Raghu, and M.N.A. Rao. 1992. Synthesis and anti-inflammatory activity of 3-(benzylideneamino)coumarins in rodents. *Journal of Pharmaceutical Sciences* 81: 964-966.

Malik, M.S., J.K. Norsworthy, A.S. Culpepper, M.B. Riley, and W. Bridges. 2008. Use of wild radish (*Raphanus raphanistrum*) and rye cover crops for weed suppression in sweet corn. *Weed Science* 56: 588-595.

Mithen, R. 2001. Glucosinolates and the degradation products. In *Advances in Botanical Research*, ed. J. Callow, 214-262. New York: Academic Press.

Molisch, H. 1937. *Der Einfluss einer Pflanze auf die andere-Allelopathie*. Fischer, Jena.

Netzly, D.H., J.L. Riople, G. Ejeta, and L.G. Butler. 1988. Germination stimulants of witchweed (*Striga asiatica*) from hydrophobic root exudate of sorghum (*Sorghum bicolor*).Weed Science 36: 441-446.

Nimbal, C.I., J.F. Pedersen, C.N. Yerkes, L.A. Weston, S.C. Weller. 1996. Phytotoxicty and distribution of sorgoleone in grain sorghum germplasm. *Journal of Agriculture and Food Chemistry* 44: 1343-1347.

Norsworthy, J.K. and J.T. Meehan. 2005. Herbicidal activity of eight isothiocyanates onTexas panicum (*Panicum texanum*), large crabgrass (*Digitaria sanguinalis*), and sicklepod (*Senna obtusifolia*). *Weed Science* 53: 515-520.

Norsworthy, J.K., M.S. Malik, P. Jha, and M.B. Riley. 2007. Suppression of *Digitaria sanguinalis* and *Amaranthus palmeri* using autumn-sown glucosinolate-producing cover crops in organically grown bell pepper. *Weed Research* 47: 425-432.

Olofsdotter, M. 2001. Rice-a step toward use of allelopathy. *Agronomy Journal* 93: 3-8.

Patterson, D.T. 1981. Effects of allelopathic chemicals on growth and physiological response of soybean (*Glycine max*). *Weed Science* 29: 53-58.

Petersen, J., R. Belz, F. Walker, and K. Hurle. 2001. Weed suppression by release of isothiocyanates from turnip-rape mulch. *Agronomy Journal* 93: 37-43.

Pramanik, M.H.R., M. Nagal, M. Asao, and Y. Matsui. 2000. Effect of temperature and photoperiod on phytotoxic root exudates of cucumber (*Cucumis sativus*) in hydroponic culture. *Journal of Chemical Ecology* 26: 1953-1967.

Price, A. J., C. S. Charron, and C. E. Sams. 2005. Allyl isothiocyanate and carbon dioxide produced during degradation of *Brassica juncea* tissue in different soil conditions. HortSci. 40:1734-1739.

Price, A.J., D.W. Reeves, and M.G. Patterson. 2006. Evaluation of weed control provided by three winter cereals in conservation-tillage soybean. *Renewable Agriculture and Food Systems* 21: 159-164.

Price, A.J., M.E. Stoll, J.S. Bergtold, F.J. Arriaga, K.S. Balkcom, T.S. Kornecki, and R.L. Raper. 2008. Effect of cover crop extracts on cotton and radish radicle elongation. *Communications in Biometry and Crop Science* 3: 60-66.

Putnam, A.R., and W.B. Duke. 1978. Allelopathy in agroecosystems. *Annual Review of Phytopathology* 16: 431-451.

Putnam, A.R. 1985. Allelopathic research in agriculture: past highlights and potential. In *The Chemistry of Allelopathy*, ed. A.C. Thompson 1-8. Washington, D.C.: American Chemical Society.

Razavi, S.M., G.H. Imanzadeh, and M. Davari. 2010. Coumarins from *Zosima absinthifolia* seeds, with allelopathic effects. *Eurasion Journal of Biosciences* 4: 17-22.

Razavi, S.M. 2011. Plant coumarins as allelopathic agents. *International Journal of Biological Chemistry* 5: 86-90.

Rimando, A.M., F.E. Dayan, M.A. Czarnota, L.A. Weston, and S.O. Duke. 1998. A new photosystem II electron transfer inhibitor from *Sorghum bicolor*. *Journal of Natural Products* 61: 927-930.

SARE (Sustainable Agriculture Research and Education). 2007. *Managing Cover Crops Profitably*, 3rd edition. Ed. A. Clark. 244 pgs. College Park, MD: Sustainable Agriculture Research and Education.

Sarrantonio, M. and E. Gallandt. 2003. The role of cover crops in North American cropping systems. *Journal of Crop Production* 8: 53-74.

Secor, J. 1994. Inhibition of barnyardgrass 4-hydroxyphenylpyruvate dioxygenase by sulcotrione. *Plant Physiology* 106: 1429-1433.

Seigler, D.S. 1996. Chemistry and mechanisms of allelopathic interactions. *Agronomy Journal* 88: 876-885.

Sterling, T.M., R.L. Houtz, and A.R. Putnam. 1987. Phytotoxic exudates from velvetleaf (*Abutilon theophrasti*) glandular trichomes. *American Journal of Botany* 74: 543-550.

Swain, T. 1977. Secondary compounds as protective agents. *Annual Review of Plant Physiology* 28: 479-501.

Truman, C.C., D.W. Reeves, J.N. Shaw, A.C. Motta, C.H. Burmester, R.L. Raper, and E.B. Schwab. 2003. Tillage impacts on soil property, runoff, and soil loss variations of a Rhodic Paleudult under simulated rainfall. *Journal of Soil and Water Conservation* 58: 258-267.

Uren, N.C. 2000. Types, amounts, and possible functions of compounds released into the rhizosphere by soil-grown plants. In *The Rhizosphere: Biochemistry and Organic Substances at the Soil-Plant Interface*, eds. R. Pinton, Z. Varanini and P. Nannipieri, 19-40. New York: Marcel Dekker, Inc.

Vyvyan, J.R. 2002. Allelochemicals as leads for new herbicides and agrochemicals. *Tetrahedron*. 58: 1631-1646.

Walters, S.A. and B.G. Young. 2008. Utility of winter rye living mulch for weed management in zucchini squash production. *Weed Technology* 22: 724-728.

Weidenhamer, J.D. 1996. Distinguishing resource competition and chemical interference: Overcoming the methodological impasse: Allelopathy in cropping systems. *Agronomy Journal* 88: 866-875.

Weir, T.L., S. Park, and J.M. Vivanco. 2004. Biochemical and physiological mechanisms mediated by allelochemicals. *Plant Biology* 7: 472-479.

Westoby, M., M. Leishman, and J. Lord. 1996. Comparative ecology of seed size and dispersal. *Philosophical Transactions of the Royal Society of London B. Biological Sciences* 351: 1309-1318.

Weston, L.A. 1996. Utilization of allelopathy for weed management in agroecosystems: Allelopathy in cropping systems. *Agronomy Journal* 88: 860-866.

Willis, R.J. 1985. The historical bases of the concept of allelopathy. *Journal of the History of Biology* 18: 71-102.

Wingard, C. 1996. Cover crops in integrated vegetable production systems. SARE Project report # PG95-033. Southern Region SARE. Griffin, GA. www.sare.org/projects.

Wolf, R.B., G.F. Spencer, and W.F. Kwolek. 1984. Inhibition of velvetleaf (*Abutilon theophrasti*) germination and growth by benzyl isothiocyanate, a natural toxicant. *Weed Science* 32: 612-615.

Wu, L., X. Guo, M.A. Harivandi. 1998. Allelopathic effects of phenolic acids detected in buffalograss (*Buchloe dactyloides*) clippings on growth of annual bluegrass (*Poa annua*) and buffalograss seedlings. *Environmental and Experimental Botany* 39: 159-167.

Yamamoto, Y. 2008. Movement of allelopathic compound coumarin from plant residue of Sweet vernalgrass (*Anthoxanthum odoratum* L.) to soil. *Japanese Society of Grassland Science* 55: 36-40.

Yu, J.Q., S.F. Ye, M.F. Zhan, and W.H. Hu. 2003. Effects of root exudates and aqueous root extracts of cucumber (*Cucumis sativus*) and allelochemicals, on photosynthesis and antioxidant enzymes in cucumber. *Biochemical Systematics and Ecology* 31: 129-139.

Part 2

Integrated Chemical Weed Management

7

Weeds in Forestry and Possibilities of Their Control

Verica Vasic[1], Branko Konstantinovic[2] and Sasa Orlovic[1]
[1]University of Novi Sad, Institute of Lowland Forestry and Environment, Novi Sad,
[2]University of Novi Sad, Faculty of Agriculture, Novi Sad,
Serbia

1. Introduction

Thanks to wide inter-row spaces and open canopy in the early phases of establishment, forest nurseries and plantations represent ideal places of floristically rich and diverse weed flora. Weeds have an exceptional capacity of adaptation to environmental conditions because most produce vast quantities of seeds which enable great expansion.

Although the geographic weed distribution and composition depends mainly on climate factors, the vegetation within each climate region is differentiated under the effect of edaphic factors. The soil physical and chemical properties, as well as climate conditions, have the primary significance for both cultivated plants and weeds.

However, all weeds do not have equal significance. When considering weed control attibutes, perennial weeds present are far greater challange due to difficulties employing mechanical means, because perennials are often stimulated to grow and disperse even more intensively. Perennial weed species, such as *Sorghum halepense, Convolvulus arvensis*, and *Cynodon dactylon,* have well-developed underground organs and are great problems not only in agriculture, but also in nursery production of forest planting materials.

The problem of forestry weeds came to the fore in recent years as more and more attention has been paid to establishing and restoring forests. In afforested areas, luxuriate development of weed vegetation, can imperil the survival and development of young seedlings. Harmful effects of weeds are reflected not only in the subtraction of basic living conditions such as humidity, light and nutrients already undergo a poor growth and receiving of seedlings.

2. The concept of weeds in forestry

When defining a weed, it should be emphasized that there is no simple and precise definition. Kojic et al., (1996) are of the opinion that weeds are plants growing among cultivated plants and interfere with man's activities; Zekic (1983) stated that weeds are plants growing in places they are not wanted. The concept of a weed is relative in nature, i.e., there is no sharp boundary between weeds and cultivated plants. While in some regions, one plant species is considered a weed, the same species is cultivated in others.

Vajda (1973) considered as forest weeds those plant species interfering with germination and growth of young forest plants and Konstantinovic (1999) those which were unfavourable under certain circumstances in the forest and interfere with forest management. According to Kovacevic (1979) the weeds in forestry are all herbaceous plants, shrubs, and trees which, in forest nurseries, stands, and clear felled areas weaken or prevent the growth and development of cultivated trees.

Evidence of the benefits of weed control for enhanced tree growth is widespread however weeds in forestry are not always harmful. Herbaceous weeds in forest plantations represent food for livestock (DiTomaso, 1997; Papachristou et al., 2009) and dry weights of some weeds are used as bedding for livestock. The fruits of some weeds are edible and weeds while they are somewhere used for human consumption. Some weeds have medicinal properties and are used as medicinal plants (Stepp & Moerman, 2001; Stepp, 2004; Dhole et al., 2009). Weeds prevent soil erosion and can be a shelter for wild animals and birds. However it should be noted that the benefits of weeds significantly less than the damage caused.

3. The properties of weed species

The knowledge gained from the study weed biology under given agro ecological conditions represents the basis in choosing the appropriate measures for their control. Compared to cultivated plants, weedy plants show considerable plasticity in relation to numerous ecological factors. One of the most important weed traits is the expressed adaptation ability. Another important weed trait is the pronounced resistance to unfavourable environmental conditions (drought, moisture, wind etc.). Many weeds are resistant to plant diseases and pests. Also, one of the weed traits is the periodicity of germination. Very often weed seeds do not germinate at once, but rather in different time periods, and it is hard to control weeds simultaneously. In addition, many weeds produce an enormous quantity of seed, which makes it easier for them to spread and expand in space.

4. Propagation of weeds

According to reproduction method, weed plants may be divided into those propagating only sexually, i.e. from seed, and those which are also vegetatively propagated. Sexual weed reproduction results in the formation of seed - reproductive organs by which the weeds are dispersed. All annual weedy plants belong to the group of weeds reproducing only by seed. Under favorable conditions, weeds produce an enormous quantity of seed, even several million of seeds per individual plant. After ripening a great part of the weed seed will end up on the soil surface and subsequently incorporated into the soil by tillage or other means. According to Wilson et al. (1985), under conditions of great weed infestations, some 300 million to 3.5 billion seeds can be found per one hectare of soil. Presence of weed seed in the soil depends on many factors and varies from field to field and region to region (Lutman, 2002).

Vegetative reproduction of perennial weeds represents a very efficient tool for their quick regeneration and distribution. Vegetative reproduction, i.e. regeneration ability of ground vegetative organs, depends first of all on their physiological state. In other words, there is a correlation between intensity of rooting of the rhizome and other ground vegetative organs and phenophases of plant development which is expressed differently in individual phases. In most cases rooting intensity is very low or completely absent during summer months. The time prior to or at the end of a vegetation period is the most favorable for vegetative regeneration and maximum regeneration ability occurs during spring and autumn (Vrbvnicanin & Kojic, 2000). Regeneration ability of vegetative ground organs depends greatly on environmental factors. Favorable climatic conditions, and in particular, the optimal condition of soil moisture and temperature, affect the regeneration process of vegetative organs.

5. The spread of weeds

Large quantities of seed produced by weeds would not be able to establish and develop in the immediate vicinity of a mother plant. Therefore, the fact that weeds are able to find different ways of quick and efficient seed and fruit distribution is fully justified from a biological point of view (Konstantinovic et al., 2005). In order for weed control to be successful, the means of their distribution must be known. Weeds can be spread by the plant itself, i.e. by self-distribution, and by other factors:

- Wind
- Water
- Animals and
- Man

6. Damage from weeds

Weeds present a large challenge both in agriculture and in forestry. They form a large mass of aboveground and inground organs engaged in a competitive relationship with cultivated plants for light, water and nutritive components in the soil (Kojic et al., 1972). Damages caused to cultivated plants by weeds can be great. According to some opinions, damages caused by weeds are greater than those caused by diseases and pests together (Kojic et al., 1996). Weedy plants grow relatively faster in forest settings and displace young forest plants living space, overshadow and stifle them, and water and nutritive matters are taken at the expense of cultivated plants.

Far less favorable impacts of weeds are found in nursery production. Due to the presence of weeds, nursery plants can experience retardation of growth, chlorosis, reduced resistance to plant diseases and pests, and death of individual parts of branches or crowns; if the weed is abundant, it often leads to drying and deterioration of the entire plant (Zekic, 1983). If weed control in nurseries is lacking, nursery plants of poor quality and fewer total plants are obtained.

7. Distribution of weeds in forestry

Competitive division of weeds in forestry is often made according to the degree of harmfulness of weeds to the trees. According to Vajda (1983) weeds in forestry are classified as either useful or harmful; Konstantinovic (1999) categorizes weeds into useful, harmful, or indifferent. According to this classification:

- Harmful weeds are plants that hinder tree development, and form thick cover
- Indifferent weeds are plants that grow individually, form weak coverage and do not hinder development of cultivated plants
- Useful weeds are plants with medical properties and plants that form fruits

The role of light has been of particular importance for emergence of weeds. In relation to light regime, weeds may be classified into **sciophytes** – plants developing in the shadow in weakly thinned forest stands or in dense forest stands and represent no threat to tree development; **semisciophytes** – semi-shadow plants that develop in thinned stands and can do a lot of harm; and **heliophytes** – plants of open habitats such as clearings, strips, burnt areas, etc., and represent a big threat to renovation and development of trees. There are a number of other weed classifications due to their adaptation to abiotic factors such as water regime, temperature, physico-chemical soil characteristics, etc. during their evolutionary development. However, very important weed classifications in forestry, which would have practical significance from the aspect of weed control, are the following weeds of forest nurseries and weeds of forest plantations and forest stands.

8. The most important weeds in forestry

8.1 Weeds in forest nurseries

Weed flora in forest nurseries differ from those found in forest plantations and forest stands. Given the extent of care measures applied, weeds in forest nurseries are very similar to those found in cultivated crops (Konstantinović, 1999). They are mostly annual and perennial herbaceous weedy species. The most common grass weed species present in the forest nurseries include: *Sorghum halepense, Cynodon dactylon, Alopecurus myosuroides, Digitaria sanquinalis, Echinochloa crus-galli, Poa annua,* and *Setaria spp.* Dominant broadleaf species include: *Amaranthus retroflexus, Ambrosia artemisiifolia, Chenopodium album, Cirsium arvense, Convolvulus arvensis, Erigeron canadensis, Datura stramonium, Galium aparine, Solanum nigrum, Sinapis arvensis,* and *Poligonum spp..*

Control of weediness in forest nurseries is very important and quality planting material is the basic prerequisite for success in forest stand establishment. Since weeds are one of the most limiting factors for the success of nursery production, their control should be approached very seriously (Vasic & Konstantinovic, 2008).

8.2 Weeds in forest plantations and forest stands

Weeds in forest plantations and forest stands differ from those in forest nurseries, because, in addition to different care measures applied in plantations and stands, the conditions in habitats also differ. Apart from ferns, herbaceous annual and perennial weeds, woody weeds such as shrubs, bushes, and shoots from the stumps of different tree types may also be present in forest plantations and stands. Woody weeds are very hardy and have a great power of regeneration; it is practically impossible to destroy them completely by mechanical means. The most common weed species present in forest plantations and stands are: *Ambrosia artemisiifolia, Amorpha fruticosa, Asclepias syriaca, Erigeron canadensis, Solidago gigantea, Sorghum halepense, Sambucus nigra, Stenactis annua, Pteridium aquilinum, Rubus caesius and etc.*

9. Weed control in forestry

There are numerous measures and procedures for weed control in forestry today, but, in order to fight weeds successfully, they should consist of different care and control measures. Described below are the six classifications of weed control measures.

9.1 Preventive measures

The main goal of preventive measures is to prevent weed distribution. All measures used to protect any surface from weeds, i.e. to prevent weed seed growth in the field are considered preventive measures (Kovacevic & Momirovic, 2004). Preventive measures in forestry weed control include:

- Control aided by sowing only pure crop seeds, which prevents spreading of weeds over sown surfaces
- Destruction of weeds on non-agricultural areas; weeds that present a constant source of weediness and transportation of seeds to arable lands are developed on such areas
- Prevention of the spread of weed seed by human activities by keeping agricultural and forest machinery and objects clean.
- Allelopathy is a phenomenon, where cultivated plants secrete exudates affecting the suppression of weeds (Pratley et al., 1999). It is manifested in such a way that in the presence of certain plant species, many others are not able to thrive, or are slowly developed (Janjic et al., 2008).

9.2 Mechanical measures

Mechanical measures for combating weeds include basic treatment such as ploughing, disking, tilling and etc. Also regular measures in forest nurseries and plantations are hoeing and farrowing, undertaken during the greatest part of the vegetation period and especially emphasised during the entire spring and in early summer.

One of the ways of suppressing the already growing weeds and preventing their seed dispersal is mowing. Multiple repetitions exhaust the stored substances in the root and the plant is killed. In addition to mowing, one of the methods of weed suppression in forestry is also the pruning of shoots and stump shoots. However, this weed suppression method is relatively expensive due to intense labor and if repeated pruning is required depending on the weed species present (Vasic et al., 2009). Concerns about increasing pesticide use have been major factors for research in physical weed control methods in Europe (Melander et al., 2005).

9.3 Physical measures

Physical weed control measures applied in forestry involve the use of flame and superheated steam. Destruction of weeds by flame can be applied in forest plantations with wider spaces between the rows, provided that the crops are previously protected by metal shields. Burning weeds is carried out on non-productive areas such as forest railways, roads, and canals. Destruction of weeds using steam is applied in forest nurseries in preparation of substrates used for sowing or planting. This is also a form of sterilization which destroys weed seeds in addition to plant diseases and noxious insects. Orloff & Cudney (1993) believe that the use of flame for the reduction of weeds is the best at the end of growing seasons, because in this way destroy most weed seeds that are dispersed on the soil surface.

9.4 Mulches

The covering of soil with a variety of materials such as straw, stubble, polyethyelene films, and others, to prevent the emergence of weeds is utilzed on smaller areas, mostly in forest nurseries. Polyethylene films of varying colors and thickness are most often used. This type of weed control is efficient for annual weeds but has no effect on control of many perennial weeds, and can be expensive compared to other methods used to fight weeds.

Many types of mulches have been tried including: sheets of plastic, newspaper, plywood, various thicknesses of bark, sawdust, sand, straw, sprayed-on petroleum resin, and even large plastic buckets. Most have proven to be ineffective, costly or both. Early trials tended to use small, short-lived materials that aided conifer seedling survival but not growth. Compared to other weed control techniques available in previous years, mulches were rather expensive. Current trends are to apply longer-lived, somewhat larger mulches of mostly sheet materials made of reinforced paper, polyester, or polypropylene (McDonald & Helgerson, 1990).

9.5 Biological weed control

Biological measures of weed control are based on the application of natural weed enemies such as insects, fungi, viruses, and bacteria in order to prevent their dissemination, and thus spreading. There are numerous examples of successful biological weed control. Application of pathogenic fungus, *Chondrostereum purpureum*, is used to control beech, yellow birch, red maple, sugar maple, trembling aspen, paper birch, and pin cherry (Wall, 1990). Exotic leaf pathogens, *Phaeoramularia* sp. and *Entyloma ageratinae*, were used for control of *Ageratina adenophora* and *Ageratina riparia* (Morris, 1991) in South Africa. Gordon & Kluge (1991) mentioned that control of *Hypericum perforatum* can be done by using insects *Chrysolina quadrigemina* and *Zeuxidiplosis giardi*. For control of *Acacia longifolia*, the widely spread invasive plant species in Portugal, the bee wasp *Trichilogaster acaciaelongifoliae* was used (Marchante et al., 2011). In those parts of the world where *Eucalyptus sp.* presents a problem the pathogen, *Cryphonectria eucalypti*, may be used for its suppression (Gryzenhout et al., 2003).

Application of biological measures in weed suppression has its limitations, though it has several advantages. Cultivated plants can be protected from some weeds, but not from all of them. It is impossible to destroy weeds completely because the biological agent depends upon the weed for survival; moreover, it is difficult to program biological protection for numerous cultivated plants from weeds with certainty since there are many similarities between weed species and cultivated plants (Konstantinovic, 1999).

9.6. Herbicides

Herbicides are used in forestry to manage tree-species composition, reduce competition from shrubs and herbaceous vegetation, manipulate wildlife habitat, and control invasive exotics (Shepard et al., 2004).

Unlike agriculture, the use of herbicides in forestry began much later and generally the application of herbicides in forestry was based on experiences from intensive agricultural production. The results of research in agriculture are applied in forestry with major or minor delays. Due to the lack of labour, high labour costs, and large areas, producers are more often interestedin the use of herbicides. Use of herbicides in forestry decreases weediness, particularly at the initial stages of development of forest nursery plants, when the effect of weeds on plants is the greatest; at the same time, much better economic efficiency in the production process is achieved. Also, possible mechanical damages to the nursery plants can be avoided, and it happens very often that any kind of mechanical treatment is prevented in early stages of plant development due to high soil humidity. Use of herbicides to control competing vegetation in young forests can increase wood volume yields by 50–150% (Guynn et al., 2004).

10. Division of herbicides

According to the type of action herbicides may be divided into the herbicides with contact action and herbicides of translocation. Contact herbicides destroy above ground parts of plant, only the parts of the plant they touch. Translocation or systemic herbicides absorbed by leaves are transferred through the whole plant.

According to the mode of action herbicides may be divided into the total and selective herbicides. Total herbicides kill all plants and selective herbicides kill weeds, and are not harmful to cultivated plants.

According to the time of application herbicides may be divided into the herbicides applied before sowing or planting, herbicides applied after sowing, and before emergence of weeds and cultivated plants and herbicides applied after emergence of weeds and cultivated plants.

11. Mechanisms of action herbicides

For contact herbicides, action is manifested at the site of penetration. Contact herbicides penetrate quickly through cuticle and epidermal cells of plants and the toxic effects on weeds are quickly observed. In systemic herbicides, absorption may occur through the root, stem, and leaf. If an herbicide is absorbed through the roots it can move via the xylem to the above ground parts; more often, it moves to leaves where disturbed respiration and photosynthesis occur. Herbicides absorbed by the leaves and stem cause harmful effects in absorbed plant parts and by spreading via the phloem to reach the root and ground plant organs (rhizomes). Whether the plant will absorb a higher quantity of the herbicide through above ground parts or the root system depends on herbicide application, type of herbicide, and several other factors (Janjic, 2005).

Herbicides exhibit different mechanisms of action. Some herbicides inhibit synthesis of amino acids in plants, and others help the formation of free radicals in plants. Lipid synthesis is the site of herbicide action used to control monocot weeds, and a great number of different herbicides inhibit the process of photosynthesis (Duke, 1990). While some herbicides act only on one process, others act on multiple processes in plants. If only one process in the cell is disturbed, the whole range of processes is affected. Due to that it is difficult to determine the primary herbicide action and consequences.

12. Selectivity of herbicides

Selectivity is a property of herbicides to destroy weeds effectively without harming cultivated plants. Selectivity is not an absolute property of any herbicide. There is a whole range of factors such as morphological, biological and physiological plant traits, chemical composition, and herbicide structure, quantity, mode and time of herbicide application, and translocation of herbicides on which selectivity of some herbicides depend (Kojic & Janjic, 1994; Owen, 1990). Cudney (1996) mentioned that herbicide selectivity is a dynamic process with complex interactions between plant, herbicide and environment.

The main mechanism of herbicide selectivity is the differential metabolism between weeds and crop species, by which susceptible weeds are less able to metabolize selective herbicides (Cole, 1994). The importance of understanding the main stages of differential metabolism-based selectivity derives from the elucidation that plants, to an extent, use cell energy to process and detoxify herbicides (Carvalho et al., 2009).

There is also morphological selectivity based on plant structure. Leaf structure such as vertical and narrow leaf, waxy coat, and a protected vegetation cone contribute to the fact that some grasses are resistant to 2,4-D herbicide. Greater resistance of coniferous species to herbicides compared to broadleaves is also based on leaf structure (Zekic, 1983).

13. Degradation of herbicides

A great part of the total quantity of herbicides applied in agriculture and in forestry is found in the soil. After introduction into soils, several processes affect the vertical and horizontal distribution of the herbicides including transport by water flow, sorption to soil components and various degradation processes. Degradation can involve biotic and abiotic processes, where microbially facilitated biodegradation is especially interesting, as it is a major process in the complete mineralisation of compounds to harmless inorganic products (Alexander, 1981; Kojic & Janjic, 1994).

Decomposition of herbicides in the soil is a complex process taking place in several stages such as photodegradation, chemical degradation, and microbiological degradation.

Photodegradation is means that some herbicide molecules such as trifluralin, dinitroaniline are degraded by the influence of ultraviolet rays, and these herbicides should be incorporated into soil after application (Konstantinovic, 1999). Chemical degradation of herbicides in the soil is done through processes of oxidation, hydrolisis, hidratation and reduction during which the herbicides are completely or partly degraded. Microbial degradation plays an important role in herbicide breakdown, and of their toxic material found in soil. Herbicides not only influence the activity of microorganisms, but the fate of herbicide in the soil depends on the activity of microorganisms (Janjc, 1996). Abilty of microorganisms to carry out herbicide biodegradation depends on the type af applied herbicide (Govedarica & Mrkovacki, 1993; Dordevic et al., 1994; Milosevic & Govedarica 2000), herbicide chemical properties (Poppell et al., 2002; Martins et al., 2001), applied herbicide concentrations (Gigliotti & Allievi, 2001), and great number of biotic and abiotic factors.

14. The toxicity of herbicides

Toxicity is the capacity of a substance to harm or disturb the health of an organism (Sovljanski, 2003). Toxic effects may be immediate (acute) or accumulative (chronic), depending upon the exposure duration, the dose, and the herbicide. The toxicity of a substance varies with the animal species, age, sex, and nutritional status and with the route of exposure – through the stomach (orally), the lungs (by inhalation), or the skin (dermally). The skin and eyes are also subject to irritation caused by chemicals.

A common way to document toxicity is by oral LD_{50} values. LD_{50} is the amount of chemical required to provide a "lethal dose" to 50% of the test population. LD_{50} is measured in mg of chemical administered per kg of body weight (Fishel et al., 2006). Toxicity tests are conducted on experimental animals, such as white rats, mice, and rabbits. Due to different ways of herbicide action when estimating its toxicity the whole range of data should be known and therefore it is difficult to express herbicide toxicity (Janjic, 2005).

15. Impact of herbicides on the environment

The impact of herbicides on the environment (water, soil, biodiversity, etc.) may have diverse effects depending on the whole range of factors and, initially, on the evironment in which it is found after application. In general, herbicides most commonly used for vegetation management in forestry (glyphosate, triclopyr, imazapyr, sulfometuron and etc.) degrade quickly once they enter the environment and thus are neither persistent nor bioaccumulative (Tatum, 2004). Forest herbicides persist short term in the environment, and have few toxic effects when operationally applied following herbicide labels (Guynn et al., 2004). Single applications of forestry herbicides at stand initiation have minor and temporary impacts on plant communities and wildlife habitat conditions (Miller & Miller, 2004). Studies carried out on the effect of herbicides hexazinone fosamine ammonium and glyphosate in forestry have revealed that these herbicides have minimal effects on soil microorganisms and exhibit little or no potential for bioaccumulation (Ghassemi et al., 1982). If herbicides are properly used, current research indicates that the negative effects on wildlife usually are short-term and that herbicides can be used to meet wildlife habitat objectives (Wagner et al., 2004).

16. Possibilities of weed control in forest nurseries

16.1 Mechanical weed control

In addition to irrigation, fertilization, and pruning of branches and tender shoots, hoeing and dusting are also very significant care measures in forest nurseries for production of planting material. Hoeing and dusting are regular measures applied during most of the vegetation period, and are particularly pronounced during the spring and at the beginning of summer. The purpose and objective of hoeing and dusting are, in addition to destruction of weeds, the maintenance of such soil structure that provides the optimum water-air regime of soil layers in which the root system develops (Roncevic et al., 2002).

The number of hoeing and farrowing applications required depends on the soil preparation, climate conditions, and on weed emergence. Markovic et al., (1995) claimed that first hoeing is a very significant measure that must be paid attention to particularly around cuttings and roots in order to avoid damages of buds and young shoots.

However mechanical measures have no long lasting effect in weed control due to relatively fast regeneration of weed flora (Table 1). Combined chemical and mechanical measures applied in forest nurseries are very effective in weed control. Us of herbicides decrease weediness in the early stages of development of cultivated plants when negative influences of weeds are the most dangerous. Mechanical injuries of nursery plants can be avoided in that way, and very often they are prevented due to high soil moisture. Mechanical measures are aimed at maintaining soil water-air regime and control of weeds that may subsequently have emerged.

Weed types	Treatment alternatives	Cost Euro/ha	Effectiveness	Potential Environmental Impacts
Perennial broadleaf and grass weeds	herbicides	50 - 100	Very effective	They can be potential polluters depending on the applied herbicide and the environment in which the herbicide occurs (water, soil)
	cutting	100 - 200	Not effective	No adverse effect on the environment
	cultivation	150 - 190	Effectiveness varies with weed and site	No adverse effect on the environment
	mulches	600 - 820	Only effective on annual weeds	No adverse effect on the environment
Annual broadleaf and grass weeds	herbicides	30 - 100	Very effective	They can be potential polluters depending on the applied herbicide and the environment in which the herbicide occurs (water, soil)
	cutting	100 - 200	Only effective on annual weeds	No adverse effect on the environment
	cultivation	150 - 190	Effectiveness varies with weed and site Very effective	No adverse effect on the environment
	mulches	600 - 820	Only effective on annual weeds	No adverse effect on the environment

Weed types	Treatment alternatives	Cost Euro/ha	Effectiveness	Potential Environmental Impacts
Woody weeds	herbicides	60 - 120	Very effective	They can be potential polluters depending on the applied herbicide and the environment in which the herbicide occurs (water, soil)
	cutting	100 - 200	Not effective	No adverse effect on the environment
Bracken	herbicides	60 – 120	Very effective	They can be potential polluters depending on the applied herbicide and the environment in which the herbicide occurs (water, soil)
	cutting	100 - 200	Weakness effect	No adverse effect on the environment

Table 1. Comments on control methods adopted and impacts in forestry

16.2 Application of herbicides

More attention is being paid to the application of herbicides as one of the control measures against weeds in forest nurseries. Due to the lack of labour, high labour costs, and large areas, the application of herbicides might be considered as the only possible way of weed control (Table 1).

When choosing herbicides, it is important to take care of several factors such as: weed composition, range of herbicide action, phenophase of development of cultivated plants, and time and manner of herbicide application.

16.2.1 Herbicides that can be used in forest nurseries

Application of herbicides in forest nurseries is that the biological and economic point of view is fully justified. Doses of herbicide application are low and there is no danger to wildlife, crops and watercourses (Zekic, 1979). For weed control in nurseries is necessary to provide a lot of labour and for mitigate this problem and increase the productivity application of herbicide is necessary. Chemical weed control in forest nurseries represents a complex job. In order to obtain the expected effect should take into account number of factors such as composition of the weed, the spectrum of action of herbicides, soil type, rainfall, temperature and etc. Herbicides that can be used in nursery production of some forest tree species are as follows:

Dasomet - is applied 2 - 5 weeks prior to sowing or planting in quantity of $30 – 60 \text{ g}/\text{m}^2$ incorporated into the soil up to the depth of $8 – 10$ cm. It is used for soil treatment in nurseries before sowing or planting.

Trifluralin – registered rate is 1,5 – 2,5 l/ha depending on soil with mandatory incorporation at a depth of 5 – 8 cm. It is used for soil treatment in nurseries before sowing or planting.

Azafenidin - is applied in quantity of 100 – 125 g/ha after sowing or planting, and before emergence of cultivated plants. It is used in production of poplar (*Populus euramericana, Populus deltoides.*) for control of a great number of broadleaf weeds (Photo 1).

Acetochlor – is applied in quantity of 2 l/ha after sowing or planting, and before emergence of cultivated plants. It is used in production of poplar (*Populus euramericana, Populus deltoides*), oak (*Quercus robur*), and black locust (*Robinia pseudoacacia*) nursery plants for control of a great number of grass and broad leaved weeds.

Dimethenamid - is applied in quantity of 1,2 – 1,4 l/ha after sowing or planting, and before emergence of cultivated plants. It is used in production of poplar (*Populus euramericana, Populus deltoides*), oak (*Quercus robur*), and black locust (*Robinia pseudoacacia*) nursery plants for control of great number of grass and broadleave weeds.

Linuron – is used for soil treatment after sowing or planting, and before emergence of cultivated plants. It is applied in quantity of 2 l/ha in production of poplar (*Populus euramericana, Populus deltoides*), willow (*Salix* sp.) and oak (*Quercus robur*).

Photo 1. Efficiency ofcombination of herbicids Acetochlor (Relay plus) and Azafenidinin (Evolus 80-WG) in nursery production of poplar (*Populus euramericana, Populus deltoides*)

Metribuzin – registered rate is 0,500 – 0,750 kg/ha after planting, and before emergence of poplars (*Populus euramericana, Populus deltoides*) and weeds. It is used for control with a greater number of annual broadleaved weeds. If applied on sandy soil with a lighter mechanical composition, it can have phytotoxic effects on poplar seedling (Photo 2).

Promethrin - is used for soil treatment after sowing or planting, and before emergence of cultivated plants and weeds. It is applied in quantity of 2 l/ha in production of poplar (*Populus euramericana, Populus deltoides*) and willow (*Salix* sp.) nursery plants (Photo 3).

Photo 2. Phytotoxic effect of n poplar seedling (*Populus euramericana, Populus deltoides*)

Photo 3. Efficiency of herbicides in nursery production of poplar plants (*Populus euramericana, Populus deltoides*)

Pendimethalin – is used for control of many annual grass weeds in production of poplar (*Populus euramericana, Populus deltoides*) nursery plants. It is applied after sowing or planting in quantity of 4 – 6 l/ha depending on soil.

Photo 4. Control plot

S-metolachlor – registered rate is 1,2 – 1,5 l/ha after sowing or planting, and before emergence of weeds and cultivated plants. It used for control of annual grass and broadleave weeds in production of poplar (*Populus euramericana, Populus deltoides*), black locust (*Robinia pseudoacacia*), willow (*Salix* sp.) and oak (*Quercus robur*) nursery plants.

Oxyifluorfen – registered rate is 1 l/ha after sowing or planting, and before emergence of cultivated plants. It is used in production of poplar (*Populus euramericana, Populus deltoides*) nursery plants for control of large number of grass and broadleave weeds.

Cycloxydim – is used for foliar control of annual and perennial grass weeds in production of poplar (*Populus euramericana, Populus deltoides*), black locust (*Robinia pseudoacacia*), oak (*Quercus robur*), maple tree (*Acer* sp.) and bee tree (*Evodia hupehensis*) nursery plants. It is applied in quantity of 3 l/ha at the stage of intensive weeds growth.

Glyphostate – is used for total control of emerged weeds on areas planned for sowing or planting, and on areas where sowing and planting have already been performed, and prior to appearance of cultivated plants or nursery plants. It is applied in quantity of 2 – 12 l/ha with water consumption of 200-400 l/ha.

Glufosinate - ammonium – is used as non-selective, contact herbicide for control of weeds on areas planed for sowing or planting, and on areas where sowing or planting have already been done, and before to the appearance of cultivated or nursery plants. It is applied in quantity of 4 -7,5 l/ha with water consumption of 400-600 l/ha.

Hexazinone – registered rate is 1,5 – 2 kg/ha in pine germination chambers 1,5 to 2 months after sowing and 2 – 3 kg/ha in nurseries. Pines are highly resistant to herbicide hexazinone, except *Pinus strobes* and *Pinus contorta*. Brown, necrotic spots appear on Norway spruce and larch is completely destroyed. It has a large range of action, long-lasting, so the soil remains clear throughout vegetation period.

Fluazifop-p-butyl – is used for foliar control of annual and perennial grass weeds in production of willow, poplar, oak, maple tree, birch, and beech. It is applied in quantity of 1,3 l/ha when weeds are at intensive growth phase.

Haloxyfop-p-methyl – is used for foliar control of annual and perennial grass weeds in production of poplar, acacia, oak and maple tree nursery plants. It is applied in quantity of 1 – 1,5 l/ha when weeds are at intensive growth phase.

Imazethapyr – is used in production of acacia nursery plants. It is applied at a quantity of 1 l/ha after black locust emergence and controls a great number of grass and broadleaved weeds.

Diquat – is used for total control of emerged weeds on areas planed for sowing or planting, and on areas where sowing or planting have already been done, and before to the appearance of cultivated or nursery plants. It is applied in a quantity of 4 – 6 l/ha.

Quizalofop-p-ethyl – is used for foliar control of annual and perennial grass weeds in production of willow, poplar, oak, maple tree, birch, and beech nursery plants. It is applied in a quantity of 1 – 1,5 l/ha at the intensive growth phase.

17. Possibilities of weed control in forest plantations

17.1 Mechanical weed control

Control of weed vegetation in forest stands is performed by mowing, cutting, pulling, etc. However, this way of weed control today is slow, inefficient and expensive and it must be repeated. Weed control between the rows is most often by mowing or by weed cutters, and within the row, or around the plants weed control is by application of herbicides.

17.2 Application of herbicides

Due to the presence of both herbaceous and woody weeds, application of herbicides in forest stands is of great significance. Woody weeds are very resistant and have great power of regeneration, therefore difficult to kill completely by mechanical means. It is very important to perform weed control in stands in a timely manner (at the stage of intensive growth) and in the right manner in order to use mininal herbicides inputs with effective results. Costs of weed control in stands by application of herbicides are much lower since reapplications are generally not required and it can be performed with less labour. Weed control in stands can be performed over the entire surface (broadcast application), within rows, or just around the nursery plants. The aim of weed control in forest stands is not complete destruction of weed flora, but prevention of competitive relationships with nursery plants and termination of growth.

However, their application is sometimes impossible in some systems because, very often, agricultural crops such as maize, soya, wheat, etc. are sown between the rows in order to use that space (Photo 5 and 6). In that case, selective herbicides for use in maize and efficient for control of weeds present in stand should be used. The most often used herbicides are those based on fluroxypyr or mesotrione.

Photo 5. Inter-row sown corn in the poplar plantation

Photo 6. Inter-row sown soybeans in the poplar plantation

17.2.1 Herbicides that can be used for weed control in forest plantations

Weed control in forest plantations should be carried out until the moment when the seedlings provided normal development and growth acording when seedlings grow beyond the zone of herbaceous weeds and shoots from the stumps. This moment the performance of various plantations in the different age because it depends on many factors such as species and age of seedlings, soil type, ground preparation prior to afforestation and etc. Weed control in forest plantations is not intended to completely destroy weed seedlings than the release from weed competition acording to stop weed growth and development. Application of herbicides in forest plantations can be done on the whole surface in rows or around trees. We should take into account that the applied herbicides do not reach the leaves of seedlings. For this purpose they may use the following herbicides:

Glyphosate – is used for complete weed control during the stage of intensive growth and concentration of 2-3%. Its efficacy on weeds is high due to good translocation into root and rhizomes. Side shoots, if present, should be removed prior to treatment with glyphosate because the preparations based on glyphoste should not reach the leaves of nursery plants (Photo 7 and 8). Glyphostate is used for treating stumps in order to prevent emergence of shoots from stumps. Concentration of 10-15% is applied immediately after cutting, but the treatment may be applied until shoots appear from May through October (Photo 9 and 10).

Photo 7. Efficiency of herbicide glyphosate in poplar plantation

Photo 8. Inter-row application of herbicide glyphosate in the plantation

Glufosinate - ammonium – is used as non-selective, contact herbicide for control of weeds in forest plantations. It is applied in quantity of 4 -7,5 l/ha with water consumption of 400-600 l/ha.

Triclopyr – is used for oak stump shoots control in the ration of 1:5 or 1:10 for other broadleaved species. The best way is to treat stumps immediately after cutting, but it can be done until the shoots emerge. It can be performed during entire year, except during freezing.

Photo 9. Treated stumps by glyphosate

Photo 10. Non treated stumps

18. Possibilities of weed control in the natural and artificial regeneration of pedunculate oak forest

In addition to harmful insects and diseases, weed vegetation represents a great problem in renovated pedunculate oak (*Quercus robur* L.) forests. Presence of weeds and great number of shrubby species per area unit is a basic limiting factor for continual spontaneous rejuvenation and offspring survival (Bobinac et al., 1991). Natural renovation is often poor or completely missing due to the presence of a great number of weeds. Due to the impossibility of preserving seedlings, and formation of quality offspring as well as the decline and slow development, the weed vegetation in young pedunculate oak forest should be suppressed. Prior to acorn planting it is important to perform the site preparation in order to provide the most favourable conditions for oak development. The acorn is planted in the autumn or spring in the soil prepared for the reception of the seed. If renovating surfaces are prepared prior to sowing then the problems with weeds encountered later when maintaining the surfaces are much smaller. But despite the completion of site preparation in the first and the second year after renovation, the occurrence of some weeds that emerged subsequently or were not affected by the preparation treatment before planting, may be expected.

18.1 Mechanical weed control

Removal of debris in the form of wood chips obtained by stump grinding is done completely or partially by collecting heaps, and spreading them on skid trails and then burning. Suppression of shrubs and shoots from the stumps may be accoplished manually by cutting with scythes, scissors, axes and etc. However, such weed suppression is inefficient and expensive, and is being replaced by faster and more efficient ways of suppression.

18.2 Application of herbicides

The most often encountered problem in pedunculate oak forest renovation is *Rubus caesius* L. (European dewberry) forming impenetrable thickets. Besides blackberry, *Crataegus monogyna, C.oxyacantha*, and *Rosa arvensis* may also pose a problem although significantly smaller. Also, if the shoots from stumps are not suppressed they can reach the height of up to 1,5 m, and blackberry and other weeds form thick cover, then the chemical control is difficult to perform due to impenetrability and requires a much higher expenditure of funds.

Photo 11. Application herbicides glyphosate after sowing and before emergence of cultivated plants (*Quercus robur*)

Photo 12. Emergence of oak plants after application of total herbicide

18.3 Herbicides that can be used for weed control in regeneration of pedunculate oak (*Quercus robur* L.) forest

To work in weed control in regeneration of pedunculate oak (*Quercus robur* L.) forest were successful among series of measures that are applied herbicide application is necessary. Herbicides that can be used for weed control in regeneration of penduculate oak (*Quercus robur* L.) forest as:

Cycloxydim – is used for foliar control of annual and perennial grass weeds. It is applied in quantity of 3 l/ha or applied twice at half-dosage (1,5 + 1,5 l/ha) when weeds are at the intensive growth stage.

Clopyralid – is used for control of a great number of broad-leaved weed species such as *Cirsium arvense, Ambrosia artemisifolia, Solanum nigrum, Erigeron canadensis* and etc. in renovated pedunculate oak forests. It causes transient symptoms of phytotoxicity in annual pedunculate oak plants (creating "a spoon like appearance"), while it is selective toward two-, and three-year oak plants, and causes no symptoms of phytotoxicity. It is applied in quantity of 1 l/ha with water consumption of 300 l/ha.

Nicosulfuron – is selective for pedunculate oak plants. It is used for control of a great number of annual broad leaf weeds, and for some perennial weeds. It is applied in the quantity of 1 – 1,2 l/ha.

Fluazifop-p-butyl – is used for foliar control of annual and perennial grass weeds in renovated pedunculate oak trees. It is applied in the quantity of 1,3 l/ha at the stage of intensive weed growth.

Glyphostate – is used for total control of weeds emerged on areas planed for sowing or planting, and on areas where sowing or planting have already been done, and before appearance of cultivated or nursery plants (Photo 11 and 12). It is applied in quantity of 2 – 12 l/ha, with water consumption of 200-400 l/ha.

Glufosinate - ammonium – is used as a nonselective, contact herbicide for weed control on areas planed for sowing or planting, and on areas where sowing or planting have already been done, and prior to appearance of cultivated or nursery plants. It is applied in quantity of 4 – 7,5 l/ha, with water consumption of 400-600 l/ha.

Photo 13. Efficiency of herbicides in artificial regeneration of pedunculate oak forest

Photo 14. Efficiency of herbicide clopyralid (Lontrel-100) in regeneration of pedunculate oak forest

19. Acknowledgment

This paper was realized as a part of the project "Studying climate change and its influence on the environment: impacts, adaptation and mitigation" (43007) financed by the Ministry of Education and Science of the Republic of Serbia within the framework of integrated and interdisciplinary research over the period 2011-2014.

20. References

Alexander, M. (1981). Biodegradation of chemicals of environmental concern. *Science* 211, 132–138

Bobinac, M.; Karadzic, D. & Cvjeticanin, R. (1991). Application of herbicides in the preparation of pedunculate oak stend for natural regeneration. Bulletin of the Faculty of Forestry, Belgrade. 73, 220-229

Carvalho, S. J. P de.; Nicolai, M.; Ferreira Rodrigues, R.; De Oliveira Figueira, A.V. & Christoffoleti, P.J. (2009). Herbicide selectivity by differential metabolism: considerations for reducing crop damages. Scientia Agricola, Vol.66, No.1, 136-142

Cole, D. J. (1994). Detoxification and activation of agrochemicals in plants. Pesticide Science, Vol.42, 209 - 222

Cudney, D. W. (1996). Why Herbicides Are Selective. California Exotic Pest Plant Council, Symposium Proceedings

Dhole, J. A.; Dhole, N. A. & Bodke, S. S. (2009). Ethnomedicinal Studies of Some Weeds in Crop Fields of Marathwada Region, India. Ethnobotanical Leaflets, Vol.2009, No.12, Article 3

DiTomaso, J. M. (1997). Risk analysis of various weed control methods. Symposium Proceedings, California Exotic Pest Plant Council

Dordevic, S., Govedarica, M., Ajder, S., Stefanovic, L. (1994): Uticaj nekih herbicida na biološku aktivnost i mikroorganizme u zemljištu. Contemporary agriculture, Vol. 42, No.3, 125-133

Duke, S. O. (1990). Overview of herbicide mechanisms of action. Environmental Health Perspectives, Vol.87, 263–271

Fishel, F.; Ferrell, J.; Mac Donald, G. & Sellers, B. (2006). Herbicides: How Toxic Are They?, PI-133, Florida Cooperative Extension Service. Institute of Food and Agricultural Sciences, University of Florida, EDIS web site at http://edis.ifas.ufl.edu.

Gigliotti, C. & Allievi, L. (2001): Differential effect of zhe herbicides bensulfuron and cinosulfuron on soil microorganisms. Journal of Environmental Science and Health, Part B-Pesticides, Food Contaminants and Agricultural Wastes, Vol.36, No.6, 775-782

Gordon, J. A. & Kluge, L. R. (1991). Biological control of St. John's Wort, *Hypericum perforatum* (Clusiaceae), in South Africa. Agriculture Ecosystems & Environment. Vol.37, No.1-3, 77 - 99

Govedarica, M. & Mrkovacki, N. (1993). Effect of different herbicides on the frequency of microorganisms under soybean. Mikrobiologija, Vol.30, No.1, 37-45

Ghassemi, M.; Quinlivan, S. & Dellarco, M. (1982). Environmental effects of new herbicides for vegetation control in forestry. Environment International, Vol.7, No. 6, 389-401

Gryzenhout, M.; Eisenberg B. E.; Coutinho, T. A.; Wingfield, B. D. & Wingfield, M. J. (2003). Pathogenicity of Cryphonectria eucalypti to Eucalyptus clones in South Africa. Forest Ecology and Management, Vol.176, No.1-3, 427-437

Guynn Jr., D. C.; Guynn, T. S.; Wigley, T. B. & Miller, D. A. (2004). Herbicides and forest biodiversity—what do we know and where do we go from here?. Wildlife Society Bulletin, Vol. 32, No.4, 1085-1092

Janjic, V. (2005). Phytopharmacy. Plant Protection Society of Serbia, ARI Serbia – Pesticide and Environmental Research Centre, Faculty of Agriculture Banja Luka, Beograd, Banja Luka

Janjic, V.; Stankovic-Kalezic, R. & Radivojevic, Lj. (2008). Natural products with allelopathic, herbicidal and toxic effects. Acta herbologica, Vol.17, No.1, 1-22

Kojic, M.; Stankovic, A. & Canak, M. (1972): Korovi biologija i suzbijanje Institut za zaštitu bilja. Poljoprivredni fakultet, Novi Sad

Kojic, M. & Janjic, V. (1994). Basis of Herbology. Institute of Agricultural Research "Serbia", Belgrade

Konstantinovic, B. (1999). Recognizing and weed control. University of Novi Sad, Faculty of Agriculture, Novi Sad

Konstantinovic, B.; Stojanovic, S. & Meseldzija, M. (2005). Biology, ecology and weed contro. University of Novi Sad, Faculty of Agriculture, Novi Sad

Kovacevic, J. (1979). Die Wald als ekologisches-phitocenologisches Phanomen. 1st Yugoslav conference on weed control in forestry, Sarajevo, 7-11

Kovacevic, D. & Momirovic, N. (2004). Weed management in organic agriculture. Acta herbologica, Vol. 13, No.2, 261-276

Lutman, W. J. P.; Cussans, W. G.; Wright, J. K.; Wilson, J. B.; Wright, McN. G. & Lawson, M. H. (2002). The persistence of seeds of 16 weed species over six years in two arable fields. Weed Research, 42, 231-241

Marchante, H.; Freitas, H. & Hoffmann, H. J. (2011). Assessing the suitability and safety of a well-known bud-galling wasp. Trichilogaster acaciaelongifoliae, for biological control of Acacia longifolia in Portugal. Biological Control, Vol.56, No.2, 193-201

Martins, J. M. F., Chevre, N., Spack, L., tarradelas, J., Mermoud, A. (2001): Degradation in soil and water and ecotoxicity of rimsulfuron and its metabolites, Chemosphere, 45, 515-522.

Marković, J., Rončević, S., Ivanišević, P. (1995) Mere nege i uzgoja u procesu proizvodnje sadnica topola i vrba. Seminar "Proizvodnja sadnog materijala vegetativnim putem" Novi Sad

McDonald, P. M. & Helgerson, O. T. (1990). Mulches aid in regenerating California and Oregon forests: past, present, and future. General Technical Reports, PSW-123, Berkeley, CA: U.S. Department of Agriculture, Forest Service, Pacific Southwest Forest and Range Experiment Station, pp. 19

Melander, B.; Rasmussen, I. A. & Barberi, P. (2005). Integrating Physical and Cultural Methods of Weed Control: Examples from European Research. Weed Science, Vol. 53, No.3, 369-381

Miller, K. V. & Miller, J.H. (2004). Forestry herbicide influences on biodiversity and wildlife habitat in southern forests. Wildlife Society Bulletin, Vol.32,No. 4, 1049-1060

Milosevic, N. & Govedarica, M. (2000). Effect of some herbicides on microbial properties of soil. Proceeding of 1st European Conference on Pesticides and Related Organic micropollutants in the environment, 61-63 Crfu, Greece

Morris, J.M. (1991). The use of plant pathogens for biological weed control in South Africa. Agriculture Ecosystems & Environment, Vol.37, No.1-3, 239-255

Orloff, S. B. & Cudney, D. W. (1993). Controlling dodder in alfalfa hay calls for an integrated procedure. California Agriculture, Vol.47, No.6, 32 - 35

Owen, W. J. (1990). Metabolism of herbicides – detoxification as a basis of selectivity. Herbicides and Plant Metabolism, Edited by Dodge, A.D., University of Bath, Society for Experimental Biology Seminar Series, pp. 171-198

Papachristou, G. T.; Spanos, A. I. & Platis, D. P. (2009). Forest vegetation management in Europe, Forest vegetation and management – Greece. European Science Foundation, Brussels, pp. 51 – 60

Poppell, C. A.; Hayes, R. M. & Mueller, T. C. (2002). Dissipation of Nicosulfuron and Rimosulfuron in Surface Soil. Journal of Agriculture and Food Chemistry, 50, 4581-4585

Pratley, J.W.H.; Lemerled, D. & Haig, T. (1999). Crop cultivars with allelopathic capability. Weed Research, Vol.39, No.3, 171-180

Shepard, J. P.; Creighton, J., & Duzan, H. (2004). Forestry herbicides in the United States: an overview. Wildlife Society Bulletin, Vol.32, No.4, 1020-1027

Sovljanski, R. (2003). Pesticidi simptomatologija i terapija trovanja. II dopunjeno i izmenjeno izdanje, Poljoprivredni fakultet Novi Sad

Stepp, J. R. & Moerman, D. E. (2001). The importance of weeds in ethnopharmacology. Journal of Ethnopharmacology, Vol.75, No.1, 19-23

Stepp, J. R. (2004). The role of weeds as sources of pharmaceuticals. Journal of Ethnopharmacology, Vol.92, No.2-3, 163-166

Roncevic, S.; Andrasev, S. & Ivanisevic, P. (2002). Production of poplar and willow reproductive and planting stock. Poplar, 169/170, 3-20

Tatum, V. (2004). Toxicity, transport, and fate of forest herbicides. Wildlife Society Bulletin, Vol.32, No.4, 1042-1048

Vajda, Z. (1973). Schadliche Waldunkrauter und deren Bekampfung. Fragmenta herbologica Jugoslavica XXV, pp. 1-8

Vasic, V. & Konstantinovic, B. (2008). Weed control in poplar nurseries using herbicides. Acta herbologica, Vol.17, No.2, 145-154

Vasic, V.; Orlovic, S. & Galic, Z. (2009). Forest vegetation management in Europe, Forest vegetation and management – Serbia. European Science Foundation, Brussels, pp. 117 – 122

Vrbnicanin, S. & Kojic, M. (2000). Biological and ecological research of weeds in Serbia development, current status perspectives. Acta herbologica, Vol.9, No.1, 41-59

Wagner, R. G.; Newton, M.; Cole, E. C.; Miller, J. H. & Shiver, B. D. (2004). The role of herbicides for enhancing forest productivity and conserving land for biodiversity in North America. Wildlife Society Bulletin, Vol.32, No.4, 1028-1041

Wall, R. E. (1990). The fungus Chondrostereumpurpureum as a silvicide to control stump sprouting in hardwoods. North. Journal of Applied Ecology, 7, 17-19

Wilson, R. G.; Kerr, E. D. & Nelson, L. A. (1985). Potential for using weed seed content in the soil to predict future weed problems. Weed Science, 33, 171-175

Zekic, N. (1979). The possibility of the application of Velpar in forest nurseries. 1st Yugoslav conference on weed control in forestry, Sarajevo, 63 - 69

Zekic, N. (1983). Weeds in forestry and weed control. Savez inzenjera i tehnicara sumarstva i industrije za preradu drveta Bosne i Herzegovine, Sarajevo

8

The Relationship Between Patch Spraying Cost and Target Weed Distribution

John Carroll[1] and Nicholas Holden[2]
[1]*Teagasc Crops Research Centre,*
[2]*University College Dublin,*
Ireland

1. Introduction

Carroll and Holden (2005) defined a method for quantifying weed distributions using distance transform analysis as a first-step in relating the distribution of weeds in a field to the type and cost of equipment used to spray the field (Carroll and Holden, 2009). The method was developed because in much of Europe, fields are sprayed at a fixed application rate determined by the average weed density of weed patches in the field, despite the fact that some areas of the field are below the economic threshold (ET) for intervention and do not require spraying (Mortensen et al., 1995). Targeted application of herbicides to weed patches, known as patch spraying, has the potential to significantly reduce herbicide use, which has both economic and environmental advantages (Lutman et al., 1998).

Patch spraying of herbicide is only viable if: (i) there is a distinct pattern of within-field variability; (ii) the variability identified can be reliably mapped; (iii) the variability has a known biological or environmental effect once managed; (iv) there is a suitable theoretical means of dealing with the weed that accounts for chemical efficacy and weed reproduction; (v) the mechanical equipment exists that can target the within-field variability in an accurate and precise manner; and (vi) the operation can be undertaken at an acceptable cost and return on investment.

The work of Carroll and Holden (2005, 2009) provides a method of quantifying weed distributions. It is documented that weeds are clustered and can be mapped (Godwin and Miller, 2003), and there is reasonable evidence to suggest that patch spraying can be a theoretically effective management approach that has an agronomic and environmental advantage (Lutman et al., 1998 and Wilkerson et al., 2004). Ford et al. (2011) showed that less herbicide could be used through variable rate application, when comparing a conventional broadcast herbicide sprayer to a variable spray weed sensing sprayer It is also known that within certain spatial constraints patch spraying can be accurate (Paice et al., 1997), but the question remains as to whether an acceptable cost and return on investment can be achieved. Carroll and Holden (2009) developed generalized relationships between field weed patterns, patch sprayer specifications and the spray quality achieved with the view to use these relationships to specify the most appropriate equipment for a certain field weed pattern based on required spray quality and cost.

The focus of the research presented within this chapter is to define the actual costs associated with using sprayer specifications, derived by analyzing weed distributions, and to compare this cost with that of uniform spraying. Any potential environmental costs or benefits associated with applying excess or precise herbicide amount in whole field or site specific applications are not considered. The analysis was undertaken to quantify the economic benefit of precision agriculture herbicide application technology.

2. Economic thresholds

Some research has been reported on the economic analysis of the benefits of patch spraying. The first requirement of any economic weed control analysis is the allocation of an ET, which is defined as the weed density at which the control cost equals the crop loss value if no control action is taken (Bauer and Mortensen, 1992) or the weed population at which the cost of control is equal to the crop value increase from control of the weeds present (Coble and Mortensen, 1992).

Coble and Mortensen (1992) wrote that the economic return associated with a crop production practice and the sustainability of that practice is of greatest immediate concern to the producer. As both biological and economic effects and costs are considered, an economic threshold offers a method by which profitable and sustainable weed management decisions can be made. The ET can be estimated by:

$$T_e = \frac{C_h + C_a}{YPLH} \tag{1}$$

Where

T_e = economic threshold
C_h = herbicide cost
C_a = application cost
Y = weed free crop yield
P = value per unit of crop
L = proportional loss per unit weed density
H = proportional reduction in weed density by the herbicide treatment.

Equation 1 reveals that any increase in herbicide or application cost will increase ET with other factors being constant. Increase in crop yield, value, degree of weed control or crop loss per unit weed density will lower ET. Three of the factors involved in ET calculations, herbicide cost, application cost and crop value can be estimated fairly accurately by individual growers. However, other factors including potential crop yield, proportional loss per unit weed density and herbicide efficacy are more difficult to estimate because of the variability associated with weather, weed species composition, weed size and cropping system effects on these variables. The focus of this research is on the application cost (C_a). This consists of depreciating sprayer value, cost of operation (including fuel and maintenance) and labor costs. All costs will relate to the size, segmentation and the control system of the sprayer.

Bauer and Mortensen (1992) discussed the Economic Optimum Threshold (EOT) concept. Economic thresholds generally refer to in-season decisions during a single crop year, and do not include a cost factor associated with possible increases in the soil seedbank due to lack of weed control. However, the term EOT is used to include the impact of seedbank dynamics on long-term profitability of weed management decisions. Not treating a near threshold weed population may affect whether or not a threshold will be exceeded in subsequent years due to increases in weed seedbank. Weaver (1996) discussed the importance of seed production by threshold density weed populations, emphasizing the importance of seed viability, dormancy and longevity. Long-term weed management programs must consider weed seed production, as well as yield losses to permit accurate cost estimation and to increase the likelihood that the threshold criterion is adopted by producers. Zanin et al. (1993) determined ET for winter wheat weed control using different herbicides or mixtures effective against individual weed species:

- *Avena Sterilis* L. subsp. Ludoviciana (Durieu) Nyman = 7 – 12 plants per m².
- *Alopecurus myosuroides* Hudson = 25 – 35 plants per m².
- *Lolium multiflorum* Lam = 25 – 35 plants per m².
- *Bromus sterilis* L. = just under 40 per m².
- *Galium aparine* L. = 2/ m².
- *Vicia sativa* L. = 2 – 10 / m².

Within the same species the different values of ET result from the different costs and efficacy of the herbicides.

Black and Dyson (1993) developed a model for calculating the economic benefit of early spraying of herbicides in wheat and barley crops, using data derived from routine herbicide evaluation field experiments. The absolute yield benefit (kg/ha) from controlling a forecast proportion of estimated weed units present when the crop is sprayed at or before the early tillering stage has 3 determinants: the weed-free yield potential of the crop (kg/ha); the number of weeds /m² at spraying; and relative growth stages of weeds and crop at spraying. The weight of evidence from the data used indicates that there is an approximately linear relationship between weed density after spraying and grain yield.

3. Economic simulations

Barroso et al. (2004) simulated the effects of weed spatial pattern and resolution of mapping and spraying on economics of site-specific weed management (SSWM). They concluded that the economic benefits of using SSWM are related to the proportion of the field that is weed-infested, the number of weed patches and the spatial resolution of sampling and spraying technologies. Different combinations of these factors were simulated using parameter values obtained for *Avena sterilis ludoviciana* growing in Spanish winter barley crops. The profitability of SSWM systems increased as the proportion of the field infested decreased and when patch distribution was more concentrated. Positive net returns for SSWM were obtained when the weed-infested area was smaller than 30% with the highest return occurring at a 12 m X 12 m mapping and spraying resolution.

Paice et al. (1998) evaluated patch spraying using a stochastic simulation model incorporating Lloyd's Patchiness Index to quantify the patchiness of the weed distribution and the negative binomial distribution to measure distribution shape. They concluded that the long-term economic benefits of patch spraying are likely to be related to the initial spatial distribution, the demographic characteristics of the weed species and the weed control and crop husbandry practices to which they are subjected and that for a system conforming to their very exact specifications, patch spraying of *Alopecurus myosuroides* Huds would not be profitable in the long term if the control area was greater than 6m X 6m. The method was not developed for field application and focuses on the agronomic rather than the mechanization aspects of weed control.

The work presented in this chapter, and the preceding papers (Carroll and Holden, 2005, 2009), provides a practical, readily applied method of selecting the most appropriate spray technology for accurate patch spraying based on the weed distribution to be targeted, and to evaluate whether an economic benefit will arise from using the technology in preference to uniform application of herbicide over the whole field.

4. Materials and methods

4.1 Weed maps and pattern quantification

The weed maps used to develop the economic analysis were the same as those used by Carroll and Holden (2005) (all scanned with a resolution of 1 m per pixel and subject to a 2 pixel radius filter after thresholding to remove noise): (i) 23 maps of blackgrass (*Alopecurus myosuroides* Huds.) in cereal fields published in a report for the Home Grown Cereals Authority in the United Kingdom (Lutman et al., 1998) with a critical density of >3 plant/m^2 and a minimum patch size of 25 m^2; (ii) 14 maps of 'scutch grass' (*Elymus repens* L. Gould) delineated by field scouting after harvest from a farm in Ireland located in Co. Kilkenny (52.6 degrees north, 7.1 degrees west) with a critical density of c. 10 plants/m^2; (iii) 9 maps delineating blackgrass (*Alopecurus myosuroides* Huds.) in cereals and sugar beet published by Gerhards and Christensen (2003), with a critical density of >5 plants/m^2; and (iv) 4 maps published by Barroso et al. (2001) delineating sterile wild oat (*Avena Sterilis*) infestations mapped by 4 different methods: counting panicle contacts, scoring panicle density from the ground, scoring panicle density from a combine and counting seed rain on the ground. 5 plants/m^2 was defined as the critical density.

The pattern of weeds as shown in each map was quantified by inward (subscript i) and outward (subscript o) distance transform analysis (Carroll and Holden, 2005) and summarised by an exponential association function fitted to the cumulative area probability distribution derived from the transformed image histogram:

$$y = a\left(b - e^{-c_n x}\right) \tag{2}$$

where a and b values account for the error of the fitted curve from the data (1 − ab indicates the deviation of the fitted curve from the data), and the c_n parameter represents the steepness of the curve (where n can be inward or outward). Values of c_n were collected for each field and were used to group fields with similar weed distribution patterns (Table 1, Figure 1).

4.2 Required resolution

Carroll and Holden (2009) defined the minimum control requirements for each on the nine weed map classes (Figure 1) in terms of boom segmentation (BS) and control distance (CD) for acceptable patch spraying (based on Spray Quality Index, SQI) for both untreated weed maps (Table 2), and after a dilation and erosion image processing algorithm had been applied to consolidate many small patches into larger patches of lower average weed density (Table 3).

Class	Description	C_i	C_o
1	Widely distributed large patches	< 0.05	< 0.025
2	Large patches closer together than in class 1	< 0.05	0.025 – 0.1
3	Large spatially aggregated patches	< 0.05	> 0.1
4	Medium, widely distributed patches	0.05 – 0.14	< 0.025
5	Medium patches closer together	0.05 - 0.14	0.025 – 0.1
6	Spatially aggregated medium sized patches	0.05 - 0.14	> 0.1
7	Small, widely distributed patches	> 0.14	< 0.025
8	Small patches closer together	> 0.14	0.025 – 0.1
9	Small spatially aggregated patches.	> 0.14	> 0.1

Table 1. Qualitative wed map class descriptions derived from quantified distance transform analysis (Carroll and Holden, 2009)

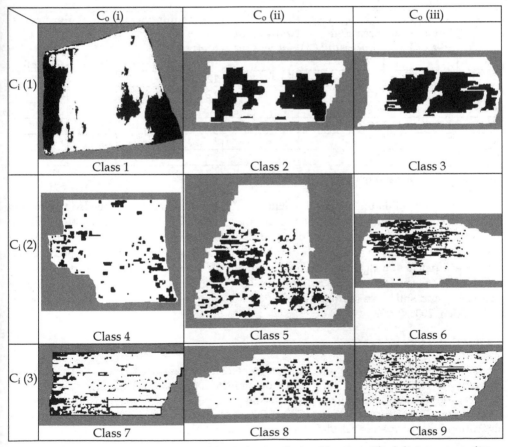

Fig. 1. Example weed maps in each of the 9 classes derived by quantified distance transform analysis.

Class	SQI @ Min Control	SQI @ Max Available Control	Min Requirements for 75% SQI	
	(BS = 30, CD = 20)	(BS = 3, CD = 2)	BS	CD
1	78	97	30	20
2	72	95	30	13
3	80	96	30	20
4	45	91	12	5
5	50	90	12	6
6	60	88	10	6
7	15	80	4	2
8	27	79	3	3
9	25	80	3	3

Table 2. SQIs at different boom segment length (BS, m) and control distance (CD, m) combinations

Class	Average initial weed area (%)	Average weed area after processing (%)	% error induced by processing	Min requirements for 75% SQI (before)		Min requirements for 75% SQI (after)	
				BS (m)	CD (m)	BS (m)	CD (m)
4	13	18	5	12	5	12	5
5	33	41	8	12	6	12	6
6	61	72	11	10	6	10	6
7	9	15	6	4	2	12	6
8	22	34	12	3	3	10	6
9	35	65	30	3	3	10	6

Table 3. Results of weed map erosion and dilation.

No change in the minimum technology requirements was predicted to be needed for Classes 1 to 6 (i.e. large to medium sized patches from widely distributed to spatially aggregated). This was due to the fact that the erosion and dilation process had very little effect in these situations. Only maps in classes 7 to 9 really benefited from this processing because the small weed patches amalgamate and produce maps classified as class 5 or 6. Pre-processing provides a means to specify readily available and relatively inexpensive spraying technology and still make savings in herbicide use compared to uniform spraying (Carroll and Holden, 2005, 2009).

4.3 Calculation of costs

For each of the nine classes, costs were calculated for an assumed model tillage farm of 100ha under winter wheat for three situations: uniform application with cheapest possible combination of sprayer and tractor; patch spraying for 75% SQI with no pre-processing algorithm; and patch spraying for 75% SQI after pre-processing with erosion/dilation algorithm. Total costs were calculated using five sub-sections:

- Sprayer: Manufacturers list prices and data from O' Mahony (2010) were used to determine the price of sprayers that could satisfy the boom length and control distance specified in Tables 2 and 3. Average costs per hectare over 10 years were found using ASAE Standard 497.4 as a template.

- Tractor: The size of tractor needed to power each sprayer was determined using ASAE Standards 2002a, EP496.2 (Table 4). List prices for these tractors were found in O' Mahony (2010). Using ASAE Standards 2002b, EP497.4 as a guide, depreciation, repair and maintenance and interest costs were calculated. The costs were then averaged over 10 years to give a cost of tractor use per hectare. Using average values of tractor work schedules (Forristal, 2005), 20% of the tractors yearly work was allocated to spraying. The Hardi Window 3.00 (Drouin, 1989) computer program was used to calculate fuel consumption and hence diesel costs based on tractor and sprayer size, distance to field and distance traveled within the field (Table 5).
- Positioning/Control System (only needed for patch spraying): The costs of a Global Positioning System with the required accuracy (e.g. 1-5 m Carroll and Holden 2009), Geographic Information System to process data and create the various maps required (e.g. ArcView GIS and AgLeader SMS) and control system for patch sprayer operation (e.g. AgLeader Insight) were determined as per manufacturers list price and again allocated per unit area using ASAE Standard 497.4.
- Labor: The Hardi Window 3.00 (Drouin, 1989) computer program was used to calculate the work rates for each type sprayer in ha/hr over the model 100 ha farm (Table 6). This was then converted to a labor cost by multiplying by the Irish national agricultural wage of €7.50/hr and allowing for three herbicide applications per year.
- Herbicide Costs: A herbicide cost of €66/ha for contact herbicide application in winter wheat crops was reported by O' Mahony (2010). There are typically three herbicide applications per year in Irish conditions. The first is a glyphosate spray on stubbled ground post harvest for control of grass weeds including scutch grass (*Elymus repens* L. Gould) and rye grasses (Lolium perenne spp.). The second spray (sulfonylureas) is applied post emergence for control of the common grass and broadleaved weeds including chickweed (Stellaria media spp.), speedwell (Veronica arvensis L.), charlock (Sinapsis arvensis L.) and knotgrass (Paspalum distichum L.). The third spray (amidosulfuron and Fenoxaprop-P (ethyl)) is applied for control of cleavers (Galium aparine) and wild oats (Avena fatua L.). For herbicide costs it was assumed that if a non-weed area is not sprayed, this will have no effect on future weed populations.

Total costs were calculated using equation 3.

$$T_c = C_t + C_s + C_r + C_l + C_h \qquad (3)$$

Where T_c = total cost
C_t = cost of tractor
C_s = cost of sprayer
C_r = cost of resolution (function of mapping/GPS/control system combination).
C_l = cost of labor
C_h = cost of herbicide

5. Results and discussion

5.1 Uniform application

Costs were calculated under the following headings

1. Sprayer: It was assumed that a simple 15m sprayer with manual valve operation could be used for uniform application of herbicide over the entire 100 ha model farm. This is a fairly typical sprayer used for these operations on Irish farms (Rice, 2005). A typical sprayer of this type has a list price of €12,000 and, allocated over 10 years, gives an average cost per hectare of €20.73. For the largest model available (30m) the extra costs of purchasing the equipment (€49,000) and the larger tractor (€90,000) were not found to be justified in this situation. However for a larger operation economies of scale may lead to larger sprayers being much more economically viable.

2. Tractor: Using the data from ASAE Standard EP496.2 it was determined that a tractor of 45 kW is required to operate this 15 m sprayer. Allowing for adverse field conditions and the use of a slightly larger tractor also used for many other farm operations, it was decided that a 65kW tractor at cost of €40,000 (O' Mahony, 2010) would be used in this situation to give an allocation of €13.82/ha.

Sprayer Width (m)	PTO Power (kW)	Required Tractor Power (kW)	Actual Tractor Power (kW)
10	24.6	29.6	50
12	29.5	35.6	55
15	36.9	44.5	65
18	44.2	53.4	75
21	51.7	62.2	85
24	59.1	71.1	90
27	66.4	80.1	100
30	73.8	88.9	110

Table 4. PTO and Tractor power requirements for different sprayer boom lengths

The costs were calculated as per {Table 7} and 20% of yearly tractor work was allocated to spraying. Diesel costs at €0.40/l (Table 5) were obtained using Hardi Window 3.00 program, which calculates sprayer use based on tractor and sprayer size, distance from field and distance traveled within the field and found to be €2.80/ha over three applications for this situation.

Boom Length (m)	Tank Size (l)	Tractor size (kW)	Work Rate (ha/hr)	Diesel Cost (€/ha)
10	800	50	4.1	2.97
12	800	55	4.5	2.91
15	1000	65	5.5	2.80
18	1200	75	6.4	2.76
21	1500	85	7.4	2.76
24	1500	90	8.1	2.67
27	2500	100	9.6	2.46
30	2500	110	10.2	2.55

Table 5. Diesel Cost based on different sprayer and tractor combinations.

3. Positioning/control system: for uniform application no mapping, positioning or control systems are used so none of these costs are incurred in this situation.
4. Labor: labor costs were calculated based on a sprayer work rate as shown in Table 6 and the agricultural minimum wage of €7.50 per hour to give a value of €4.09/ha for the 15 m sprayer over the 3 herbicide sprays.
5. Herbicide: From O' Mahony (2010) a herbicide cost of €66/ha was allocated for the three contact herbicide applications.

Boom Length (m)	Work Rate (ha/hr)	Labor cost (€/hr)	Number of runs	Labor (€/ha)
10	4.1	7.50	3	5.63
12	4.5	7.50	3	5.01
15	5.5	7.50	3	4.09
18	6.4	7.50	3	3.52
21	7.4	7.50	3	3.04
24	8.1	7.50	3	2.81
27	9.6	7.50	3	2.34
30	10.2	7.50	3	2.21

Table 6. Labor costs/ha calculations

5.2 Patch spraying

Costs were calculated under the same headings as for uniform for systems required for 75% SQI and best available technology in Ireland at the current time before and after processing with the erosion/dilation algorithm. At 75% SQI efficacy of spray is at or near 100%.

1. Sprayer: For each group the cost of sprayers with a resolution required for 75% SQI and best available technology (BAT) from table 2 were calculated. For 75% SQI it was found that the sprayer used in the uniform application had the necessary resolution for groups 1 to 3 to give a cost of €20.73/ha. For groups 4 to 6 a 15 m sprayer with control over 3 x 5 m segments and a control distance of less than 6 m was required. A sprayer with this resolution retailed at €15,000 to give a cost/ha of €25.91. For groups 7 to 9 a 15 m sprayer with control over 5 x 3m segments and a control distance less than 3 m was required. A retail price of €17,000 led to an allocation of €29.36/ha over 10 years. After pre-processing the required resolution of the sprayer remained the same for groups 1 to 6 so the same costs were incurred. For groups 7 to 9 the required resolution was decreased so the same sprayer as used for groups 1 to 6 could be used to give a cost of €25.91/ha, a decrease of €3.45/ha.

From table 2 the best available technology has control over 3m boom sections at a control distance of approximately 2m on a 15m boom. A retail price of €19,000 led to an allocation of €32.82/ha over 10 years to each group. Other emerging technologies may in the future lead to a much higher accuracy but as yet are not suited for herbicide application at high resolutions in cereal crops.

2. Tractor: For both 75% and best available technology with and without pre-processing, the 15m boom was used for each group as described above to give a cost of €13.82/ha

using the method as shown in Table 7. The first column shows the amount of depreciation (at 15% cumulative per annum) in each of the 10 years that the tractor is used. The second column shows repair and maintenance costs, which will naturally increase, as the tractor gets older. The 3rd column shows interest on capital expenditure at 5% per annum. The costs were then averaged over 10 years per hectare. Diesel costs came to €2.80/ha.

Year	Depreciation	R&M	Interest	Cost/ha
1	7500	166	2125	97
2	6375	833	1806	90
3	5418	1166	1535	81
4	4605	1666	1305	75
5	3915	3333	1109	83
6	3327	5000	942	92
7	2828	5833	801	94
8	2404	7500	681	105
9	2040	9166	579	117
10	1737	10000	492	122

Table 7. Allocation of costs over 10 years in €/ha for a €50,000 tractor

Group	% weed		Herbicide Cost (€)				
	Before processing	After processing	Uniform	Before processing		After processing	
				75% SQI	B.A.T.	75% SQI	B.A.T.
1	20	20	66	16.50	13.60	16.50	13.60
2	28	28	66	23.10	19.40	23.10	19.40
3	37	37	66	30.53	25.40	30.53	25.40
4	13	18	66	10.73	9.35	14.85	12.95
5	33	41	66	27.73	23.96	33.83	29.77
6	61	72	66	50.33	45.09	59.40	53.22
7	9	15	66	7.43	7.13	12.38	11.88
8	22	34	66	18.15	17.57	28.05	27.15
9	35	65	66	28.88	27.72	53.63	51.48

Table 8. Herbicide cost at different accuracy levels

3. GPS/Mapping/Control System: mapping costs of €15/ha for resolution required for 75% SQI and €18/ha for best available technology were allocated (Barroso et al, 2004). The AgLeader Insight system, which provides positioning, control, and analysis components at a retail price of €5,000 was used as the base for the control system. This gave a cost of €9.52/ha. For best available technology a retail price of €6,000 was assumed to give a cost of €11.43/ha over 10 years.

4. Labor: For both 75% SQI and best available technology the 15 m boom gave a work rate of 5.5 ha/hr to give a labor cost of €4.09/ha over the three spray applications from Table 6.
5. Herbicide: Calculated based on average percent weed in each class before and after pre-processing from Carroll and Holden (2009) as shown in table 8. As can be seen for the 75% and best available technology categories the cost of herbicide will depend on the percent weed present in the field.

Once all these costs had been calculated they were collected and combined to give an overall cost for spraying with the three different methods for each group. Examples of 3 groups are shown in Table 9 and the final figures for all groups are shown in Table 10.

Cost	Group 1			Group 5			Group 9		
	Uni-form (€/ha)	At control requirements for (€/ha)		Uni-form (€/ha)	At control requirements for (€/ha)		Uni-form (€/ha)	At control requirements for (€/ha)	
		75%	B.A.T.		75%	B.A.T.		75%	B.A.T.
Sprayer	20.73	20.73	32.82	20.73	25.91	32.82	20.73	29.36	32.82
GPS/ Mapping/ Control System	0	24.52	29.43	0	24.52	29.43	0	24.52	29.43
Tractor	16.62	16.62	16.62	16.62	16.62	16.62	16.62	16.62	16.62
Labor	4.09	4.09	4.09	4.09	4.09	4.09	4.09	4.09	4.09
Herbicide	66	16.50	13.60	66	27.23	23.96	66	28.88	27.7
Total	107.44	82.46	96.56	107.44	98.73	106.92	107.44	103.47	110.68

Table 9. Total cost of uniform versus patch spraying for some sample groups.

Group	Total costs (€/ha) @				
	Uniform Spraying	Before processing		After processing	
		75% SQI	B.A.T.	75% SQI	B.A.T.
1	107.44	82.46	96.56	-	-
2	107.44	89.06	102.36	-	-
3	107.44	96.49	108.36	-	-
4	107.44	81.87	92.31	85.99	95.91
5	107.44	98.73	106.92	104.97	112.73
6	107.44	121.47	128.05	130.54	136.18
7	107.44	82.02	90.09	83.52	94.84
8	107.44	92.74	100.53	99.12	110.11
9	107.44	103.47	110.68	124.77	134.44

Table 10. Total costs at different resolution levels

By analyzing the above data, the cost benefit from patch spraying can be described using the following function.

Cost benefit = f (group, percent weed, required resolution)

The required resolution will be determined by the group to which the field is allocated and the resolution cost can be described using the function.

Resolution cost = f(sprayer, tractor, mapping, positioning, control system)

It is clear from the above data that patch spraying at 75% SQI would give a reduction in costs in most cases. This reduction is very much related to the percent weed present with a Pearson's Correlation Coefficient of 0.975 ($p < 0.0001$). As percentage weed increases the benefits derived from patch spraying will decrease linearly. In the groups with well spread out weed patches (1,4 and 7) cost benefits of up to €25/ha could be achieved. Even though the costs were greater in group 7 due to the increased resolution of the spraying system a large benefit could still be achieved due to the major reduction in herbicides. Cost benefits are least in the spatially aggregated patch groups due mostly to the fact that in these groups the percent weed is almost always greater than in other groups. The dilation/erosion pre-processing, while reducing the equipment costs in groups 7 to 9 actually led to an increased patch spraying cost in all cases. This was due to the production of more weed pixels by the process and hence an increase in percent weed, which led to greater herbicide costs.

While these results focus only on the sprayers and tractors required for specific use on a model 100 ha winter wheat farm, the data from Tables 4,5 and 6 could be used to allocate costs based on different sized systems.

6. Conclusion

Using the above method it is clear that patch spraying using basic, readily available equipment should be economically advantageous in certain situations. For many of the weed map classes containing medium to large weed patches (1 to 6) there should be economic benefits (up to €25/ha) from patch spraying. Some benefits are also expected in fields with smaller, more aggregated weed patches but at higher weed populations the extra cost of more sophisticated equipment may outweigh the savings from reduced herbicide usage. For patch spraying to become a more attractive option to farmers a cheap, standardized mapping method must be maintained and control systems that can adapt normal sprayers for site specific application must become more readily available and cost effective. If these conditions are met and the methods described by Carroll and Holden (2005, 2009) are used to allocate the correct sprayer to the correct field and weed distribution, patch spraying may be of great economic benefit to a large number of farmers as well as decreasing pesticide introduced into the agro-environment.

7. References

ASAE Standards, 49th Edition, 2002a EP496.2 Agricultural Machinery Management. St. Joseph, Michigan: ASAE.

ASAE Standards, 49th Edition, 2002b EP497.4 Agricultural Machinery Management. St. Joseph, Michigan: ASAE.

Barroso, J., Fernandez-Quintilla, C., Maxwell, B. and Rew, L., 2004. Simulating the effects of weed spatial pattern and resolution of mapping and spraying on economics of site-specific management. *Weed Research*, 44, 460 – 468.

Barroso, J., Ruiz, D., Fernandez-Quintanilla, C., Ribeiro, A. and Diaz, B., 2000. Comparison of various sampling methodologies for site-specific sterile wild oat management. *Proceedings of the 3rd European Conference on Precision Agriculture*, Montpellier Ecole Nationale Superieure Agronomique, 578 – 581.

Bauer, T.A. and Mortensen, D.A., 1992. A comparison of economic and economic optimum thresholds for two annual weeds in soybean. *Weed technology*, 6, 228 – 235.

Black, I.D. and Dyson, C.B., 1993. An economic threshold model for spraying herbicides in cereals. *Weed Research*, 33, 279 – 290.

Carroll, J.P. and Holden, N.M., 2005. A method to quantify weed distribution for relating to patch spraying systems. *Transactions of the ASAE* 48(1), 27 – 35.

Carroll, J.P. and Holden, H. H., 2009. Modeling the relationship between patch sprayer performance and weed distribution. *Transactions of the ASABE*, 52:1051-1056.

Coble, H.D. and Mortensen, D.A., 1992. The threshold concept and its application to weed science. *Weed Technology*, 6, 191 – 195.

Drouin, B., 1989. Developer of Hardi Window Version 3.00. Application Technology Group for Hardi International A/S. Copyright Baler Software Corporation.

Ford, A., Dotray, P., Keeling, J., Wilkerson, J., Wilcut, J. and Gilbert, L., 2011. Site-specific weed management in cotton using WebHADSS. *Weed Technology*, 25, 107 – 112.

Forristal, D., 2005. Department of Engineering. Teagasc, OakPark, Ireland. Personal Communication.

Gerhards, R. and Christensen, S., 2003. Real-time weed detection, decision making and patch spraying in maize, sugar beet, winter wheat and winter barley. *Weed Research*, 43, 385 – 392.

Godwin, R.J. and Miller, P.C.H., 2003. A review of the technologies for mapping within-field variability. *Biosystems Engineering*, 84, 393 – 407.

Lutman, P.J.W., Rew, L.J., Cussans, G.W., Miller, P.C.H., Paice, M.E.R., & Stafford, J.E. 1998. Development of a 'Patch Spraying' system to control weeds in winter wheat. *Home Grown Cereals Authority Project Report No.* 158, London, U.K.

Mortensen, D.A., Johnson, G.A., Wyse, W.Y. and Martin, A.R. 1995. Managing Spatially Variable Weed Populations. *Site-Specific Management for Agricultural Systems, 2nd International Conference*, 397 – 415, ASA, CSSA, CSSA, Madison, WI.

O' Mahony, J., 2010. Crop Costs and Returns. Teagasc, Oak Park Carlow.

Paice, M.E.R., Miller, P.C.H. and Lark, A.G., 1997. The response characteristics of a patch spraying system based on injection metering. *Aspects of Applied Biology*, 48, 41 – 48.

Paice, M.E.R., Day, W., Rew, L.J. and Howard, A., 1998. A stochastic simulation model for evaluating the concept of patch spraying. *Weed Research*, 38, 373-388.

Rice, B., 2005. Head of Engineering Department, Teagasc, OakPark, Ireland. Personal Communication.

Weaver, S.E., 1996. Simulation of crop-weed competition models and their applications. *Phytoprotection*, 77, 3 – 11.

Wilkerson, G., Price, A., Bennett, A., Krueger, D., Roberson, G. and Robinson, B., 2004. Evaluating the potential for site-specific herbicide application in soybean. *Weed Technology*, 18, 1101 – 1110.

Zanin, G., Berti, A. and Taniolo, L., 1993. Estimation of economic thresholds for weed control in winter wheat. *Weed Research*, 33, 459 – 467.

9

Potential Use of Tembotrione (HPPD-Inhibitor Herbicides) in Grain Sorghum

Hugo De Almeida Dan[1], Alberto Leão De Lemos Barroso[2],
Rubem Silvério De Oliveira Junior [1], Lilian Gomes De Moraes Dan[1],
Sergio De Oliveira Procópio[3], Jamil Constantin[1]
and Guilherme Braga Pereira Braz[1]

[1]*Center for Advanced Studies in Weed Research, Agronomy Department,
State University of Maringá, Paraná,*
[2]*University of Rio Verde, Goiás,*
[3]*Brazilian Agricultural Research Corporation, Londrina, PR,
Brazil*

1. Introduction

Sorghum [*Sorghum bicolor* (L.) Moench] is a very important cultivated species in India, United States and some countries in Africa due to its high nutritional value both for food (grains) and feed (forage and grains) (Dahlberg *et al.*, 2004). In Brazil, sorghum has increasingly attained a level of recognition mainly as an option for the second crop cycle known as "safrinha". It has also been considered a viable alternative to replace crops such as cotton [*Gossypium hirsutum* (L.) Moench], corn (*Zea mays* (L.) Moench] and millet [*Pennisentum glauco* (L.) Moench] in crop rotations, serving not only for straw residue in conservation agriculture systems but also for the production of grains and forage as well (Gontijo Neto *et al.*, 2002).

Grown in tropical and subtropical climate regions, grain sorghum presents upright growing habit, mid-range height and uniform development even under limited water availability (Kismann, 2007). Despite its rusticity, grain sorghum has a slow initial growth, becoming vulnerable to the interference caused by weed competition. In this context, weeds may become a limiting factor for the development of the crop. It is estimated that the coexistence of weeds along with grain sorghum during the four first weeks after crop emergence may cause reductions ranging from 40 to 97% in grain yield (Tamado *et al.*, 2002).

In spite of being a remarkable crop on grain production worldwide, there are a limited number of studies on the selectivity of herbicides for this species, making weed control options more limited, mainly in large areas (Abit *et al.*, 2009). One of the major obstacles that has limited sorghum expansion is the difficulty to manage weeds due to the crop sensitivity to grass herbicides currently available (Archangelo *et al.*, 2002). Since the aggravating factor is the difficulty to control grass weeds, new research on this issue must be considered.

The identification of post-emergence herbicides capable of controlling grasses, with suitable crop selectivity, is crucially important to keep sorghum cultivated areas expanding. Most of the registered herbicides used on sorghum farming were initially developed to be used on other large scale crops, particularly on corn and sweet corn (Stahlman & Wicks, 2000).

In this regard, the objective of this study is to gather information concerning the actual status of reported effects of weed interference on grain sorghum and also discuss options of chemical weed control through post-emergence herbicides including tembotrione.

2. Importance of weed control in sorghum

Sorghum, as well as other agricultural plant species, is subjected to a series of biotic and abiotic factors, which directly or indirectly influence its growth and development (Magalhães et al., 2000). Among these factors, weed-imposed interference on crops is one of the most remarkable. Low sorghum yields have been correlated both to the absence and inefficient weed control (Erasmo & Pitelli, 1997).

The negative effects of weeds on sorghum agrosystems occur mainly due to the competition for crops' vital resources, such as water, light and nutrients. Furthermore, weeds can host pests and diseases, raising the cost of production, not to mention the depreciation of the product's quality (Grichar et al., 2005; Andres et al., 2009).

Initial development of sorghum is slow when compared to other cultivated species, which ensures that weeds, mainly those with a more aggressive growth habit, are more advantaged in the competition for resources, making sorghum more susceptible to interference exerted by weed community (Rizzardi et al., 2004). Even showing a slow initial growth, sorghum utilizes a C4 photosynthetic pathway and is able to grow under low soil moisture conditions (Rodrigues et al., 2010). Noteworthy for fodder sorghum, an annual crop used for feeding during dry periods, dense sowing increases this crop's competitive efficiency in relation to weeds, due to a faster land cover and, therefore, to the limited available spaces for weed emergence and growth.

For the majority of cultivated species, most troublesome weeds are those with a similar morphophysiology and life cycle, such as Echinochloa crus-galli and Brachiaria plantaginea, in areas cultivated with corn, sorghum and pear millet (Andres et al., 2009; Rodrigues et al., 2010; Dan et al., 2011a).

However, in the United States and Mexico, the biggest problems to weed competition are related to the presence of broadleaves. These species have caused steep yield reductions, encouraging research focused on such weeds (Grichar et al., 2005; Rosales-Robles et al., 2005). Weed density increases of one single plant of Amaranthus palmeri per square meter have caused a 1.8% reduction on grain yield (Moore et al., 2004). In subtropical areas, some grasses are still considered even more aggressive. According to Norris (1980), the presence of 175 Echinochloa crus-galli plants per square meter was enough to cause a 52% reduction on grain sorghum yield.

Weed management on sorghum crops in small properties has been carried out during the first 40 to 50 days after emergence, and two to three manual weedings are required. From this point on, the sorghum canopy will contribute to reduced favorable conditions for weed germination, growth and development, mainly by reducing the incidence of radiation (Rizzardi et al., 2001). In larger areas, weed control is usually accomplished by herbicide applications. Despite the method used for weed control, it is also important to observe that the period within the crop cycle when weed interference is prevented may also be a determinant for crop success. The stage of the crop cycle when weed control is established strongly influences competition levels, bringing about impacts on crop growth, development and grain yield (Silva et al., 2009).

At the start of crop development, sorghum and weeds can coexist for a given period without the latter affecting either quantitatively or qualitatively crop production. This phase is called 'period prior to interference' (PPI). By determining interference periods in sorghum crops in tropical regions, Rodrigues et al. (2010) concluded that sorghum and the weed community could coexist for 42 days (PPI) with no yield reduction (Figure 1). On the other hand, this interference could occur earlier in the crop cycle depending on the density and species of weeds.

The period of time after sorghum emergence during which it must be free from weed competition is called 'total period of interference prevention' (TPIP).

By definition, the period in which weeds effectively interfere with the crop and the period during which competition must not exist is called 'critical period of competition' (PCPI) (Pitelli & Durigan, 1984). During this period, there has been observed a drastic crop yield reduction (54%) when the control was achieved late in time. Silva et al. (1986) observed that the absence of weed control on the first four weeks after sorghum emergence can lead to a reduction in grain production of 35% and that, without any control during the entire crop cycle, the reduction can be as high as 70%.

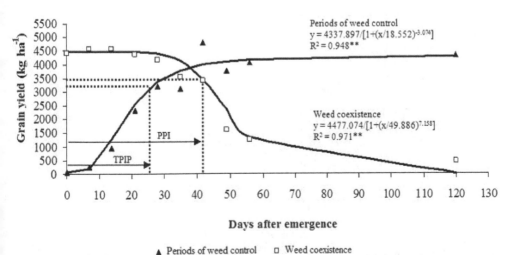

Fig. 1. Sorghum grain yield as a function of periods of weed control and weed coexistence in tropical regions (Rodrigues et al., 2010).

However, total period of interference prevention (TPIP) was 26 days (Rodrigues *et al.*, 2010). On this basis, it can be concluded that PPI was longer than TPIP, and, in this case, there was no PCPI. Under this scenario, accomplishing weed control just once during the crop cycle would be enough to preserve yield crop potential, as long as it is carried out between the end of PPI and the end of TPIP. Nevertheless, it must be understood that those periods may vary, mainly in relation to the intensity of competitive potential of weeds and to the density range as well as the predominant environmental conditions which may be more or less favorable to weeds. *Abutilon theophrasti* is noted to be more competitive than *Ipomoea purpurea* and *I. hederacea* in relation to sorghum, but the period of competition varies according to soil moisture level, exposure to solar radiation and nitrogen fertilization (Feltner *et al.*, 1973). Further studies should be carried out to determine critical periods of weed interference under different environmental and soil conditions.

Another approach to study weed interference on crops is based on crop development stage. For sorghum, the plant's phenological stage is usually a better indicator than the number of days after crop emergence due to both biotic and abiotic factors affecting crop growth (Larcher, 2000).

Losses can reach 80% of grain production under no weed control method (Andres *et al.*, 2009). Weed control on fodder sorghum crop should be accomplished along the period of the crop cycle between third and seventh leaf emission. Proper weed control during this period ensures no significant damage to the crop's grain yield. Figure 2 represents sorghum grain yield in relation to the phase of crop cycle in which weed control was accomplished (Andres *et al.*, 2009).

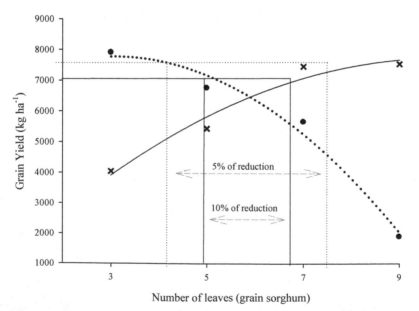

Fig. 2. Sorghum grain yield in relation to periods of initial control and coexistence of weeds in sorghum crop cv. BRS 305 in temperate climate lowlands. (x) periods of initial control; (•) periods of initial coexistence (Andres *et al.*, 2009).

Lack of adoption of weed control measures may affect sorghum quality and/or productivity, and, as a result, decrease a farmer's profitability. However, management of the weed community at specific periods of time ensures lower damages because sorghum can exert the crop's control as well as express its full productive potential.

Local variations on the critical period of weed interference are due to differences in crop genotype, sowing and emergence timing, water and nutrients availability, and density and composition of the weed community.

3. Selectivity of herbicides to grain sorghum

Traditionally, sorghum is more susceptible to herbicides than corn, mainly for graminicides applied postemergence. This response limits the utilization of chemical control as the main tool for weed management in sorghum areas.

To date, most studies have focused on the selectivity of herbicides applied pre-emergence such as s-metolachlor, dimethenamid and atrazine. However, the use of s-metolachlor has always been limited to the utilization of protective agents known as "safeners". Seed treatment using protectors such as fluxofenim, oxabetrinil, benoxacor, cyometrinil and naphthalic anhydride improves selectivity of s-metolachlor for sorghum (Horky & Martin, 2005).

It is estimated that approximately 95% of sorghum area is treated with post-emergence herbicides, particularly with atrazine. In Brazil, little attention has been given to pre-emergent herbicides in sorghum, due to the fact that most areas are cultivated in no or minimum-tillage areas. Therefore, sorghum sowing is often associated with the presence of a variable amount of straw (ranging from 2 to 8 ton dry matter per hectare) from the previous cropping cycle, usually following soybeans. With the increasing area of no-till farming and the growing problems of herbicide-resistant weeds, there has been a growing demand for herbicides with different mechanisms of action, mainly those applied post-emergence. Table 1 summarizes main current post-emergence options studied and utilized in weed management for grain sorghum.

Common Name	Level of selectivity	Author
atrazine	Good	Martin (2004)
bentazon	Good	Ferrell et al (2008)
bromoxynil	Good	Rosales-Robles et al. (2005)
2,4-D (amine)	Inter.	Dan et al. (2010b)
carfentrazone	Good	Ferrell et al. (2008)
dicamba	Good	Smith & Scott (2006)
halosulfuron	Inter.	Ferrell et al. (2008)
mesotrione	Inter.	Abit et al. (2009)
prosulfuron	Good	Rosales-Robles et al. (2005)

Inter: Intermediate (Some restrictions); Good: (No restrictions)

Table 1. Compilation of results related to herbicide selectivity in post-emergence application in sorghum

One of the most commonly used herbicides to control weeds post-emergence in sorghum is atrazine. Atrazine has been the basis of chemical weed control in corn for the last 50 years and its mechanism of action inhibits the electron flow in photosystem II; other than its know selectivity to corn, it has been considered selective to other grass crops such as pear millet and sorghum (Dan *et al.*, 2011a). In contrast, one of the main limitations of this herbicide is its low effectiveness on grasses. Previous reports confirm the limited effectiveness of atrazine postemergence applications to control grass weeds like *Cenchrus echinatus* and *Digitaria horizontalis* in corn and sorghum (Dan *et al.*, 2011a,b).

Herbicides like 2,4-D, carfentrazone and dicamba have also been considered excellent alternatives for the control of broad-leaved weeds. However, they present limitations regarding grass control. Furthermore, additional caution concerning the use of synthetic auxins like 2,4-D and dicamba, should be taken since the combination of late applications and high doses of these chemicals can cause foliar and root dymorphism, which in some cases, leads to yield reduction (Dan *et al.*, 2010b).

Among the graminicides and broadleaf herbicides with potential post-emergence use in sorghum, carotenoid biosynthesis inhibitor herbicides, particularly those that inhibit the enzyme 4-hydroxyphenylpyruvate dioxygenase (HPPD) are noteworthy (Miller & Regehr, 2002). The inhibition of HPPD blocks the pathway of prenylquinone biosynthesis in plants. Early effects, prior to the appearance of visible phytotoxicity symptoms, are decreased levels of tocopherols and plastoquinone in the plant tissue and a reduced photosynthetic yield. Indirect inhibition of phytoene desaturase as an effect of blocked plastoquinone biosynthesis leads to a decrease in carotenoid levels particularly in young, still expanding leaves. This causes typical foliar bleaching symptoms because the photosynthetic apparatus is no longer stabilized by these pigments. Under high light intensity, excess energy is not quenched and chlorophyll molecules are destroyed (Wichert *et al.*, 1999). Since carotenoids play an important role in dissipating the oxidative energy of singlet O_2, bleaching occurs due to the loss of the protection provided these pigments, leading to a chlorophyll oxidative degradation and, in some extreme cases, to cell membrane oxidation (Mitchell *et al.*, 2001; Armel *et al.*, 2003; Grossmann & Ehrhardt, 2007). Current carotenoid biosynthesis inhibitors registered for use in Brazil include clomazone, isoxaflutole, mesotrione and tembotrione, but clomazone and isoxaflutole have been limited to pre-emergence applications.

Some crops, such as corn, show good tolerance to these herbicides. It has been suggested that selectivity of HPPD inhibitors occur due to a rapid metabolism of herbicide molecules, mainly caused by the action of cytochrome P450 hemoprotein. The cytochrome P450 enzyme, responsible for this metabolism, is likely encoded by the active allele, *Nsf1* (Pataky *et al.*, 2008). Sweet corn hybrids, homozygous for the inactive allele (*Nsf1*), are highly sensitive to mesotrione (Pataky *et al.*, 2008).

Recent studies have demonstrated the possibility of using mesotrione in sorghum as post-emergence applications. Mesotrione is a HPPD inhibitor and belongs to triketone chemical family. It is derived from a natural phytotoxin (callistemone) obtained from the *Callistemon citrinus* plants. A large variability of crop response in the 85 sorghum hybrids treated with 0, 52, 105, 210, and 315 g ha[-1] mesotrione was found when plants were sprayed at the 3 to 4-leaf stage (Abit *et al.*, 2009). From the total number of hybrids tested, 23 were classified as susceptible, 45 as intermediate, and 17 as tolerant. From the 17 hybrids classified as tolerant, four were grown in the field. In field, the level of injury symptoms did not correlate to yield reduction. Since sorghum hybrids were able to recover from injury as the growing season progressed, injury symptoms were not good predictors of yield loss. This study demonstrated that post-emergence applications of mesotrione to sorghum grain hybrids caused a differential crop injury response ranging from susceptible to tolerant. To develop mesotrione as a good alternative for post-emergence weed grass management in sorghum, it may be crucially important for regionalized studies to understand the diversity of genotype tolerance across different producing regions throughout the world.

3.1 Selectivity of tembotrione to grain sorghum

Tembotrione was discovered in 1997 and launched as a commercial herbicide in 2007/2008 in Austria, Hungary, USA and Brazil. When tembotrione is applied to the foliage, a very high percentage of the applied compound is rapidly absorbed. In cases where the herbicide comes in contact with the soil, only small amounts enter the plants via the roots. Accordingly, this herbicide acts after post-emergence application predominantly via the foliage. Tembotrione is mobile both in the plant symplast (phloem) and in the apoplast (xylem). The mobility in the phloem is of particular importance, since it ensures that after a post-emergence spray application the herbicide will be distributed in the stream of assimilates from the mature leaves (metabolic sources) to the developing, highly susceptible leaves (metabolic sinks) at the shoot apex. In accordance with the translocation data obtained with ^{14}C-labeled tembotrione, it can be demonstrated that after controlled foliar placement of the herbicide on susceptible weed species new shoot growth is inhibited due to phloem systemicity (Van Almsick et al., 2009).

As a member of the triketone family of active ingredients, tembotrione shows properties of a weak acid (pKa = 3.18), resulting in high water solubility and low lipophilicity, e.g. a low octanol/water partition coefficient. These properties are pH-dependent in the environmentally relevant pH range between pH 5 to 9 (log Pow = –1.09 at pH 7 and –1.37 at pH 9). Consequently, it can be assumed that the behavior of tembotrione in soil and aqueous systems is also influenced by pH. This expectation was confirmed by the differences in the water solubility of tembotrione. Solubility is low at pH 4 (0.22 g L^{-1}) and significantly higher at pH 7 and 9 (28-29 g L^{-1}). The high solubility in water at neutral to weakly alkaline pH correlates favorably with the low logPow. Therefore, under environmentally relevant pH conditions, tembotrione is mainly present in its ionic form indicating a very low potential for accumulation in biological systems and a tendency to form salts in the environment. In addition, with the values determined for vapor pressure and the Henry's law constant it is estimated that no significant volatilization from soil or water surfaces will occur (Tarara et al., 2009). Typical bleaching caused by tembotrione applications in sorghum occurs in leaves that develop after spraying (Figure 3).

Tembotrione is currently registered for post-emergence use in corn in the United States and Brazil and has showed quite satisfactory results on weed control, particularly for grasses. Commercial formulations of this herbicide include the safener isoxadifen-ethyl, granting higher selectivity to corn and popcorn crops (Waddington & Young, 2006). Field evaluations of crop tolerance provided by mesotrione, topramesone and tembotrione applications in corn, lead to the conclusion that tembotrione caused the least crop injury when compared to topramesone and mesotrione (Bollman et al., 2008).

Fig. 3. Simptoms of tembotrione (200 g ha-1) injuries in late post-emergence application in sorghum.

When assessing selectivity of tembotrione applied to 4-leaf stage in five sorghum cultivars, different levels of crop tolerance were found (Dan *et al.*, 2009a). Results from evaluation performed seven days after application (DAA) of tembotrione demonstrated typical injuries of carotenoid pigment biosynthesis inhibitor herbicides (Figure 3). Throughout the post-application evaluation period, all cultivars showed intoxication (0 to 23% crop injury) when compared to those plants with no herbicide treatment (Table 2). Although there have been visible injuries in all cultivars at 7 DAA, progressive recovery of sorghum plants lead to less than 5% of visual injuries and no bleaching at 21 DAA (Table 2).

Cultivar AG-1020 was the most susceptible genotype among cultivars, and its shoot dry biomass was severely (~30%) affected when plants were harvested 28 DAA. Cultivars have not differed concerning the extent to herbicide sensitivity after 75.5 ha-1 tembotrione application to sorghum crop in tropical regions.

Based on the effect of crop dose-response in relation to stages when the herbicide application was performed, results so far indicate that earlier applications are more harmful to grain sorghum development (Dan *et al.*, 2010a). In this study, they evaluated the effect of tembotrione (0, 42, 88, 126, and 168 g ha-1) applied to three phenological stages of sorghum (S1: 3-leaf stage, 15 days after emergence; S2: 5-leaf stage, 23 days after emergence; S3: 8-leaf stage, 31 days after emergence). Cultivar AG-1040 presented the greatest injury levels (59, 46 and 38% at 7 DAA), respectively for the highest dose of 168 g ha^{-1}. Results are shown below (Figure 4).

Cultivar	Visual crop injury (%)			SDW (g plot $^{-1}$)			
				Dose (g ha^{-1})			
	7 DAA	14 DAA	21 DAA	0.0		75.5	
DKB 599	17.0	6.5	4.4	13.2	abA	11.2	abA
AG 1020	23.0	19.4	2.3	12.9	abA	9.1	bB
BRS 308	13.5	4.3	1.5	10.3	bA	9.3	bA
AG 1040	14.7	8.6	3.2	15.3	aA	13.4	aA
AGN-8040	11.3	10.3	4.3	12.6	bA	10.2	aA
CV%				13.12			
DMS				3.23			

Means followed by the same letter (low case letter in the column and capital letter in the row) do not differ from each other by Tukey $p \geq 0.05$ test (Dan *et al.*, 2009a).

Table 2. Visual rating of crop injury and shoot dry weight (SDW) of five sorghum cultivars after application of two doses of tembotrione.

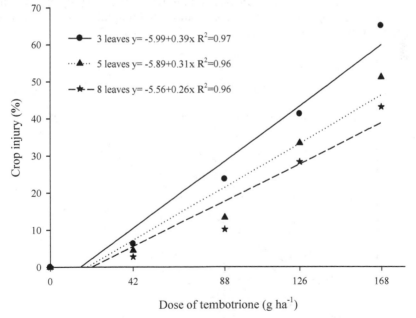

Fig. 4. Sorghum visual injury seven days after application for different doses of tembotrione in three crop growth stages. Source: Dan *et al.* (2010a).

Despite the rapid injuries recovery at 21 DAA, the authors have reported that trends evidenced at 7 DAA were maintained, indicating that applications accomplished in the earlier stages of sorghum crop development have provided the highest levels of crop injury, implying that herbicide tolerance increases as plants get older. Similar effects related to tembotrione applications in pearl millet have also been described (Dan *et al.*, 2010c). In pear millet, higher tolerance occurred when tembotrione (75 g ha⁻¹) was applied at the beginning of tillering, as compared to prior-tillering.

Increasing doses of tembotrione can trigger significant reductions on the amount of shoot dry weight and final plant height. More evident reductions of sorghum growth were observed when the herbicide application was carried out at earlier growth stages (3 leaves stage) (Dan *et al.*, 2010a). Injury reduction was twice as much more pronounced when compared to applications at 5- and 8-leaf stage. Nevertheless, effects on dry weight are directly related to crop stage at herbicide spraying. Abit *et al.* (2009) observed that all 85 sorghum hybrids evaluated showed significant reductions in the amount of dry weight after exposure to mesotrione, an herbicide which exhibits a very close chemical structure and similar mechanism of action to that of tembotrione.

Results lead to the conclusion that younger plants are less able to recover from injuries caused by tembotrione and that this fact directly reflects on dry weight accumulation, which may represent a negative factor for sorghum crops destined to forage production. For this reason, proper care should be taken concerning the dose and time of application of this herbicide.

In relation to grain yield, intoxication caused by tembotrione can cause significant reductions due to dose increment. Studies carried out with doses ranging from 0 to 168 g ha⁻¹, demonstrated grain yield reductions of 25, 16 and 15% for applications performed at 3, 5 and 8 expanded leaves stages, respectively (Figure 5) (Dan *et al.*, 2010a).

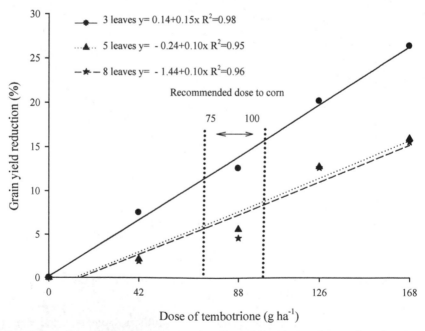

Fig. 5. Sorghum grain yield reduction as a function of increasing doses of tembotrione applied in three crop growth stages (Dan *et al.*, 2010a).

Currently, doses ranging from 75.6 to 100.8 g ha^{-1} of tembotrione are recommended for weed control in corn in Brazil. Taking into account the lowest recommended dose (75.6 g ha^{-1}), for instance, the greatest reduction observed for sorghum grain yield was about 11% when applications were carried out at the 3-leaf stage. Applications performed in other crop stages reached 7.3% and 6.1% in 5- and 8- leaf stages, respectively. These results indicate a potential use of this herbicide on grain sorghum, however, further studies evaluating other cultivars are required to supplement information on the selectivity of this herbicide.

Despite the different levels of crop injury, it is important to highlight that interference caused by weeds could pose a much more important risk due to losses up to 97% on grain sorghum yield (Tamado et al., 2001), justifying the need for weed control.

The tolerance of corn to tembotrione in combination with the safener isoxadifenethyl has been attributed to a much faster metabolic degradation of the herbicide than in susceptible dicotyledonous and grass weed species. Herbicide metabolism studies in corn, with and without a safener, reveal that isoxadifen-ethyl enhances tembotrione metabolism resulting in non-phytotoxic products. Corresponding to the specificity of safener action in corn, no significant enhancement of herbicide metabolism is found in Brachiaria plantaginea as one example of a representative target weed species (Tarara et al., 2009).

3.2 Weed control by tembotrione in grain sorghum

Besides selectivity, another decisive factor leading to the adoption of a certain herbicide is related to the spectrum of weed control. The list of weeds controlled by tembotrione in Brazil comprises important grasses like Brachiaria decumbens, Cenchrus echinatus, Digitaria horizontalis, D. ciliaris and Brachiaria plantaginea and broad leaf species like Alternanthera tenella, Commelina benghalensis, Ipomoea nil, I. purpurea, I. acuminate, Sida rhombifolia, Nicandra physaloides, Euphorbia heterophylla, Raphanus raphanistrum, Bidens pilosa, B. subalternans, Richardia brasiliensis and Leonurus sibiricus, but the registration is limited to corn. However, the control on broad leaf species such as A. tenella, B. pilosa and Ageratum conyzoides is usually extended by using the combined use of atrazine and tembotrione (Barroso et al., 2009). The efficiency of tembotrione alone is clearly limited when it is applied to weeds in a more advanced growth stage.

Among main grass species that are present in areas cultivated with sorghum in Midwestern Brazil, post-emergence applications of tembotrione may have a differential level of efficacy. D. horizontalis is more sensitive than Cenchrus echinatus; control of both species becomes more evident (>80%) in doses ≥88 g ha^{-1} for D. horizontalis in applications carried out before tillering. However, similar levels of control of C. echinatus are obtained only by using doses of 126 g ha^{-1} (Figure 6).

Other studies have also evaluated the spectrum of weeds controlled by tembotrione. Applied at 92 g ha^{-1}, control of broadleaves and grass species was reported (Hinz et al., 2005; Lamore et al., 2006), including redroot pigweed (Amaranthus retroflexus L.), common lambsquarters (Chenopodium album L.), common ragweed (Ambrosia artemisifolia L.), velvetleaf (Abutilon theophrasti Medic.), giant foxtail (Setaria faberi Herrm), barnyardgrass [Echinochloa crusgalli (L.) Beauv], and woolly cupgrass [Eriochloa villosa (Thunb.) Kunth].

Further work on this issue must investigate the possibility of using mixtures with other herbicides such as atrazine, among others. In addition, it is equally important to highlight other cropping techniques targeted to reduce the infestation in order to reduce pressure by making the control easier to ensure a more successful tillage management.

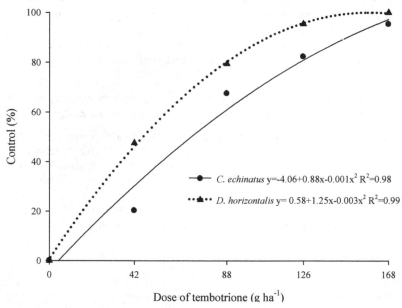

Fig. 6. Weed control for sorghum crop at 21 days after applying increasing doses of tembotrione (Dan et al., 2009b).

4. Concluding remarks

Weeds present a great competitive potential with grain sorghum. However, effects are converged by a number of factors such as weed species and densities, moment of crop cycle when control is imposed and farming practices such as tillage system. Results have demonstrated that more intense interference occurs, in most cases, starting at the 4-leaf stage, weed free period. Although sorghum cropping is widespread in a great variety of regions throughout the world, current selective herbicides have not been sufficiently evaluated and the options available so far are not enough. Studies have provided results that confirm HPPD-inhibitor herbicides potential, mainly for mesotrione and tembotrione, assisting mainly in post-mergence grass weed control. Nevertheless, regionalized studies on different genotypes of sorghum must be conducted to supplement information regarding the selectivity of this herbicide for grain sorghum and to support recommendations.

5. References

Andres, A. *et al.* Períodos de interferência de plantas daninhas na cultura do sorgo forrageiro em terras baixas. Planta Daninha, v.27, n.2, p.229-234, 2009.

Abit, J.M. *et al.* Differential response of grain sorghum hybrids to foliar-applied mesotrione. Weed Technology, v. 23, n.1, p.28-33, 2009.

Archangelo, E.R. *et al.* Tolerância do sorgo forrageiro ao herbicida Primestra SC. Revista Brasileira de Milho e Sorgo, v.1, n.2, p.59-66, 2002.

Armel, G.R.; Wilson, H.P.; Richardson, R.J. Mesotrione combinations in no-till corn (*Zea mays*). Weed Technology, v.17, n.3, p.111-116, 2003.

Barroso, A.L.L. *et al.* Controle de plantas daninhas na cultura do milho In: Worshop Comigo 2009, Rio Verde, GO. Anais... Resultados CTC Comigo 2009, 2009. p.90-96.

Bollman, J.D.; Boerboom, C.M.; Becker, R.L. Efficacy and tolerance to HPPD-inhibiting herbicides in sweet corn. Weed Technology, v.22, n.4, p.666-674, 2008.

Dahlberg, J.A.; Burke, J.J.; Rosenow, D.T. Development of a sorghum core collection: refinement and evaluation of a subset from sudan. Economic Botany, v.58, n.4, p.556-567, 2004.

Dan, H.A. *et al.* Efeito do herbicida atrazine no controle do capim-carrapicho (*Cenchus echinatus*) e capim-colchão (*Digitaria horizontalis*). In: Worshop Comigo 2009, Rio Rerde, GO. Anais... Resultados CTC Comigo 2009, 2009a. p.77-80.

Dan, H.A. *et al.* Seletividade de Herbicidas para a cultura do sorgo granífero. In: Worshop Comigo 2009, Rio Rerde, GO. Anais. Resultados CTC Comigo 2009, 2009b. p.80-85.

Dan, H.A. *et al.* Tolerância do sorgo granífero ao herbicida tembotrione. Planta Daninha, v.28, n.3, p.615-620, 2010a.

Dan, H.A *et al.* Tolerância do sorgo granífero ao 2,4-D aplicado em pós-emergência. Planta Daninha, v.28, n.4, p.785-792, 2010b.

Dan, H.A *et al.* Seletividade do herbicida tembotrione à cultura do milheto. Planta Daninha, v.28, n.4, p.793-799, 2010c.

Dan, H.A. *et al.* Influência do estádio de desenvolvimento de *Cenchrus echinatus* na supressão imposta por atrazine. Planta Daninha, v.29, n.1, p.179-184, 2011a.

Dan, H.A. *et al.* Supressão imposta pelo atrazine a *Digitaria horizontalis* em função do estádio de desenvolvimento. Revista Caatinga, v.24, n.1, p. 27-33, 2011b.

Erasmo, E.A.L.; Pitelli, R.A. Efeitos da adubação fosfatada nas relações de interferência entre sorgo granífero e tiririca. I. Crescimento inicial. Planta Daninha, v.15, n.2, p.114-121, 1997.

Feltner, K. C.; Vanderlip, R. L.; Hurst, H. R. Velvetleaf and morningglory competition in grain sorghum. Kansas Academia Science, v.76, n.4, p.282-288, 1973.

Ferrell, J.A.; Macdonald, G.E.; Brecke, B.J. Weed management in sorghum – 2008. Series of the Agronomy Department. 2008. 5p

Gontijo Neto, M.M.G. et al. Híbridos de sorgo (*Sorghum bicolor* (L.) Moench) cultivados sob níveis crescentes de adubação. Rendimento, proteína bruta e digestibilidade *in Vitro*. Revista Brasileira de Zootecnia, v.31, n.4, p.1640-1647, 2002

Grichar, W.J. *et al.* Weed control and grain sorghum (*Sorghum bicolor*) response to postemergence applications of atrazine, pendimethalin, and trifluralin. Weed Technology, v.19, n.4, p.999-1003, 2005.

Grossmann, K.; Ehrhardt, T. On the mechanism of action and selectivity of the corn herbicide topramezone: a new inhibitor of 4- hydroxyphenylpyruvate dioxygenase. Pest Management, v.63, n.3, p.429-439, 2007.

Hinz, J.; Wollam, J.; Allen, J. Weed control with AE 0172747 in corn. Proc. North Central Weed Science Society, v.60, p.90, 2005.

Horky, K.T.; Martin, A.R. Evaluation of preemergence weed control programs in grain sorghum. In: Weed Control in Specialty Crops. Lincoln, NE: 2005 NCWSS Research Report. v.62. 2005, p.30-32.

Lamore, D.; Simkins, G.; Watteyne, K.; Allen, J. Weed control programs with tembotrione in corn. Proc. North Central Weed Science Society, v.61, n.2, p.119, 2006.

Larcher, W. Ecofisiologia vegetal. São Carlos: RIMA, 2000. 531 p.

Kissmann, K.G. Plantas infestantes e nocivas. TOMO I. 3° ed. São Paulo: Basf Brasileira S. A., 2007. CD-ROM.

Magalhães, P.C. *et al.* Fitotoxicidade causada por herbicidas na fase inicial de desenvolvimento da cultura do sorgo. Planta Daninha, v.18, n.3, p.483-490, 2000.

Miller, J.N.; Regehr, D.L. Grain sorghum tolerance to postemergence mesotrione applications. Weed Science, v.57, n.2, 136-143, 2002.

Mitchell, G.D.W. *et al.* Mesotrione: a new selective herbicide for use in maize. Pest Management, v.57, n.4, p.120-128, 2001.

Moore, J.W.; Murray, D.S.; Westerman, R.B. Palmer amaranth (*Amaranthus palmeri*) effects on the harvest and yield of grain sorghum (*Sorghum bicolor*). Weed Technology, v.18, n.5, p.23-29, 2004.

Norris, R.F. Barnyardgrass [*Echinochloa crus-galli* (L.) Beauv] competition and seed production. Weed Science, v.5, n.20, p.143-149, 1980.

Pataky, J.K. *et al.* Genetic basis for varied levels of injury to sweet corn hybrids from three cytochrome P450-metabolized herbicides. Journal of the American Society Horticultural Science, v.133, n.1, p.438-447, 2008.

Pitelli, R.A.; Durigan, J.C. Terminologia para períodos de controle e convivência das plantas daninhas em culturas anuais e bianuais. In: Congresso Brasileiro de Herbicidas e Plantas Daninhas, 15., 1984, Belo Horizonte. Resumos. Belo Horizonte: SBHED, p. 37, 1984.

Rizzardi, M.A. *et al.* Competição por recursos do solo entre ervas daninhas e culturas. Ciência Rural, v.31, n.4, p.707-714, 2001.

Rizzardi, M.A.; KARAM, D.; CRUZ, M.B. Manejo e controle de plantas daninhas em milho e sorgo. In. VARGAS, L.; ROMAN, E. S (Eds.). Manual de Manejo e Controle de Plantas Daninhas. Bento Gonçalves: Embrapa Uva e Vinho, 2004. p.571-594.

Rodrigues, A.C.P. *et al.* Períodos de interferência de plantas daninhas na cultura do sorgo. Planta Daninha, v.28, n.1, p.23-31, 2010.

Rosales-Robles, E. *et al.* Broadleaf weed management in grain sorghum with reduced rates of postemergence herbicides. Weed Technology, v.19, n.2, p.385-390, 2005.

Silva, A.F. *et al.* Período anterior à interferência na cultura da soja-RR em condições de baixa, média e alta infestação. Planta Daninha, v.27, n.1, p.57-66, 2009.

Silva, J.B.; Passini, T.; Viana, A.C. Controle de plantas daninhas na cultura do sorgo. Informe Agropecuário, v.12, n.144, p.43-45, 1986.

Smith, K.; Scott, B. Grain Sorghum Production Handbook: Weed control in grain sorghum. Little Rock, AR: University of Arkansas Cooperative Extension Service. 2006. 74 p.

Stahlman, P.W.; Wicks, G.A. Weeds and their control in grain sorghum. In: Smith, C.W.; Frederiksen, R.A. (eds.). Sorghum: Origin, History, Technology, and Production. New York, NY: John Wiley and Sons. 2000. pp. 535-690.

Tamado, T.; Schu¨tz, W.; Milberg, P. Germination ecology of the weed *Parthenium hysterophorus* in eastern Ethiopia. Annals of Applied of Biology, v.140, n.3, p.263-270, 2002.

Tarara, G.; Fliege, R.; Kley, C.; Peters, B. Environmental fate of tembotrione. Bayer CropScience Journal, v.62, p.63-78, 2009.

Van Almsick, A.; Benet-Buchholz, J.; Olenik, B.; Willms, L. Tembotrione, a new exceptionally safe cross-spectrum herbicide for corn production. Bayer CropScience Journal, v.62, p.5-15, 2009.

Waddington, M.A.; Young, B.G. Interactions of herbicides and adjuvants with AE 0172747 on postemergence grass control. Weed Science, v.61, n.4, p.108-115, 2006.

Wichert R.; Townson J.K.; Bartlett D.W.; Foxon G.A. Technical review of mesotrione, a new maize herbicide. The BCPC Conference – Weeds, v.1: p.105–110, 1999.

Part 3

Integrated Weed Management and Soil Fertility/Quality

10

Managing Competition for Nutrients in Agro-Ecosystems

Moses Imo

Department of Forestry and Wood Science,
Chepkoilel University College, Moi University, Eldoret,
Kenya

1. Introduction

Knowledge of competitive plant interactions is important in designing more productive cropping systems in both agriculture and forestry. These interactions are often variable in nature, and may be competitive, synergistic or complementary depending on several factors such as species mixture, environmental conditions and management practices, which are also influenced by prevailing socioeconomic factors. Most of these interactions often involve primarily competition for the major plant growth resources namely: light, moisture, nutrients and space. Unfortunately, segregating the specific mechanisms involved at any time in the competition process has often been a major problem for many agro-ecologists because of the complex interactive nature of the requirements by plants for these growth resources. Although significant attempts and gains have been made with respect to understanding mechanisms for competition for a single aboveground resource (i.e. light), little progress has been made with respect to competition for a broad range of belowground resources (i.e. nutrients and moisture). This is mainly because of the multiple belowground interactions involving complex processes and mechanisms of availability, uptake and utilization by plants. In the case of nutrients, plants compete for a broad range of essential plant mineral elements that differ in molecular size, valence, oxidation state and mobility within the soil. Unfortunately, and leess understood, belowground competition often reduces plant performance more than aboveground competition (Wilson 1988), and it is the principal form of competition occurring in ecosystems with extremely low plant densities such as arid lands and low fertility sites (Fowler, 1986).

This chapter reviews the mechanisms and ecological importance of nutrient competition, emphasizing methodologies for measuring nutrient compeition in cropping systems and their advantages and limitations. This is particularly important in understanding the roles of plant competition for nutrients in the productivity of agro-ecosystems, and provide guidelines for their management. The approach is to combine knowledge in soil fertility and plant nutrition with physiological ecology in order to merge various diagnostic tools for decision making at farm level. The goal is to illustrate a simple graphical diagnostic model for identifying overall nutrient interaction effects and how to optimize various factors affecting nutrient competition in different agro-ecosystems. To be useful, such tools must help determine the benefits and consequences of crop and weed management strategies in any give system, and facilitate determination of the relative importance of various interaction types and the associated specific mechanisms.

2. Definition and importance of nutrient competition

Competition for soil nutrients is one of the major factors that structure plant communities (Grace and Tilman 1990). Understanding the mechanisms that control plant competition for soil nutrients is therefore an essential step in predicting the outcome of interspecific competition in many plant communities, and in designing effective cultural practices in agro-ecosystems. Generally, plant competition is broadly defined as the process by which two or more individual plants or populations of plants interact, such that at least one exerts a negative effect on its neighbor in terms of reduced survivorship, growth or reproduction. Competition for nutrients therefore can be said to occur when a plant depletes a soil nutrient and negatively impacts availability of the nutrient element to which another plant shows a positive response. This definition, which is derived from that of Goldberg (1990), essentially means that the reduced level of the nutrient (an intermediary resource), has a negative effect on the performance of the competing plants measured per individual or per unit size. The advantage of this physiologically based definition identifies how nutrient uptake by one individual plant can affect the quantity of the nutrient taken up by another, and often help determine the consequences for plant performance as shown in Figure 1.

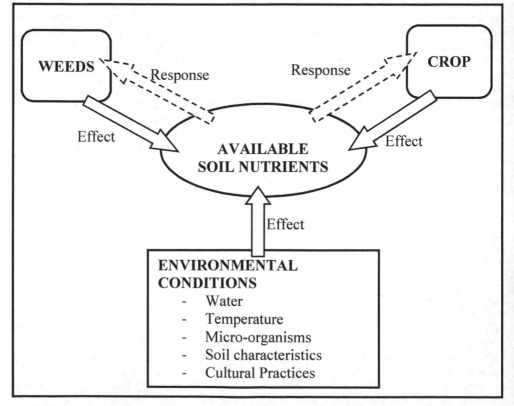

Fig. 1. Plant competition for nutrients showing the effects and responses of competing plants to changing nutrient availability. Both the effect and the response should have the appropriate sign for competition to occur.

Characteristically, like competition for all the other growth resources, nutrient competition is reciprocal, i.e., it occurs only when nutrient resources are in limited supply. The competing plants can either be of the same species (*intraspecific* competition) or of different species (*interspecific* competition). These responses are usually described by yield-density relationships that follow the 'law of constant final yield' (Begon *et al.* 1996). That is, at low density, total resource availability to each individual plant is high resulting in a few large individuals, and total growth will respond to small changes in density. At high densities, however, resource availability to each individual is low resulting in many small individuals, and total production is less responsive to changes in density and attains a final constant value reflecting complete utilization of available growth resources. When one species has a negative effect on the second species, yet both can utilize more efficiently available resources when in mixture than in monoculture, it is referred to as the *interference*, but it is *facilitation* if one species has a positive effect on the other (Vandermeer, 1984). The term *interaction* is also often used to simply mean mutual or reciprocal effects in situations where species performance in mixtures is not equal to the sum of their performances when they are grown separately. Although plant interactions may also be due to other effects such as allelopathy through production of toxins, parasitism by natural enemies and mutualisms, the focus in this chapter, however, is on interactions involving soil nutrients.

3. Mechanisms of plant competition for nutrients

3.1 Nutrient uptake

Soil nutrients reach the root surfaces for uptake through three general processes (Marschner, 1995): *root interception* (the capture of nutrients as the root grows through the soil, physically displacing soil particles and clay surfaces), *mass flow* (the movement of dissolved mineral nutrients in water driven by plant transpiration and is a function of the rate of water movement to the root and the concentration of dissolved nutrients), and *diffusion* of nutrients toward the root surfaces when nutrient uptake exceeds the supply by mass flow thereby creating localized nutrient concentration gradients. Overall, root interception is the least important of the three processes, while diffusion occurs only when mass flow and root interception are inadequate to meet plant requirements. However, these mechanisms almost always work together. Various mechanisms have been proposed to explain partitioning of soil nutrients among neighboring competing plants, and corresponding theories are often linked to a specific theoretical framework developed for a specific type of ecological system under study. Overall, the competitive ability of a plant is determined by its capacity to capture and exploit resources rapidly and the ability to utilize shared nutrient resources in shortest supply by two or more species (Tilman 1988). Thus, understanding mechanisms of nutrient partitioning requires knowledge of factors related to the plant's ability to acquire a greater proportion of nutrients, utilize nutrients more efficiently, and allocate assimilates in ways that maximize the capacity of an individual for survival and growth (Berendse and Elberse 1990; Nambiar and Sands 1993). These factors are discussed below.

3.2 Nutrient acquisition strategies by plants

The characteristics of plant root systems (e.g. root length, density, surface area and diameter), rates of nutrient diffusion in the soil and uptake by plants, morphological and physiological plasticity, and spatial and temporal soil partitioning are the major factors determining the nutrient competitive ability of most plant nutrient competition (Gillespie 1989; Neary *et al.* 1990; Smethurst and Comerford 1993). Generally, these factors have been used to predict nutrient uptake of competing plants as a function of the nutrient concentration in solution at the soil-root interface, which is determined by the balance between plant demand for nutrients and the ability of the soil to supply that demand. As roots absorb nutrients, concentrations around the root surface declines, thus creating 'nutrient depletion zones' around the root surface. Nutrient competition then occurs when the depletion zones for adjacent roots overlap, thus interfering with nutrient availability for each plant and resulting in reduced uptake.

Root morphology plays a major role in determining nutrient depletion. Regardless of nutrient mobility, competition for all nutrients increases with root length or density (Barber 1984; Gillespie 1989). In addition, thicker roots have steeper depletion gradients and wider depletion zones than thinner roots. Hence, thinner and longer roots are less likely to compete than thicker and shorter roots because finer roots will be able to absorb nutrients at much lower nutrient concentration in solution (Sands and Mulligan 1990). Spatial segregation of roots of different species may reduce interspecific competition. For example, the ability of woody plants to develop deep rooting systems (Eastham and Rose 1990; Stone and Kalisz 1991) may be an important strategy to avoid competition with shallow rooted herbaceous plants. However, roots of most woody plants are also concentrated in the surface soil (Nambiar 1990), thus making direct competition with herbaceous plants inevitable.

3.3 Nutrient use by plants

Plant growth usually increases with the amount of nutrient present in biomass. However, there are considerable species differences in the amount of nutrients required to produce biomass (Wang *et al.* 1991), indicating the differential ability in species to utilize nutrients for growth (i.e. nutrient-use-efficiency [NUE]). Generally, trees produce more biomass per unit of nutrients (i.e. higher NUE) than herbaceous weedy species, probably because trees increasingly produce woody tissue which is low in nutrients as it is not active photosynthetic tissue (Chapin 1990; van den Driessche 1991; Nambiar and Fife, 1991). This mechanism partially explains why the capacity of trees to dominate a site accelerates with increasing age of the trees.

Nutrient losses from a plant (litter fall, leaching, root decay and herbivory) also determine the plant's total nutrient requirements, i.e. the amount of nutrient that must be absorbed by an individual or population just to maintain or replace its biomass (Berendse and Elberse 1990). A species with low nutrient loss rate and high uptake rate and/or higher NUE will have a relatively low demand for external nutrients. According to Tilman (1988), such a species is predicted to be able to meet its nutrient requirements at a lower soil nutrient supply rate, and its total biomass will increase only if it absorbs more nutrients than its demand, but declines if it absorbs less. Hence, partitioning of a limiting nutrient between competing species is expected to be proportional to their demand relative to uptake; therefore, a species with a low demand relative to uptake will have a higher competitive ability.

3.4 Carbon allocation

Relative allocation of assimilates to root and shoot growth modifies root-shoot ratios of plants and influences the ability of plants to acquire below- and above-ground resources. Thus, one critical point in understanding plant strategies in overcoming nutrient competition is the allocation of growth to below-ground during nutrient limitations (Chapin 1980). Root-shoot ratios increase with reduced nutrient availability (Axelsson and Axelsson 1986; Boot and Mensink 1990) perhaps as a means of enhancing nutrient acquisition from the soil. However, such data must be analyzed with caution since the positive effects of nutrient status (e.g. higher nutrient concentration in plant tissue) on plant growth can overshadow the effect of preferred growth allocation to roots. For example, high soil nutrient availability may enhance both root and shoot growth resulting in a larger root system, but root-shoot ratio may decline.

3.5 Nutrient interactions with other resources

Since plant growth involves acquisition of multiple resources, interactions among these resources in the environment (particularly in the soil) can complicate interpretation of competition effects. As discussed earlier, soil moisture can greatly affect nutrient availability and uptake, while light can alter demand for the nutrients thus influencing the outcome of competition for the nutrients in short supply. Also, removal of one species may increase availability of all resources together because of reduced uptake, or the removal may indirectly modify resource availability through microclimate modification, as has been demonstrated in other weed control experiments (Smesthurst and Nambiar 1989; Woods *et al.* 1990). In these studies, weed removal by herbicide application increased available moisture and nutrients, and altered soil temperature that favored faster N mineralization. Weed removal (especially in young plantations) also increases light availability that favors faster growth and further creates a larger demand for soil moisture and nutrients. This evidence demonstrates the need for a systematic and integrated approach to resource competition studies to allow a better segregation of the competition processes involved. Figure 2 is a conceptual model for resource competition by plants that provides a comprehensive starting point for any such approach. Since interspecific interactions may result in either reduced growth (competition) or enhanced productivity (synergism), understanding processes of resource partitioning is necessary to increasing productivity in these ecosystems, and requires knowledge of soil and plant processes related to nutrient availability, uptake and utilization by plants.

4. Framework for understanding plant competition for nutrients

4.1 Ecological relationships

The model is based on Berkowitz's conceptual model of plant competition for resources (Berkowitz 1988), simplified here to illustrate the key processes involved as shown in Figure 2. This diagram distinguishes three major variables: state, rate and intermediate variables. State variables (represented by boxes in Figure 2) are measurable quantities (biomass, nutrient content, soil water, soil nutrients and solar radiation). Each state variable is characterized by a rate or efficiency variable (represented as valves in Figure 2) that determines the rate of flow in material or energy between state variables due to specific processes (represented as hexagons in Figure 2). The intermediate variables (represented as circles in Figure 2) define plant characteristics that directly determine the capture of available resources. The other variables not included in this model are deriving variables that characterize the effect of environmental conditions on the whole system.

The amount of resource acquired by a plant is determined by resource availability and the rate of resource uptake or resource capture efficiency (*LCE, WCE* and *NCE* for light, water and nutrient capture efficiency, respectively, in Figure 2). Plant biomass is considered a product of its cumulative total resource uptake and resource use efficiency (i.e. rate of biomass production per unit of resource acquired or *LUE, WUE* and *NUE* for light, water and nutrient use efficiencies, respectively, in Figure 2). Biomass is composed of carbon (*W*) and nutrient content (*U*), and the ratio *U/W* gives a measure of nutrient concentration (*C*). Carbon is allocated to different plant components (leaves, stem, fruits and roots) that determine the plant's physiological characteristics that further influence resource capture (Berendse and Elberse 1990). Greater proportional allocation to roots increases root density, hence improve efficiency or rate of capturing soil resources, while preferred allocation to shoots enhances growth rate because of higher leaf area index (*LAI*), thus photosynthesis. The loss rate (*LR*) determines the rate of nutrient return to the soil as mulch or root decay. Application efficiency (*NA*) determines the rate by which added nutrients are made available for uptake.

Resource acquisition and use involve three major processes: nutrient uptake, moisture absorption, and photosynthesis. Moisture availability influences growth through photosynthesis because of its effect on plant water status and hence leaf conductance (Burdett 1990), and nutrient absorption since it is required for diffusion and absorption of nutrient ions (Gillespie 1989; Smesthurst and Comerford 1993). In ecosystems in which at least one resource is limiting, biomass production can be regarded as a function of the amount of limiting resource(s) and their utilization efficiency. Interspecific plant interactions occur when one species affects availability of one or more resources to the other species, or simply alters environmental conditions favorable for the flow of material between state variables. Competition will reduce availability and flow of resources, while synergism will increase resource availability to the neighboring species. The mechanisms associated with this model are discussed in the following section.

4.2 Partitioning of soil nutrients

The conceptual framework in Figure 2 shows that there is mutual interdependence between nutrient uptake, moisture absorption and photosynthesis (Kropff *et al.* 1984). For example, increased nutrient availability and uptake accelerates photosynthesis that, in turn, promotes uptake and use of nutrients, and *vice versa*. Competition for soil nutrients may also reduce nutrient content and leaf area, thus resulting in reduced photosynthesis and growth that, in part, has a negative effect on nutrient uptake. Also, competition for moisture reduces plant water status and leaf conductance (thus reduced photosynthesis and growth), and may impair nutrient uptake since moisture is required for nutrient absorption by the roots.

The question as to which of these resources is the primary (direct) and secondary (indirect) growth limiting factor in competitive situations often poses considerable difficulty in interpreting observed effects of competition (Nambiar and Sands 1993). For example, most

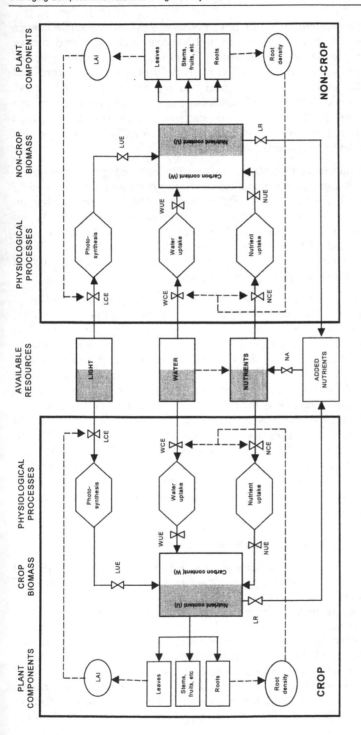

Fig. 2. Conceptual processes of resource competition between a crop and neighboring non-crop. Boxes represent state variables; valves represent rate variables; circles represent intermediate variables; and hexagons represent physiological processes. Solid arrows represent flow of material, while broken arrows are flow of information or signals, and indicate occurrence of feed-back processes. Available resources are partitioned between competing plant species: shaded for the crop, and unshaded for the non-crop. LCE, WCE and NCE are light, water and nutrient capture efficiency, respectively. LUE, WUE and NUE are light-, water- and nutrient-use- efficiency, respectively. Biomass is composed of carbon (W) and nutrients (U), and nutrients are returned to the soil as mulch as determined by the loss rate (LR). Nutrient application efficiency (NA) determines the rate of conversion of added nutrients to available nutrients. Adapted from Imo (1999), and based on Kropff *et. al* 1984).

studies reporting on moisture competition hardly have measurements on tree nutrition even when treatments may have substantial impacts on soil nutrient availability and uptake (Coates *et al.* 1991). Hence, the relative significance of competition for moisture versus nutrients is usually ignored, especially in environments in which both moisture and nutrients are limiting. Unlike moisture, however, nutrients accumulate in plant biomass. Since moisture affects both photosynthesis and nutrient absorption, the approach in this chapter is to evaluate the effects of competition on carbon accumulation and nutrient content as shown in Figure 2. The fundamental question, therefore, is how competition affects carbon assimilation relative to nutrient uptake.

In Figure 2, carbon (W) and nutrient content (U) are assumed to be an integrated measure of availability, uptake and use of light, moisture and nutrients by plants, and are regulated by other environmental factors (i.e. deriving variables). When competition is primarily for soil nutrients, a reduction in U will be large relative to a corresponding reduction in W, and U/W (i.e. nutrient concentration) will also decline. When competition is primarily for light, then reduction in U will be small relative to reduction in W, and U/W will increase. When nutrients are non-limiting and competition is primarily for moisture, reduction in U will also be small relative to reduction in W, and U/W will increase.

In environments where both water and nutrients are limiting, moisture competition is expected to have overriding effects because water molecules are more mobile than nutrient ions in the soil, hence would have larger and greater overlapping depletion zones than nutrients (Gillespie 1989; Smesthurst and Comerford 1993). Thus, reduction in U will be small relative to reduction in W, and U/W will increase. Also, increased growth associated with (1) elevated nutrient levels (both concentration and uptake) reflects positive fertility effects, and (2) increased uptake but decreased nutrient concentration exemplifies improved moisture, light and or microclimate favorable for crop growth without significant effects on nutrient availability. These principles have been illustrated using interactions involving N responses in order to confirm these interpretations (Imo and Timmer, 1999a; 1999b). The treatments in these studies were selected to represent competition-free status and the other three interaction types (antagonistic, synergistic and compensatory). Although this technique may be applicable to all other nutrients when the same resources are removed, total resource availability to the other species should increase resulting in maximum growth and nutrient uptake potential (Imo 1999).

4.3 Nutrient uptake and plant growth relationships

Since nutrient concentration and content of a plant gives an integrated estimate of both total uptake and use by a plant (Imo 1999), studying the relationships between the two fundamental processes involved (i.e. nutrient accumulation and biomass production) can provide insight into the mechanisms involved. Chemical analysis of plants is frequently used to diagnose the nutritional status of plants since the plant itself is the object of interest, and its nutrient composition reflects many of the factors affecting its nutrition. Traditionally, plant nutrient composition is expressed either in relative term s (i.e. concentration [C], the amount of nutrient present per unit amount of biomass) or on total mass basis (i.e. absolute content [U], the total amount of nutrient present in a specific amount of plant tissue [W]). Total content is obtained by multiplying concentration by dry mass of the sample, thus $U = C(W)$. Imo and Timmer (1997) and Timmer (1991) have previously argued that using concentration alone does not reveal the mechanism on how nutrient content and dry mass are related, since changes in concentration may be caused by changes in either biomass or nutrient uptake or both, and there is no way of distinguishing between these mechanisms. Changes in concentration as a result of changes in content implies that the plant itself altered nutrient uptake and synthesis, while changes in concentration due to changes in biomass can be regarded as a growth response without any specific effects on metabolism of the nutrient.

5. Measuring plant competition for nutrients

5.1 Root exclusion experiments

Belowground competition is measured by quantifying the extent that root interactions reduce nutrient uptake and plant growth by preventing root interactions using root exclusion tubes, trenching, or neighbor removal to separate the roots of target individuals from those of neighboring plants. Root competition is determined by comparing the growth or survival of target plants inside the partitions with those having root systems that can interact freely with neighboring vegetation. Neighbors within partitions are killed by a fast-degrading herbicide, severed at the soil surface to remove shoots, or removed completely by excavating, sieving, and replacing the soil. Unfortunately, such methods often alter the soil environment and may even affect the availability of resources for which the plants are competing. To address this issue, many studies have proposed various competition indices to characterize the degree to which the growing space of a target crop is shared by weedy vegetation in agro-ecosystems and forest plantations by developing functional relationships between target crop or tree responses to some measure of non-crop (weed) proximity. Some of these indices are discussed below.

5.2 Competition indices

5.2.1 Additive and substitutive indices

The traditional approach has been to use competition indices to predict yield losses due to weeds in forestry and in agricultural systems (Morris and McDonald 1991; Wagner, 1994; 1993). Similarly, a competition index (i.e. tree-crop-interaction [TCI] equation) has been proposed to quantify the balance between competitive and beneficial effects of trees on crops in agroforestry systems (Ong 1996). Although these competition indices often demonstrate occurrence of likely competition, they fail to explain the specific processes and mechanisms involved, thus complicating interpretation of competition or beneficial effects between plants and making extrapolation of results to other situations difficult. Available mechanistic models usually focus on competition for a single resource. Physiological models of competition for light or moisture in agricultural ecosystems (e.g. Kropff 1993; Ong *et al.* 1996) and competition for nutrients in natural plant communities (e.g. Berendse *et al.* 1989) are examples of models based on single resources. However, plant growth involves acquisition, partitioning and interactions among multiple resources, making it difficult to determine if competition occurred for one or more resources. Understanding how growth resources are partitioned between neighboring species is, therefore, important in providing a scientific base for designing more productive cropping systems.

Various experimental methods have been used to elucidate competitive interactions in cropping systems, most of which have often considered only two-species mixtures as summarized by Radosevich (1988). The outcome of interspecific plant competition is influenced by factors of plant proximity such as density, spatial arrangement of plants, and the proportion of each species in mixtures. These methods rely mainly on using various regression models based on the growth-density relationship by assessing intercrop productivity in relation to performance in monoculture. To achieve this, three main methods used have been proposed to elucidate interspecific plant competition in agro-ecosystems, additive, substitutive and neighborhood experimentation. In additive experiments, two or more species are grown together with the density of one species held constant while that of the other species is varied. Hence, the additive approach is relevant to studying weed competition in agricultural systems where weeds often invade an area occupied by a fixed density of the crop and typically follows the 'law of diminishing returns' (Roadosevich 1988). Crop productivity diminishes with increasing weed density until weed density does not reduce crop productivity significantly. The main disadvantage of this approach is that both density and proportion of the species under study keep varying making it difficult to assess the relative effects of intraspecific and interspecific competition on total productivity. Some of these limitations have been addressed using the replacement series (or substitutive) experimental designs, whereby proportions of the two species vary in mixture, but total density remain constant. This approach is important where there are management interventions that are likely to have significantly different outcomes depending on which of these principles is in play under a specific set of conditions.

The yield of each species in mixture is expressed relative to their respective yields in monoculture. The sum of the relative yields is referred to as the relative yield total (RYT) or land equivalent ratio (LER), and have been used widely to assess the competitive ability of different species in mixture, and to evaluate the advantages associated with intercropping (Spitters 1983). If $RYT > 1.0$, then there is a true advantage of mixed cropping and indicates that the mixture as a total captures more resources than the respective monocultures, whereas competitive effects are indicated by $RYT < 1.0$. The main problem with this approach is that model coefficients vary with total density (Taylor and Aarssen 1989).

5.2.2 Neighborhood competition indices

To address these issues, Goldberg and Werner (1983) introduced a 'neighborhood' experimental approach, in which performance of a target individual is assessed as a function of the number, biomass or distance of its neighbors. The target species is either grown alone or is surrounded by individuals of the neighboring species. The relationship between target species and its neighbors is then expressed in terms of production of individual plants. This enables the determination of whether competition for resources is occurring since biomass is assumed to be proportional to total resource use by plants (Goldberg 1990). The basic argument is that comparing the slopes of the relationship between competing species provides a useful approach to studying competitive effects of neighboring species (Malik and Timmer 1996; Imo and Timmer 1997). Thus, lack of relationship indicates no interaction (neutral), and that resource use efficiency is constant with increasing neighbor biomass, while positive relationships would show synergistic interactions giving rise to over-yielding at increasing density. On the other hand, negative relationships indicate competition for resources resulting in lower yields with increasing density.

5.2.3 Limitations of competition indices

However, it is difficult to determine from these regression models if competition occurred for one or more resources since competitive interactions involve partitioning of multiple resources (Nambiar and Sands 1993; Trenbath 1976). Berkowitz (1988) proposed a conceptual model of resource competition between a target plant and its neighbor that explicitly includes partitioning of the three main limiting resources (light, moisture and nutrients). In this model, competition may result from direct reduction in resource availability because of competitive uptake, or indirectly by reducing the uptake capacity of the other plant through resource-mediated alterations of the environment. These mechanisms are examined in the following section with respect to nutrient partitioning between competing plants, a major factor affecting species performance in plant communities (Berendse and Elberse 1990).

Despite general agreement that plant competition usually involves interactions between plants for different resources, most studies reporting competition fail to explain the processes and mechanisms involved (Nambiar and Sands 1993; Wilson and Tilman 1991), thus making it difficult to identify the resources which led to the interactions. According to Goldberg (1990), examining effects and responses of competing plants to resource availability can provide insight into the processes and mechanisms involved. This approach is adopted in this chapter, since both plant effects on resources and plant responses to resource availability can be quantified. As indicated earlier, the objective is to examine the concept of plant competition for nutrient resources, examine mechanisms partitioning of soil nutrients between neighboring plants, explain relationships between nutrient availability and plant growth, and to develop a theoretical basis for elucidating interspecific plant nutrient interactions. Although both light and moisture influence plant growth, the focus here is on interactions involving nutrients. However, attention is given to the role of moisture and light where these resources may directly influence nutrient availability, uptake and use by plants.

Unfortunately, these indices often focus on relating crop responses to some measure of weed proximity (Goldberg and Werner 1983; Radosevich 1988). Also, treatment impacts are often evaluated on the basis of survival, growth and yield of the target crops without considering non-crop responses, a major weakness for ecologically based assessments. Moreover, applications of these indices hardly explain the mechanisms involved. Little attention has also been given to determine whether competition is primarily for light, moisture and/or nutrients. The advantages and limitations of these indices are discussed in a review by Burton (1993) who advocated replacing static competition indices with a more site-specific phytometric approaches that feature greater accounting of neighboring non-crops and systematic local calibration and verification.

Distinguishing between these processes is important to test hypotheses related to the effect of changing nutrient supply on plant growth and nutrient composition. One way of solving this problem is by first studying the effects of nutrient supply on each of the individual plant response variables (i.e. biomass, nutrient concentration and content), and then examining their interrelationships using *Vector Nutrient Analysis* model (Timmer 1991; Imo and Timmer 1998). Traditionally applied to a single crop, this diagnostic format is unsuitable for interpreting competition effects in which growth resources are partitioned between interacting plants. Thus, the format has been modified by combining responses of competing plants to allow interpretation of interspecific interactions in a model called vector competition analysis (Imo and Timmer 1998). The focus, however, remains the same: identifying growth and nutritional interactions within the framework of vector diagnosis by characterizing the relationships between biomass production, nutrient accumulation and nutrient concentration of both species in mixture relative to their status in monoculture. These relationships are then synthesized into a diagnostic model (vector competition analysis) for elucidating interspecific plant growth and nutrient interactions.

In the following section, a graphical framework that helps discern interspecific competition effects involving nutrients and helps understanding probable mechanisms for nutrient competition in cropping systems is outlined. Functioning of the model is demonstrated using case studies from previous studies involving nutrient relationships in plants growing alone or in mixture, thus elucidating crop and weed nutrient interactions by examining the extent to which nutrients influenced the observed growth responses will be used to illustrate functioning of the model in screening impacts of different vegetation management regimes including use of herbicides, fertilization and nutrient loading as weed control measures in young forest plantations (Timmer 1997; Mead and Mansur, 1993), managing tree-crop interactions in alley cropping (Imo and Timmer 2000), and crop and tree interactions in *taungya** systems of plantation establishment (Imo 2010).

6. The vector competition analysis model

The objective of this section is to provide the theoretical background of the model and to demonstrate its function in elucidating interspecific growth and nutrient interactions in plants. Although scientists have often studied competition to understand succession patterns, as well as growth, diversity and dynamics of plant communities, such studies have focused mainly on minimizing the effects of competing weeds by developing predictive tools for yield-loss assessment, and also to minimize the use of herbicides (Altieri and Liebman 1988). Also, much research has been conducted to maximize the output of intercropping systems (Willey 1979; Vandermeer 1989; 1998). In agricultural ecosystems, concepts such as 'vegetation management' or 'weed management' are important considerations in yield improvement, and often include a broad spectrum of concerns such as biodiversity, the effects of different management practices on competitive interactions, and sustainable crop production. The general view is that any vegetation or weed management should aim to suppress the non-crop only to the extent that it significantly interferes with the target crop. More recently, similar concepts have been extended to agroforestry where biophysical benefits and consequences of including trees in farm land management (e.g. competition for resources and improvement of soil fertility) are major factors in designing mixed cropping systems (Ong and Huxley 1996). The objective here is to minimize competitive effects between plants, while taking advantage of the beneficial effects of trees on the crop. According to Nambiar and Sands (1993), the difficulty with these concepts is defining levels of 'significant' interference, as well as establishing criteria for segregating competition effects.

---

*An agroforestry system involving planting tree seedlings in combination with food crops by first growing crops with tree seedlings for 3 - 4 years, after which trees are left to grow alone. This planting sequence eliminates weed competition, while tree and crop competition is minimized.

6.1 Model theory

Vector competition analysis is modeled on regression analysis to evaluate competition effects (Goldberg and Werner 1983; Malik and Timmer 1995), and vector diagnosis is often used to assess plant nutrient status by identifying nutritional effects of nutrient dilution, deficiency, sufficiency and excess uptake (Haase and Rose 1995; Imo and Timmer 1997). The effect of non-crop weedy vegetation (V) on the target crop (T) is evaluated using the linear relationship $\{(T_W = Y_W + X(V_W)$ or $T_U = Y_U + X(V_U)\}$, where Y represents crop response in the absence of competition; X is the slope of the regression; and W and U signify biomass and nutrient uptake, respectively. According to Goldberg and Werner (1983) and Goldberg (1990), the slope of the regression has a physical meaning: it provides a measure of competition intensity. The slope indicates no interactions when it is not significantly different from zero (0); indicates beneficial or facilitative effects of the vegetation if it is positive (+); and indicates competitive effects of the vegetation if it is negative (-).

Malik and Timmer (1995; 1996) have shown the effectiveness of this approach in describing relationships between growth of interacting black spruce (*Picea mariana*) seedlings and neighboring vegetation over a two-year period, which clearly demonstrated a significant negative relationship between seedling biomass and neighboring vegetation with seedling performance suffering at the expense of weed growth (Malik and Timmer 1996). Thus, reduction in weed biomass due to herbicide application resulted in reduced weed competition and a corresponding increase in biomass or nutrient content of the target crop. Since such a response often has both direction and magnitude, it can be characterized by a vector showing the combined crop and non-crop response (i.e. vector response).

Also, the slopes of such regressions serve as indicators of competition intensity (Goldberg 1990; Goldberg and Werner 1983), and may vary depending on site quality (Weldon *et al.* 1988). Management practices can also influence competition intensity as was demonstrated with herbicide application and nutrient loading (Malik and Timmer 1995; 1996). On the basis of these results, various possible impacts of weed management practices on the competitive interactions of crops and weeds, for example, herbicide application, may reduce weed competition in favor of the crop because of weed elimination. Fertilization may favor both the crop and weeds because of stimulated growth or may favor weeds more than the crop due to rapid weed growth resulting in shading of the crops; shading under the shelterwood system may reduce growth of both species. These varying treatment responses can be classified into four quadrants by two-dimensional graphical representation of the performance (biomass, nutrient content, yield or density) of one species (i.e. crop) plotted against that of the other species (e.g. weedy vegetation).

The principle is illustrated in Figure 3, which displays some possible combinations of the crop and non-crop (i.e. weedy vegetation): crop response plotted on the y-axis, and non-crop responses on the x-axis. First, performance of the crop in the absence of non-crop competition (when there is zero biomass of the non-crop) is plotted on the y-axis, while performance of the non-crop in the absence of the crop (zero biomass of the crop) is plotted on the x-axis (Figure 3). The process of interspecific interaction can then be visualized as a change in these values along each axis when in mixtures. Competition will reduce performance of either species, while facilitation will increase species performance. Hence, performance of each species in mixture is represented by a point on the graph as summarized in Figure 3 depicting vectors of changing biomass production and nutrient uptake of interacting plants relative to competition-free status as described in the following section.

6.2 Constructing vector competition diagrams

In this model, biomass or nutrient content of the target crop is plotted on the y-axis against those of the neighboring non-crop vegetation on the x-axis (Figure 3). Although absolute data can be used in plotting vector competition diagrams, use of relative (normalized) values allows multiple comparisons among sites, treatments and nutrient elements by eliminating inherent differences in plant size and nutrient content (Timmer 1991). Normalization is achieved by dividing response parameters (biomass or nutrient content) for each treatment by corresponding values of the control or reference treatment, and expressed as percentage by multiplying by 100% to obtain relative biomass (W) and relative nutrient content (U). The choice of the reference treatment depends on the specific objective of the analysis being conducted. For studies in which interspecific interactions is the focus, species performance without competition is used as the reference for comparison with its performance in mixture. Thus, effects of non-crops on the target crop are evaluated when performance of the sole crop is used as the reference (Figure 3). Similarly, effects of the crop on non-crops are assessed when performance of the non-crop in the absence of the target crop is the reference.

After normalizing the data, plotting starts with the reference treatments (i.e. sole crop [y] = 100, and sole non-crop [x] = 100) as shown in Figure 3. This reference point represents the combined crop and non-crop response without competition. When plotting, it is important to have the same scale on both axes to form a square so as to avoid visual exaggerations. The second step is to plot crop and non-crop responses in mixture (W or U) on the same diagram, each point representing the combined crop and non-crop mixture response to different treatments within one of the quadrants A, B, C or D (as shown in Figure3). The third step is to draw vectors from the reference point to the plotted data points to show the combined response in biomass or nutrient content of the crop as well as non-crop in mixture or due to specific treatments.

The final step is to draw a vertical and horizontal dashed line across the reference point to divide the vector competition diagram into four distinct quadrants (A, B, C and D) that help define interaction types (Box I in Fig. 3). Vector shifts below the horizontal dashed line indicate competitive effects of the non-crop since responses are negative, while shifts above indicate beneficial or facilitative effects of the non-crop because responses are positive. Similarly, vector shifts to the left of the vertical dashed line indicate competitive effects of the crop on the non-crop, while shifts to the right signify facilitative effects of the crop. Interpretations of these vector relationships are discussed below.

6.3 Vector interpretations

Diagnostic interpretation of the impacts of alternative management strategies on crops as well as non-crops is based on vector direction and magnitude observed as no change (0), increase (+), or decrease (-) relative to the reference status (Box I, Figure 3). Vector shifts in quadrant A indicate treatments that inhibit both species (antagonistic [-,-]); shifts in B show treatments that favor non-crop vegetation but not the target crop (compensatory [-,+]); shifts in C exemplify treatments that favor both species (synergistic [+,+]); and shifts in D illustrate treatments that favor the crop but not non-crop (compensatory [+,-]).

Slopes of the vectors define symmetry of the interactions. If crops and non-crops influence each other such that both species change by the same magnitude and in the same direction (as shown by vectors A, B, C and D in Figure 3), the slope will be one indicating symmetric interaction. Deviations from a slope of unity represent asymmetric interactions, and can be interpreted on the basis of vector orientation and magnitude within each of the four quadrants projected horizontally and vertically from the reference point (Figure 3). Vector deviations closer to the horizontal dashed line imply the non-crop is more sensitive to the treatments than the target crop, while those closer to the vertical dashed line indicate the crop is more sensitive to treatments. Similarly, a slope of zero indicates treatments that affect the non-crop without influencing the crop, while a slope of infinity exemplifies treatments that affect the crop without affecting non-crop.

6.4 Diagnosis of plant growth and nutritional interactions

Since nutrient concentration (C) is a function of uptake (U) and biomass (W), thus $C = U/W$, interpretation of growth and nutritional interactions for each species can be determined based on the ratio between nutrient content vector (U) and biomass vector (W), thus (U/W). This approach is adapted for several reasons: first, W and C can easily be determined using standard laboratory procedures, and U is calculated for each sample by taking the product of concentration and dry mass, thus $U = C*W$. Secondly, nutrient content (U) gives a direct measure of the amount of nutrients actually absorbed by the plant (Berkowitz 1988). Finally, the relationship $C = U/W$ enables one to determine whether changes in C are associated primarily with changes in U, W or both.

Diagnostic interpretations of these interactions are summarized in Box II, Figure 3. Nutrient dilution (or decline in concentration) occurs when $W > U$ (i.e. vector ratio $U/W < 1$). Such dilution effect is antagonistic dilution if it is associated with reduced growth and nutrient uptake, or is growth dilution if it is associated with increased growth and nutrient uptake. Nutrient sufficiency occurs when $W = U$ (i.e. vector ratio $U/W = 1$, Box II in Figure 3) indicating that rate of nutrient uptake matched growth increase or steady state nutrition (Ingestad and Lund, 1986; Imo and Timmer 1997). Nutrient accumulation occurs when $W < U$ (i.e. vector ratio $U/W > 1$, Box II in Figure 3) indicating that rate of nutrient uptake was higher than growth rate. Such accumulation is a deficiency response if it is associated with increase in both W and U, but excess uptake if associated with decline in biomass and or nutrient uptake (Timmer 1991).

These interactions can also be determined graphically by plotting both biomass and nutrient uptake responses on the same vector competition diagram (illustrated later) as follows. For the crop component (y), drawing a horizontal line (parallel to the x-axis) across the point indicating W enables determination of the change in relative nutrient concentration depending on the relative position of the point indicating U. Nutrient dilution in the crop occurs when U is below the horizontal line (i.e. $U < W$); nutrient sufficiency occurs when U and W lie on the same horizontal line (i.e. $U = W$); while nutrient accumulation occurs when U is above the horizontal line (i.e. $U > W$). Similarly, changes in relative nutrient concentration for the non-crop (x) can be determined by drawing a vertical line (parallel to the y-axis) across the point indicating W. Nutrient dilution in the non-crop occurs when U is to the left (i.e. $U < W$), while accumulation occurs when U is to the right (i.e. $W < U$) of the vertical line. Sufficiency occurs when U and W lie on the same vertical line (i.e. $W = U$).

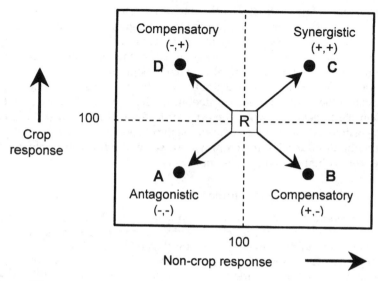

Box I: Competition responses

Vector	Relative response			
shift	Crop	Non-crop	Interaction type	Likely management options
A	-	-	Antagonistic	Weed control required; inter-cropping not feasible
B	-	+	Compensatory	Weed control required; not favoring inter-cropping
C	+	+	Synergistic	Weed control not required; inter-cropping feasible
D	+	-	Compensatory	Weed control not required; inter-cropping possible

Box II: Growth and nutritional interactions

Vector ratio (U / W)	Relative change in			
	Biomass (W)	Nutrient content (U)	Possible diagnosis	Possible interaction effects
< 1	-	-	Antagonistic dilution	Nutrient competition
< 1	+	+	Growth dilution	Improved moisture and light availability
= 1	+	+	Sufficiency	Improved nutrient, moisture and light availability
> 1	-	+ -	Excess uptake	Light / moisture competition
> 1	+	+	Deficiency	Improved nutrient availability

Fig. 3. Graphical vector competition analysis model showing total nutrient use by neighboring tree crop and non-crop weedy species. Competition-free crop or non-crop weedy vegetation status is normalized to 100% as a reference (R) for comparison with corresponding plants growing in mixture, respectively. The vertical and horizontal dashed lines divide the model into four quadrants (A, B, C and D) that characterize the type of interaction (Box I), while the associated growth and nutritional interactions are identified in terms of vector ratio in Box II. Adapted from Imo (1999); see also Imo and Timmer (2000; 2002).

For a limiting nutrient, this interaction type is associated with a deficiency response (i.e. vector ratio $U/W > 1$, Box II in Figure 3) since uptake is accelerated faster than growth in a manner similar to fertilization response. Synergistic nutrient interactions may also result from improved moisture availability, for example, mulching, reducing surface runoff and evaporation. This interaction may increase growth faster than nutrient uptake, thus resulting in growth dilution of nutrients (i.e. vector ratio $U/W < 1$, Box II Figure 3) or sufficiency if both growth and nutrient uptake are increased at the same rate (i.e. vector ratio $U/W = 1$).

Competition for nutrients reduces growth and nutrient uptake of the species in mixture, and often occurs when nutrient availability is not sufficient to support the demand by either species, resulting in antagonistic competition (Shift A in Figure 3). This type of interaction results in antagonistic dilution of nutrients (i.e. $U/W < 1$, Box II in Figure 3) indicating that competition reduced nutrient absorption more than photosynthesis. Interspecific nutrient interactions may also result in compensatory competition in favor of one species while the other is suppressed (Figure 3). For example, increasing nutrient availability through fertilization may favor growth of one species and cause preemption of other resources. Since nutrients are not limiting, this type of interaction results in excess nutrient uptake ($U/W > 1$, Box II in Figure 3) indicating that photosynthesis was reduced more than nutrient absorption presumably because of light and or moisture preemption. The function of this model is demonstrated with response data from the following study.

It is important to note that this model has been developed primarily for screening alternative strategies of integrated vegetation management in forest plantations, cropping and agroforestry systems by evaluating crop and weed interactions in a bivariate graphical model depicting vectors of changing biomass production and nutrient uptake relative to competition-free status. Conceptually, this approach has the potential to contribute to efficient nutrient management in intensively managed cropping systems by providing a systematic framework for rationalizing management prescriptions as has been demonstrated for agroforestry systems in Western Kenya (Imo and Timmer (2001) and young forest plantations (Imo and Timmer 2000) where management of competing non-crop species is an important objective and the other where complementary use of growth resources by species in mixture is an important consideration in management decisions.

7. Practical applications

Since nutrient content is often used to give an integrated measure of total nutrient uptake and use by plants, determination of nutrient content of neighboring plants can provide insight into the processes of partitioning of soil nutrient resources between them (Berkowitz 1988). Although Imo and Timmer (1997) have previously diagnosed these nutritional effects using vector competition analysis without linkage to availability of other resources required for growth, it is well-known that plant growth depends on acquisition, retention and use of multiple resources (carbon, water, nutrients and light) as illustrated in Figure 2 (Trenbath 1976). Carbon and nutrients are converted into biomass, while light and water are necessary for growth and other physiological processes (Salisbury and Ross 1992), often involving complex interactions among various resources (Neary et al. 1990; Sands and Mulligan 1990; Woods et al. 1992). Plant growth characteristics may also influence resource interactions, for example, due to trade-offs in carbon allocation between resource acquiring organs or greater growth rate and overall plant size. Trade-offs between uptake organs predict a negative relation between competitive abilities for different resources ,while accelerated growth and resource use predict a positive relation between competitive abilities for different resources (Tilman 1988). Unfortunately, most studies on plant competition often focus on effects of single resources without considering the processes involved (Nambiar and Sands 1993). This makes it difficult to determine whether interactions involved more than one resource. Figure 3 illustrates the impacts of management practices on resource partitioning between target crops and neighboring non-crop weedy vegetation as discussed below.

7.1 Compensatory competition

Compensatory competition occurs when growth of one species increases, while that of the other decreases (Shift B and C, Figure 3), and indicates re-allocation of available resources from one species for use by the other. This type of interaction occurred between natural vegetation and seedlings after fertilizer addition on the weed prone sites (Imo and Timmer 1999). Although fertilizer addition increased N availability to both species, growth and N uptake of neighboring non-crop vegetation were increased while those of seedlings declined. Nitrogen uptake and growth of the faster growing weeds were higher with carbon assimilates allocated primarily to aboveground, typical of plant responses to increased nutrient availability (Chapin 1987). With this allocation pattern, neighboring vegetation intercepted more light because of larger leaf area. Once these processes were established, a positive feedback mechanism between growth and resource capture (Grime 1979) presumably preempted available light and moisture from seedlings that in effect became suppressed. These responses confirm the hypothesis that competitive advantage of a species for one resource leads to an advantage for acquisition and use of the other resources as well (; Kropff et al. 1984).

Comparison of carbon (W) and nitrogen (U) accumulation in seedlings shows that both W and U were reduced, but proportional reduction in W was larger than the corresponding reduction in U, hence U/W increased signifying excess uptake of N (Box II in Figure 3). Thus, N was not limiting, and the suppressed seedling growth was probably because of preemption of available light and or moisture by the weeds. The proportionally lower W than U in weeds after fertilization indicates that U/W also increased signifying a deficiency response (Box II in Figure 3).

7.2 Antagonistic competition

Antagonistic competition occurs when performance of both species in mixture is reduced (Shift A, Figure 3), probably because consumption by each species reduces resource availability to each other. This interaction type has been demonstrated in conventionally fertilized seedlings planted on high competition sites without any weed control treatment (Imo and Timmer 1998). In this study, growth and N uptake of both the seedlings and neighboring vegetation were reduced exemplifying mutual antagonism due to low N availability that was not sufficient to support growth demand of both the seedlings and competing vegetation, thus reducing uptake of both species. Reduction in weed biomass was, however, small compared to decline in seedling biomass (Imo and Timmer 1999) exemplifying asymmetric competition with the trees being more sensitive to competitive effects than the weeds. Further analysis of the results from this study indicated antagonistic dilution effects (Imo and Timmer 1997), suggesting that competitive interactions reduced nutrient uptake more than photosynthesis. It was therefore concluded that competition for N between the seedlings and the weedy vegetation was more important than for light and moisture. In contrast, nutrient loading, however, improved competitive ability of the seedlings, presumably because of a build-up of pre-plant N during the nursery phase.

7.3 Synergistic interactions

Synergistic interactions occur when performance of one or both species mixture is more than their performance when grown alone (Shift C in Figure 3), and indicates beneficial effects of one species on the other. This interaction type was demonstrated in an alley agroforestry system involving *Leucaena* trees inter-cropped with maize (*Zea mays* L.) in Western Kenya (Imo and Timmer 2000). In this study, maize crop productivity was significantly higher than the sole crop, which was attributed to higher N availability mineralized from added mulch. Apparently, the beneficial effects of the mulch for the crop were higher than the negative effects of competition by the *Leucaena* hedgerows for the other resources (i.e. available light and moisture). The frequent and rapid pruning regime applied ensured return of mulch to the soil hence maintaining high N availability to the crop while ensuring minimal effects of light competition by the hedgerow trees. It was also noted that, although both biomass and nutrient content of the crop increased, the increase in the latter was larger than a corresponding increase in the former, which resulted in a typical of deficiency response (Imo and Timmer 1997) due to improved fertility from mulching.

7.4 Maximizing crop nutrient use

In view of the factors influencing nutrient partitioning between competing plants discussed in the previous sections, crop productivity can be maintained or improved by tilting the balance in resource capture in favor of the target crops. Previous developments on integrated weed management in forestry and agriculture), and tree and crop inter-cropping in agroforestry (Ong and Huxley 1996) have repeatedly emphasized the need to incorporate a combination of management approaches for different objectives. Given that such integrated methods for crop management are linked to biological, environmental and economic considerations (Nambiar and Sands 1993), the problem often encountered by many decision-makers is related to determining the level of significant interference of beneficial value of non-crop management interventions. The vector competition analysis approach presented in this thesis is a simple decision-support tool to reconcile these variable objectives, especially in relation to nutrient management. In this model, both crop and non-crop interaction types and the associated nutritional mechanism s are evaluated in a systematic manner.

Figure 3 above has shown how this simple approach can be used to determine the partitioning of soil nutrients between neighboring plant species, and how efficiently the available nutrient resources are utilized on a site-specific and management specific-regime basis. Notice that the model emphasizes total nutrient use by the neighboring plant species according to the competitive production principle or facilitation discussed earlier. The lower-left portion (*antagonistic* interactions) in Figure 3 indicates competitive interactions between the species and demonstrates combinations in which total nutrient uptake and use by the two species in mixture is less than their total nutrient uptake and use when grown separately. This portion, therefore, represents inefficient exploitation of the site by the two species if planted as monocultures. The upper-right portion (*synergistic* interactions), on the other hand, shows complementary use of nutrients (or facilitation) between the neighboring species, and indicates that total nutrient uptake and use in by the plants in mixture is greater than their total uptake when grown separately, thus represents higher efficiency in resource capture and utilization. In practical terms, management practices that aim to suppress weed competition (as demonstrated in by Imo and Timmer 2001) operate within the lower-left portion, while inter-cropping practices such those in agroforestry (as demonstrated in Imo and Timmer 2000) aim at achieving species mixture within the upper-right portion of Figure 3. The main advantage of this approach is that it provides an instant evaluation of the advantage of intercropping or of specific management practices, and the possible processes involved.

In these studies, herbicide application on young black spruce plantations on high competition forest sites (Imo and Timmer 2001) eliminated weed competition, thus maximizing the amount of available resources to the target tree seedlings whose growth was increased. In theory, supplementing the resource pool of the crop by, for example, fertilization and irrigation should reduce competitive effects of the weeds. Results from this study, however, illustrated one major difficulty in this approach as was demonstrated by Imo and Timmer (2001) after fertilizing weed prone sites. Applied fertilizer was preferentially taken up by the weeds resulting in rapid weed growth and light preemption, consistent with the well established notion that weed resource use often increases more rapidly with added nutrients than that of the target crops.

In agroforestry, tree and crop inter-crops can be managed for spatial or temporal complementary use of nutrients to reduce competitive interactions while enhancing total nutrient use in the whole system (unshaded portion of Figure 3). Here, optimizing tree density and spacing is a key factor in complementary use of nutrients in these systems, and may be explained in terms of either the competitive production principle or facilitation (Vandermeer 1989). Several mechanisms may be associated with complementary use of nutrients in these systems such as nitrogen fixation since *Leucaena* is a N-fixing species (Kang *et al.* 1985), or the ability to access different nutrient pools or use of nutrients by the other species that would otherwise be lost to deep percolation.

These processes were confirmed from results with herbicide application on the high competition sites (Imo and Timmer 1999). Herbicides eliminated weed competition resulting in increased total resource availability to the seedlings. Light availability increased because of removal of aboveground weed biomass, while moisture and nitrogen availability increased presumably due to elimination of uptake by the weeds. Further, the dead weed material was returned to the soil as residue that mineralized to increase available N as was confirmed in the field. Vegetation removal presumably increased soil temperature as well and was favorable for rapid mineralization.

In the absence of weed competition, all available resources were utilized in seedling growth, thus the trees were able to achieve maximum carbon assimilation and nutrient uptake potential. Biomass production under this competition-free status was, therefore, maximized as supported by the significantly higher growth and nitrogen content of seedlings at the end of the growing season after herbicide application (Imo and Timmer 1999). Since both carbon

(*W*) and nitrogen (*U*) content of the seedlings were maximized, seedling nutrient status can be considered sufficient for growth requirements under Compensatory competition

8. Conclusions

Several conclusions can be drawn regarding the effectiveness of the vector competition model to elucidate complex interspecific plant growth and nutrient interactions in cropping systems in a simplified graphical format. First, integrating processes of resource acquisition and use within a conceptual graphical framework provides an approach for obtaining insight into the mechanisms involved in nutrient partitioning between competing plants. Thus, appropriate management interventions can be designed to alter allocation of soil nutrients to favor targeted plant crops or to maximize total nutrient use by the competing species. This conceptual approach clearly illustrates the complex nature of resource acquisition, uptake and use by plants, yet the overall effects on growth and nutrient uptake can be elucidated in a very simplified manner using vector competition analysis. Also, these interactions vary over time, and are regulated by feedback processes within the plant itself and affected by many environmental conditions. The conceptual graphical framework may also provide a simplified framework for simulating individual processes of resource availability, uptake and use by the competing species, thus enabling the understanding of how these processes change under specific environmental conditions and management regimes.

Specifically, the model enabled identification of the nature of interspecific growth and nutritional interactions in plants competing for the same resources in terms of antagonistic, compensatory or synergistic interactions, as well as discern phenomena of symmetrical interactions by isolating the most responsive species and sites under different management regimes. By comparing biomass production, nutrient accumulation, and nutrient concentration of the competing species, vector competition analysis facilitated characterization of interspecific interactions involving nutrient competition, synergistic nutrient interactions, or non-nutrient competition responses. These interpretations were based on nutritional effects, namely: nutrient dilution, sufficiency or accumulation as summarized in Box II of Figure 3. Thus, the model is an improvement over traditional competition indices based only on morphological parameters. Since biomass and nutritional responses were normalized to a standard reference treatment (100%), it was possible to compare treatments, sites and nutrient elements simultaneously. This standardization permitted ranking of weed problem sites, the model enabled identification of the nature of interspecific growth and nutritional interactions in plants competing for the same growth resources in terms of antagonistic, compensatory or synergistic interactions as shown in Box I of Figure 3.

The model further helped identify phenomena of symmetrical interactions by isolating the most effective vegetation management practices over a wide range of ecological conditions. Thus, the model can provide farm managers with a decision-support mechanism for identifying and ranking weed problem sites, and permits recommendations regarding silvicultural treatments for specific sites. Appropriate management practices that favor resource allocation to target crops or maximize total resource use can be designed to improve productivity of the whole cropping system. The vector competition analysis approach can then be used as a decision-support tool to evaluate and rank such practices in a systematic manner.

9. Future research directions

As discussed earlier, weed or non-crop vegetation management in forestry and agriculture involves a wide range of complex issues: social, economic, ecological and their interactions. Thus, one of the most important concerns is associated with the use of different analytical criteria for the assessment of alternative crop and non-crop management alternatives. For example, assessments based entirely on yield (a biological criterion) can lead to completely different conclusions regarding desirability of alternative cropping systems than from an assessment based purely on cash value (an economic criterion). Thus, it is possible to recommend an ecologically viable cropping system that lacks economic feasibility, or vice versa. How can the different management criteria be integrated into a single vector competition model? This should allow prescription of appropriate management practices that are acceptable using both ecological and economic criteria.

Secondly, while the principles articulated in this chapter can be of general application over a wide range of interspecific plant growth and nutrient interactions, interpretations thus far are limited to only a two-species mixture, one growing season, and harvesting the total aboveground plant parts for analysis. Thus the question is whether the vector competition analysis approach can be extended to evaluate mutual or competitive effects of more than two species in mixture. An answer to this question is important especially in field studies under natural conditions where individuals of more than two species often affect each other. The other area is whether the vector competition model is able to evaluate interspecific competition responses over several seasons, thus enabling understand long-term dynamics in competition responses. The approach taken by Imo and Timmer (1997) with traditional vector nutrient diagnosis using *Prosopis chilensis* seedlings can provide a basis for modeling time-dependent competition responses in vector competition analysis.

The other issue is whether plant parts (rather than the total aboveground plant) can give an indication of competitive responses. This is an important question especially in forestry and agroforestry where monitoring competition effects will normally not involve harvesting the whole crop. Traditional methods usually applied to assess tree growth such as foliage biomass and chemistry, stem diameter or leaf area index can be useful indices if incorporated into vector competition analysis to assess competition responses. Some of these methods have been applied successfully with the traditional vector nutrient diagnosis (e.g. Imo and Timmer 2000).

Finally, the process model (Figure 2) integrating processes of acquisition, uptake and use of different resources raises a number of fundamental questions in terms of mechanisms involved in partitioning of nutrients between competing plants that need further investigation. For example: the critical 'competition-free' period during which all nutrient resources should be available for the target crops and the timing of management interventions should be investigated further to help design management interventions to maximize productivity of the target crop. Relating growth and nutritional interactions to environmental or plant variables that can be manipulated to provide further insight of limiting resources. For example, the relative effects of moisture availability on photosynthesis and nutrient uptake need to be quantified. There is a need to integrate vector competition analysis into detailed ecophysiological simulation models for the individual processes of resource availability, uptake and utilization by the competing species thus permitting understanding of how these processes change under specific environmental conditions and management regimes. Such understanding can help to predict the potential role of various environmental, plant attributes and management practices in determining the outcome of competitive interactions.

10. Acknowledgment

Funding support from the Moi University Annual Research Grant (ARG) is greatly acknowledged. I am also grateful to many colleagues who have helped review and proof read this manuscript.

11. References

Altieri, M . A. and Liebm an, M . (eds). 1988. *Weed Management in Agroecosystem s: Ecological Approaches*. CRC Press Inc., Boca Raton, Florida.

Axelsson, E. and Axelsson, B. 1986. Changes in carbon allocation patterns in spruce and pine trees following irrigation and fertilization. *Tree Physiology*, Vol. 2, pp 189-204.

Barber, S. A. 1984. *Soil Nutrient Bioavailability: A M echanistic Approach*. John Wiley and Sons, New York.

Begon, M ., Harper, J. L. and Townsend, C. R. 1996. *Ecology*, 3rd ed. Blackwell Sciences Ltd, London.

Berendse, F. and Elberse, W. Th. 1990. Competition and nutrient availability in heathland and grassland ecosystems. *In* Grace, J. B. and Tilman, D. (eds.). *Perspectives on Plant Competition*, Academic Press, New York. pp. 93-116. Berendse, F., Bobbink, R. and Rouwenhorst, G. 1989. A com parative study on nutrient cycling in wet heathlands ecosystem s. II. Litter decom position and nutrient m ineralization. *Oecologia*, Vol. 78, pp 338-348.

Berkowitz, A. R. 1988. Competition for resources in weed-crop mixtures. *In* Altieri, M . A. and Liebman, M . (eds.). *Weed Management in Agroecosystems: Ecological Approaches*. CRC Press Inc., Boca Raton, Florida. pp. 89-119.

Boot, R. G. B. and Mensik, M. 1990. Size and morphology of root system s of perennial grasses from contrasting habitats as affected by nitrogen supply. *Plant and Soil*, Vol. 129, pp 291-299.

Burdett, A. N. 1990. Physiological processes in plantation establishment and the development of specifications for forest planting stock. *Canadian Journal of Forest Research*. Vol. 20, pp 415-427.

Burton, P. J. 1993. Some limitations inherent to static indices of plant competition. *Canadian Journal of Forest Research*, Vol. 23, pp 2141-2552.

Chapin, F. S., III. 1990. The ecology and economics of storage in plants. *Annual Review of Ecological Systems*, Vol. 21, pp 423-447. Chapin, F. S., III. 1987. Adaptations and physiological responses of wild plants to nutrient stress. *In* Gabelm an, H.W. and Loughman, B.C. (eds.). *Genetic Aspects of Plant Mineral Nutrition*. Martinus Nijhoff, Boston. pp. 15-26 .

Chapin, F. S., III. 1980. The m ineral nutrition of wild plants. *Annual Review of Ecological Systems*, Vol. 11, pp 233-260.

Coates, K. D., Emmingham, W. H., Radosevich, S. R. 1991. Conifer seedling success and microclimate at different levels of herb and shrub cover in a Rhododendron-

Vaccinium-Menziensia com m unity of south central British Columbia. *Canadian Journal of Forest Research*, Vol. 21, pp 858-866.

Eastham, J. and Rose, C. W. 1990. Tree/pasture interactions in a range of tree densities in an agroforestry experiment. I: rooting patterns. *Australian Journal of Agricultural Research*, Vol. 41, pp 683-695.

Fowler N.L. 1986. The role of competition in plant communities in arid and semiarid regions. *Annual Review of Ecological Systems*, Vol. 17, pp 89–110

Gillespie, A. R. 1989. Modeling nutrient flux and interspecies root competition in agroforestry interplantings. *Agroforestry Systems*, Vol. 8, pp 257-265.

Goldberg, D. E. 1990. Components of resource competition in plant com m unities. *In* Grace, J. B. and Tilm an, D. (eds). *Perspectives on Plant Competition*, Academic Press, Inc. New York. 27-49 pp.

Goldberg, D. E. and P.A. Werner. 1983. Equivalence of com petitors in plant com m unities: A null hypothesis and a field experim ental approach. *American Journal of Botany*. Vol. 70, pp 1098-1104.

Grace, J. B. and Tilman, D. (eds). 1990. *Perspectives on Plant Competition*. Academic Press, Inc., New York.

Haase, D. L. and Rose, P. 1995. Vector analysis and its use for interpreting plant nutrient shifts in response to silvicultural treatments. *Forest Science, Vol.* 41, pp 54-66. Imo, M. 2010. Interactions amongst trees and crops in *taungya* systems of western Kenya. *Agroforestry Systems*, Vol. 76. pp 263 – 273. Online: ODI 10.1007/s104457–9164-z.

Imo, M. and Timmer, V.R. (2002) Growth and nutritional interactions of nutrient-loaded black spruce seedlings with neighboring natural vegetation under greenhouse conditions. *Forest Science*, Vol. 48, No. 1, pp 77-84.

Imo, M. and Timmer, V.R. (2000) Vector competition analysis of *Leucaena*-maize alley cropping system in western Kenya. *Forest Ecology and Management*, Vol. 126, pp 255 - 268.

Imo, M. (1999). Vector competition analysis: a model for evaluating nutritional interactions in cropping systems. PhD thesis, University of Toronto, Canada. 164p.

Imo, M. and Timmer, V.R. (1999) Vector competition analysis of black spruce seedling responses to nutrient loading and vegetation control. *Canadian Journal of Forest Research*, Vol. 29, pp 474-486.

Imo, M. and Timmer, V. R. (1998). Vector competition analysis: a new approach for evaluating vegetation control methods in young black spruce plantation. *Canadian Journal of Soil Science, Vol.* 78, pp 3-15.

Imo, M . and Timmer, V. R. (1997). Vector diagnosis of nutrient dynamics in mesquite seedlings. *Forest Science*, Vol. 43. pp 268-273. Ingestad, T. and Lund, A. B. (1986). Theory and techniques for steady-state m ineral nutrition and growth of plants. *Scandinavian Journal of. Forest. Research.* Vol. 1, pp 439-453

Kang, B. T., Grim m e, H. and Lawson, T. 1985. Alley cropping sequentially cropped m aize and cowpeas with *Leucaena* on a sandy soil in Southern Nigeria. *Plant and Soil.* Vol. 85, pp 267-277.

Kropff, M . J. (1993). Eco-physiological models for crop-weed competition. *In* Kropff, M . J. and van Laar, H. H. (ed.). *Modeling Crop-Weed Interactions*, CAB International, Wallingford, UK. pp. 25-31.

Kropff, M . J., Vossen, F. J. H., Spitters C. J. T. and de Groot, W. (1984). Competition between a maize crop and a natural population of *Echinochloa crus-galli* (L.). *Netherlands Journal of Agricultural Sciences*, Vol, 32, pp 324-327.

Malik, V. and Timmer, V. R. (1998). Biomass partitioning and nitrogen retranslocation in black spruce seedlings on competitive mixedwood sites: a bioassay study. *Canadian Journal of Forest Research*, Vol. 28, pp 206-215.

Malik, V. and Timmer, V. R. (1996). Growth, nutrient dynamics, and interspecific competition of nutrient-loaded black spruce seedlings on a boreal mixedwood site. *Canadian Journal of Forest Research*, Vol. 26, pp 1651-1659.

Malik, V. and Timmer, V. R. (1995). Interaction of nutrient-loaded black spruce seedlings with neighboring vegetation in greenhouse environments. *Canadian Journal of Forest Research*, Vol. 25, pp 1017-1023.

Marschner H. (1995). *Mineral Nutrition of Higher Plants*. London: Academic. 2nd ed.

Mead, D. J. and Mansur, I. (1993). Vector analysis of foliage data to study competition for nutrient and moisture: an agroforestry example. *New Zealand Journal of Forest Science*, Vol. 23, No. 1, pp 27-39.

Morris, D. M . and McDonald, G. B. (1991). Development of a competition index for young conifer plantations established on boreal mixedwood sites. *Forestry Chronicle*, Vol. 63, pp 1696-1703.

Nambiar, E. K. S. (1990). Interplay between nutrients, water, root growth and productivity in young plantations. *Forest Ecology and Management*, Vol. 30, pp 213-232.

Nambiar, E. K. S. and Fife, D. N. (1991). Nutrient retranslocation in temperate conifers. *Tree Physiology*, Vol. 9, pp 185-207.

Nambiar, E. K. S. and Sands, R. (1993). Competition for water and nutrients in forests. *Canadian Journal of Forest Research*, Vol. 23, pp 1955-1968.

Neary, D. G., Rockwood, D. L., Comerford, N. B., Swindel, B. F. and Cooksey, T. E. (1990). Importance of weed control, fertilization, irrigation and genetics in slash and loblolly pine early growth on poorly-drained Spodosols. *Forest Ecology and Management*, Vol. 30, pp 271-281.

Nye, P. H. and Tinker, P. B. 1977. Solute movement in the soil-root system. Blackwell, Oxford.

Ong, C. K. 1996. A framework for quantifying the various effects of tree-crop interactions. *In* Ong, C. K. and Huxley, P. (eds.). *Tree-Crop Interactions: A Physiological Approach*, CAB International, UK. pp. 1-23.

Ong, C. K. and Huxley, P. (eds.). (1996). *Tree-Crop Interactions: A Physiological Approach*. CAB International, Wellington, UK.

Ong, C. K., Black C. R., Marshall F. M . and Corlett J.E. (1996). Principles of resource capture and utilization of light and water. *In* Ong, C. K. and Huxley, P. (eds.). *Tree-Crop Interactions: A Physiological Approach*. CAB International, Wallingford, UK. pp. 73-158. Radosevich, S. R. 1988. Methods to study crop and weed interactions. *In* Altieri, M. A. and Liebm an, M. (eds.), *Weed Management in Agro-Ecosystems: Ecological Approaches*. CRC. Press Inc., Boca Raton, Florida. pp. 121-143.

Salisbury, F. B., and Ross, C. W. 1992. *Plant Physiology*. Wadsworth Publishing Company, Belmont, Carlifornia, USA.

Sands, R. and Mulligan, D. R. (1990). Water and nutrient dynamics and tree growth. *Forest Ecololgy and Management*, Vol. 30, pp 91-111

Smesthurst, P. J. and Comerford, N. B. (1993). Simulating nutrient uptake by using single or competing root systems. *Soil Science Society of America Journal*, Vol. 57, pp 1361-1367.

Smesthurst, P. J. and Nambiar, E. K. S. (1989). Role of weeds in the management of nitrogen in a young *Pinus radiata* plantation. *New Forest*, Vol. 3, pp 203-224. Spitters, C. J. T. 1983. An alternative approach to the analysis of mixed cropping experiments. 1.

Estimation of competition effects. *Netherlands Journal of Agricultural.Sciences*. Vol. 31, pp 1 - 11.

Stone, E. L. and Kalisz, P. J. (1991). On the maximum extent of tree roots. *Forest Ecology and Management*, Vol. 46, pp 59 - 102.

Taylor, D. R. and Aarssen, L. W. (1989). On the density dependence of replacement-series competition experiments. *Journal of Ecology*, Vol. 77, pp 975-988.

Tilman, D. (1988). *Plant Strategies and the Dynamics and Structure of Plant Com munities*. Monographs in Population Biology 20, Princeton University Press, Princeton, New Jersey.

Timmer, V. R. (1997). Exponential nutrient loading: a new fertilization technique to improve seedling performance on competitive sites. *New Forest*, Vol. 13, pp 279-299.

Timmer, V. R. (1991). Interpretation of seedling analysis and visual symptoms. *In* van den Driessche, R. (ed.). *Mineral Nutrition of Conifer Seedlings*. CRC Press, Boca Raton. pp. 113-134. Trenbath, B. R. 1976. Plant interactions in mixed crop communities. *In* Stelly, M . (ed.). *Multiple Cropping*. American Society of Agronomy, Special Publication No. 27. pp. 129-169.

van den Driessche, R. (1991). Effects of nutrients on stock performance in the forest. *In* van den Driessche, R. (ed.). *Mineral Nutrition of Conifer Seedlings*. CRS Press, Boca Raton, Boston.

Vandermeer, J. H. (1998). Maximizing crop yield in alley crops. *Agroforestry Systems*, Vol. 40, pp 199-206.

Vandermeer, J. H. (1989). *The Ecology of Intercopping*. Cambridge University Press, Melbourne.

Vandermeer, J. H. (1984). Plant competition and the yield-density relation. *Journal of Theoretical Biology*, Vol. 109, pp 393-399.

Wagner, R. G. (1993). Research directions to advance vegetation management in North America. *Canadian Journal of Forest Research*, Vol. 23, pp 2317-2327.

Wagner, R. G. (1994). Toward integrated forest vegetation management. *Journal of Forestry*, Vol 92, pp 26-30.

Wagner, R. G. and Radosevich, S. R. (1991). Neighborhood predictors of interspecific competition in young Douglas-fir plantations. *Canadian Journal of Forest Research*, Vol. 21, pp 821-828.

Wang, D., Bormann, F. H., Lugo, A. E. and Bowden, R. D. (1991). Comparison of nutrient use efficiency and biomass production in five tropical tree taxa. Forest Ecology and Management, Vol. 46, pp 1-21. Weldon, C. W., Slauson, W. L. and Ward, R. T. (1988). Competition and abiotic stress shrubs in northwest Colorado. *Ecology* Vol. 69, pp 1566-1577.

Willey, R. W. (1979). Intercropping - its importance and its research needs. Part I. Competition and yield advantages. *Field Crop Abstracts*, Vol. 32, pp 73-85.

Wilson, J.B. (1988). Shoot competition and root competition. *Journal of Applied Ecology*, Vol. 25, pp 279-96

Wilson, S. D. and Tilman, D. (1991). Components of plant competition along an experimental gradient of nitrogen availability. *Ecology*, Vol. 73, pp 1050-1065.

Woods, P. V., Nambiar, E. K. S. and Smethurst, P. J. (1992). Effect of annual weeds on water and nitrogen availability to *Pinus radiata* trees in a young plantation. Forest Ecology and Management, Vol. 48, pp 145-163.

11

Weed Responses to Soil Compaction and Crop Management

Endla Reintam and Jaan Kuht
Estonian University of Life Sciences,
Estonia

1. Introduction

Soil compaction first affects physical properties, as compaction occurs when soil particles are pressed together, reducing pore space between them and increasing the soil bulk density (Lipiec & Hatano, 2003; Raper, 2005; Reintam, 2006; Reintam et al., 2009). Soil compaction also influences chemical and biological processes, such as decreasing organic carbon (C) and N mineralization, the concentration of CO_2 in the soil (Conlin & Driessche, 2000), nitrification and denitrification, and activity of earthworms and other soil organisms (Ferrero et al., 2002). At high soil moisture, the difference in soil resistance between non-compacted and compacted soil is low and may be smaller than the value that limits root growth (>2 MPa). But as the soil dries, soil compaction is more observable (Hamza & Anderson, 2005). Further soil compaction effects are decreased root size, retarded root penetration, smaller rooting depth (Unger, and Kaspar, 1994), decreased plant nutrient availability and uptake (Kuchenbuch & Ingram, 2003; Reintam, 2006), and greater plant stress (Reintam et al., 2003), which are among the major reasons for reduced plant productivity and yield (Arvidsson, 1999; Reintam et al., 2009).

When estimating the decreased plant productivity in agro-ecosystems due to compaction, the greatest attention is usually paid to cultivated plant yields. On arable land, different weed species communities exist not only due to the different type of soil, but also because of cultivated plant diversity in agro-ecosystem, in response to different cultures, management intensity, and agro-ecosystems isolation from natural vegetation (van Elsen, 2000). Changing tillage practices consequently changes plant species composition, vertical distribution, and density of weed seed banks in agricultural soils (Buhler, 2002; Carter & Ivany, 2005). Pollard and Cussans (1981) reported that most weeds showed no consistent response to tillage and Derksen et al. (1993) suggested that composition changes in weed communities were influenced more by environmental factors (location and year) than by tillage systems. However, many weed species are more tolerant to poor soil conditions than cultivated plants. Because weeds are more efficient in nutrient uptake, the nutrient content of a crop decreases when competition with weeds increases (Koch & Köcher, 1968).

The composition of weed community is widely reported in intensive management systems. In experiments in Norway, there were no changes in the weed community during five years, even at the highest herbicide intensities (Fykse & Wærnhus, 1999). However, changing tillage or management intensity and soil physical parameters, following compaction, caused changes in weed flora. Without regular ploughing, selection for annual weeds decreases and selection for perennial weeds increases. On the other hand, in the experiments of Carter and Ivany (2005), direct seeding did not reduce the soil weed seed bank, but mouldboard ploughing for 14 years did reduce the weeds seed bank. Soil compaction caused by traffic (Jurik & Zhang ShuYu, 1999), or soil compaction in a first year's no-tillage system (Lampurlanés & Cantero-Martínez, 2003) changes dominant weed species in the community due to higher soil bulk density and penetration resistance. Many investigations have compared conventional tillage to reduced- or no-tillage systems and reported increasing numbers of perennial weed species, such couch grass (*Elytrigia repens* L.), Canadian thistle (*Cirsium arvense* L.), perennial sow thistle (*Sonchus arvensis* L.), and decrease of cultivated plant production (Blackshaw et al., 2001; Reintam et al., 2008) under no-tillage systems. Stevenson *et al.* (1998) reported that the reduction in midseason dry weight of 36% and seed yield of 59% of barley whole plant weight due to the chisel plough relative to the mouldboard plough treatment. Yield loss in this experiment was associated with interference from broadleaf plantain (*Plantago major* L.) and dandelion (*Taraxacum officinale* Weber in Wiggers). In central Iowa, a single wheel-tracking pass at crop sowing increased the cumulative number of seedlings of giant (*Setaria faberi* L.) and yellow foxtails (*S. glauca* L. [*S. pumila*]) by 187%, common water hemp (*Amaranthus rudis* L.) by 102% and common lambsquarter (*Chenopodium album* L.) by 30%. Researchers have suggested that compaction from wheel traffic apparently did not create a physical impediment to emergence; rather, it altered micro-environmental conditions in ways that stimulated weed germination and emergence (Jurik & Zhang ShuYu, 1999). Tillage effect on soil properties influences both number and diversity of weed populations (Hooker et al., 1997).

Most weeds have higher dry matter nutrient content than crops. Certain weed species have a lower optimal N requirement than crops, giving those weeds a competitive advantage in some situations (Di Tomaso, 1995). When growing with cereal crops, weeds can benefit from fertilizers (Bischoff & Mahn, 2000) irrespective of fertilizer placement (Salonen, 1992). On the other hand, many emerging weeds gain little advantage from fertilization when competing with established crops because of light competition. Nitrogen application rate weakly influences the weed flora (Andersson & Milberg, 1998); soil tillage influenced weeds more than the source of nutrients (McCloskey et al., 1996). Corn spurry (*Spergula arvensis* L.) is reported to be dominant on sandy soils and also clay soils where soil fertility and the competition with other plants are low (Mahn & Muslemanie, 1989). In addition, dry matter of corn spurry grown alone increased with increasing N up to 60 kg ha-1. Competition from rye (*Secale cereale* L.) severely reduced dry matter production of corn spurry and the weed itself was only weakly competitive under increasing N rates. Furthermore, common lambsquarter is reported to dominant in biomass where N was applied, while corn spurry and shepherds-purse (*Capsella bursa-pastoris* L.) dominated on experimental plots without N (Mahn & Muslemanie, 1989). Common lambsquarter and wild mustard (*Sinapis arvensis* L.) are the most widespread weed species on mouldboard ploughed, nutrient rich, neutral soil (Zanin et al., 1997). However, common lambsquarter and wild mustard are not the major species present in cases of low fertility and dense soil (Shrestha et al., 2002).

Plant age plays an essential role on nutrient uptake by weeds. Some weed species, such as corn mayweed (*Matricaria inodora* L.) and common chickweed (*Stellaria media* (L.) Vill.), grow during vegetation period 2–3 generations, but the young plants have a higher mineral content than more mature plants (Bockholt & Schnittke, 1996). Chickweed emerges continually from spring to autumn and starts flowering within one or two months after emergence. Chickweed seed germinate in response to soil disturbance rather than seasonal cues (Miura & Kusanagi, 2001). Both species, corn mayweed and chickweed, tolerate compacted soil (Reintam et al., 2006). Walter et al. (2002) found that chickweed was positively cross-correlated with clay and negatively cross-correlated with pH and potassium (K) content.

The objective of our experiment was to investigate continuous soil compaction effects on plant community composition and nutrient content in some of the most widespread weed species found in barley (*Hordeum vulgare* L.) production.

2. Material and methods

Data presented in current chapter were collected from the research field at the Estonian University of Life Sciences (58°23′N, 26°44′E) on a sandy loam soil, *Stagnic Luvisol*, at Tartu County in 2001–2004.

2.1 Experiment design

Soil compaction was accomplished using a 4.9 Mg tractor MTZ-82 before sowing time in spring 2001, 2002, 2003 and 2004. Passes of one, three and six passes with a wheeled vehicle loaded with 2.22 Mg on the first axle and 2.62 Mg on the rear axle (total load was 4.84 Mg) uniformly covered the entire experimental plot area. The inflation pressures in the wheels of the tractor were 150 kPa. An area without applied compaction served as the control, thus four compaction treatments were established on the experimental field. The compaction treatments were split to four replications and the size of each experimental plot (16 plots) was 12 x 9 m (108 m^2). Direct seeding of barley utilizing a drill (crosswise to compaction treatments) in rate of 450 germinating seeds per m^2 was accomplished in the middle of May. No fertilizers and herbicides were applied to decrease interactions during the compaction investigation on weed species and barley. Every autumn (in September) the soil was ploughed to the 0.21– 0.22 m depth.

2.2 Soil description

Soil was classified a sandy loam *Stagnic Luvisol* according to the WRB 1998 classification. From the genetic and diagnostic horizons the humus (32 cm), ferralic accumulation (8 cm), stagnic (10 cm) and argillic (29 cm) horizons were defined in the soil. The soil characteristics of the humus horizon (in beginning of experiment in 2001) are presented as follows: C 1.4%, N 0.11%, K 164 mg kg^{-1}, P 183 mg kg^{-1}, Ca 674 mg kg^{-1}, Mg 101 mg kg^{-1}, pH$_{KCl}$ 6.2, sand (2.0–0.02 mm) 67.9%, silt (0.02–0.002 mm) 22.9% and clay (<0.002 mm) 9.2%. The investigated soil formed on bisequal-textured reddish-brown till and is sensitive to soil compaction. This type of soil covers 5.9% of the total area, and 15.1% of the arable land in Estonia, mostly in southern and south-eastern part (Reintam& Köster, 2006).

2.3 Field sampling

The sampling of soil and plants were accomplished in the earing phase of barley in growth stage 75–79 by numeric code description according by BBCH Growth Scale of plants. All barley fruits reached final size in the middle of July in all experimental plots. Data regarding the content of the plant community were obtained from taking vegetation samples from a 0.25 m^2 plot (n=4). Partitioned plant part components (barley and observed weed species) were determined, counted, measured and weighed (wet weight). Parts of plants were taken to dry them in oven at 60°C temperature to calculate dry matter content and dry weight. Homogenised plant part samples from each treatment were taken for measuring nutrient content. Root samples were taken by 1131 cm^3 (h=15 cm, Ø=9.8 cm) steel cylinders in 15 cm layers down to 60 cm in 4 replications in years 2002–2004. Before root washing on 0.5 mm sieve, the soil from cylinders was weighted and soil bulk density calculated. No root measures were made in 2001. The soil bulk density was also measured with 50 cm^3 (h=5 cm, Ø=3.5 cm) cylinders in 0.1 m layers down to 0.4 m in four replications. At the each layer depth, samples were taken for measuring soil moisture, pH$_{KCl}$ and nutrient (C$_{org}$, N$_{total}$, plant available P, K, Ca, Mg) content. Penetration resistance was measured with a cone penetrometer (cone angle 60°, stick diameter 12 mm) in every 0.05 m layer down to 0.6 m in six replications from every experimental plot. Soil moisture and penetration resistance was measured also every spring after compaction.

2.4 Laboratory analyses

Soil and plant analyses were carried out at the laboratories of the Department of Soil Science and Agrochemistry, Estonian University of Life Sciences. The plant samples (aboveground and root parts separately) were dried at 60°C temperature and milled after removing the plants from a field. The Kjeldahl method was used to determine the content of total N of plants. The content of phosphorus (P) was determined colorimetrically on the basis of yellow phosphorus-molybdatic. Potassium content was determined by flame photometer in dipping solution diluted with distilled water. Air–dried soil samples were sieved through a 2 mm sieve and used to determine: soil reaction (pH) in 1M KCl 1:2.5, organic carbon (C$_{org}$) after Tjurin, calcium, magnesium, sodium in NH$_4$OAc at pH 7 and phosphorus and potassium after Melich-3 method. To determine water content in the soil, the soil samples taken from the field were weighted and dried at 105 °C to the constant weight and weighted again. After that the water content was calculated. Samples for the particle size determination were treated with sodium pyrophosphate to break down aggregates. Sands were sieved and fractions finer than 0.05 mm were determined by pipette analysis.

2.5 Weather conditions

In 2001 and 2003 the barley growing period (from May to August) was relatively rainy and cold. The precipitation totals were 373 mm and 450 mm, respectively. Average air temperature was 15.8°C in 2001 and 15°C in 2003 during the barley growing period. More precipitation occurred in May and August and less in June and July. Average air temperature was highest in July (20.1°C) and lowest in May (11.6°C). In 2002 the growing period was relatively warm and dry. During the vegetation period only 163 mm of rain fell and the average air temperature was 17.4°C. More precipitation occurred in June and in end of July, less in May and August. Average air temperature was highest in July (20.1°C) and lowest in May (13.9°C). The rainiest year was 2004, when 475 mm total precipitation fell during the growing period; average air temperature was 16.2°C. More precipitation occurred in June and less in May. Average air temperature was highest in August (17.1°C) and lowest in May (12.1°C).

2.6 Statistics

The one-way, two-way and three-way analysis of variance (ANOVA) was used to determine the impact of trial factors based on the collected data. Soil bulk density (soil compaction), and fertilization rate were considered fixed effects while year was considered random. The significance of experiment factors was calculated using the Fischer test and the level of significance $P<0.05$ was used. To compare the differences between values the standard Student's t-test was used and least significant differences (LSD) at significance $P<0.05$. Correlation analysis was also used to process the data. The program Statistica 7.0 was used for data analysis.

3. Results and discussion

3.1 Compaction effect on soil properties

The values of soil bulk density and penetration resistance (Fig. 1 and 2) changed among the experiment years due to the different weather conditions during sampling. The values of both, penetration resistance and bulk density depend on the soil moisture content (Fig. 3). However, the effect of traffic on the soil properties was significant in every experimental year between the non-compacted and the six times compacted soil. The differences in soil penetration resistance between one and three times compacted soil were significant after four years of soil compaction (Fig. 2). In average of the four years data there were no significant differences between those treatments. However, the six passes increased the soil penetration resistance by 2.0–3.0 MPa compared with to non-compacted soil.

Soil compaction did not caused significant (p<0.05) differences in soil nutrient uptake in first year (Table 1). After four years with continuous compaction and without fertilizers use, significant positive effects of soil compaction on organic carbon, total nitrogen, available phosphorus and potassium content were detected in soil. Compaction by one pass, three and six passes increased organic carbon content 13%, 32% and 39%, respectively, compared to non-compacted soil. Three and six passes increased total nitrogen content 16% and 9%, available phosphorus content 17% and 7%, and potassium 16% and 66%, respectively, compared to non-compacted soil. One pass compaction did not cause significant changes in nitrogen and phosphorus content. Without fertilizer use, the carbon phosphorus and potassium content decreased with four years in the control by one time and three pass compacted treatments. The highest decrease by was in non-compacted soil, where the decrease was 28% in case of carbon, 13% in case phosphorus and 41% in case of potassium. In six times compacted soil the content of carbon was 4%, content of nitrogen 23% and content of phosphorus 15% higher than in first year.

Compaction effect on soil properties depended on the machinery weight and number of passes, but also from soil moisture and weather conditions during compaction treatment application and vegetation period. Moist soil is more sensitive to soil compaction than dry. Low impact of soil compaction on soil bulk density and penetration resistance in the first year was caused by dry soil (110 g kg^{-1}) at the application of compaction and the subsequent rainy growing season following application. In following years, at the time of compaction application the soil was moist (200 g kg^{-1}) and compaction had a higher effect. Soils with higher clay and moisture content are more sensitive to soil compaction. Dry summers after a rainy spring make soil more susceptible to compaction.

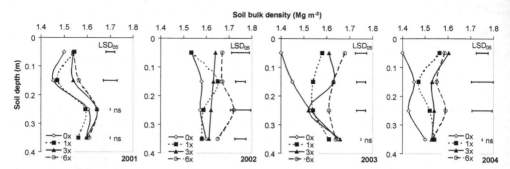

Fig. 1. Effect of soil compaction on soil bulk density in earing phase of spring barley (*Hordeum vulgare* L.) in years 2001–2004; 0x – non-compacted control, 1, 3, 6 – number of passes; LSD$_{0.05}$ – least significant differences at significance at p<0.05; ns – differences are not significant

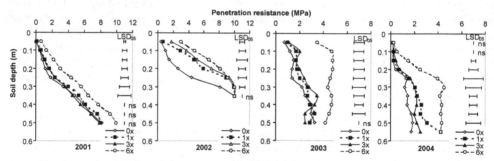

Fig. 2. Effect of soil compaction on soil penetration resistance in earing phase of spring barley (*Hordeum vulgare* L.) in years 2001–2004; 0 – non-compacted control, 1, 3, 6 – number of passes; LSD$_{0.05}$ – least significant differences at significance at p<0.05; ns – differences are not significant

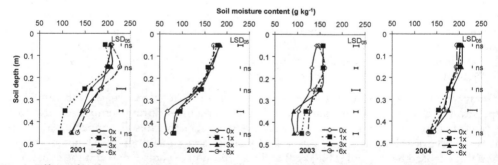

Fig. 3. Effect of soil compaction on soil moisture content in earing phase of spring barley (*Hordeum vulgare* L.) in years 2001–2004; 0 – non-compacted control, 1, 3, 6 – numbers of passes; LSD$_{0.05}$ – least significant differences at significance at p<0.05; ns – differences are not significant

Year/ Compaction variant	C_{org} (g kg^{-1})	N_{tot} (g kg^{-1})	P (mg kg^{-1})	K (mg kg^{-1})
2001				
Control	13.95	1.29	217.7	181.3
1 time compacted	14.20	1.21	238.5	156.0
3 times compacted	13.32	1.27	252.2	149.0
6 times compacted	13.37	1.20	176.4	217.2
LSD$_{0.05}$[a] (comp.)	ns[b]	ns	ns	ns
2004				
Control	10.02	1.36	189.5	107.7
1 time compacted	11.33	1.36	188.3	130.5
3 times compacted	13.25	1.58	220.8	125.2
6 times compacted	13.88	1.48	203.7	178.9
LSD$_{0.05}$ (comp.)	1.04	0.14	35.9	3.1
LSD$_{0.05}$ (year)	0.81	0.15	16.2	27.0

[a] Least significant difference at $p < 0.05$
[b] No significant differences between variants

Table 1. Changes in soil organic carbon (C_{org}), total nitrogen (N_{tot}), available phosphorus (P) and potassium (K) content due to soil compaction and experiment year of upper soil layer (0–0.3 m) in earing phase of spring barley (*Hordeum vulgare* L.)

Compaction by a 4.9 Mg tractor with tire inflation pressure 150 kPa increased soil bulk density and penetration resistance in the first and second year. However, no hardpan was formed in subsoil, likely due to deep-freezing (up to 0.5 m) in those years and because of the moderate tractor weight. However, after the third year of continuous direct compaction, a hardpan formed below the plough layer even after one tractor pass. The soils of the experiment area have a medium fine texture. They are moderately susceptible to soil compaction when moist but not particularly vulnerable when dry. The recommended maximum tire inflation pressure for medium fine textured soils is 120 to 160 kPa (van den Akker, 2002). Significant compaction has commonly been observed to a depth of about 30 cm at an axle load of 4 Mg. The natural processes of freezing/thawing, wetting/drying and bioactivity alleviate topsoil compaction. In Sweden, one pass by a 5.4 Mg tractor brought resulted in little compaction, but repeated passes led to over-compaction. In the same time one pass by the wheel-loader (9.9 Mg) increased the degree of compaction almost as much as three passes by the tractor (Etana & Håkansson, 1996). When the plough layer is severely compacted, however, the recovery of heavy clay soils may take five years in spite of annual ploughing and frost heaving. In our experiment, the highest values of soil bulk density occurred in 2002 in all compaction treatments. In other experiment years, the highest values of soil bulk density were caused from low soil moisture content (110 g kg^{-1}) and the lowest soil bulk density values occurred in 2001 and 2004 when high soil moisture content (210 g kg^{-1}) was present at compaction application. In experiments of Pickering and Veneman (1984), soil dry density increased to the soil moisture content 0.11–0.12 kg kg^{-1} and started to decrease at higher moisture contents.

Changes in soil nutrient availability due to compaction were reflected in both reduced plant growth (see Tables 3 and 4) and changed soil physical parameters (Fig. 1 and 2). Soil compaction influences both physical properties and chemical and biological processes in the soil (Ferrero et al., 2002). Higher amounts of free P and K in six times compacted soils were directly correlated with reduced nutrient removal. As the nutrient acquisition by plants was reduced, there were higher amounts of free nutrients in the soil. Phosphorus and K ions are more sensitive to soil compaction than N ions. In a rainy year the nutrients, especially P, were leached to deeper soil layers. Phosphorus is more mobile than K. The less mobile K tended to concentrate near the soil surface. A compacted soil layer, because of its high strength and low porosity, confines the crop roots to the top layer and reduces the volume of soil that can be explored by the plants for nutrients and water (Lipiec et al., 2003). There is also an interaction between compaction and soil water content. Carbon mineralization increases with increasing water content in loose soil but decreases in compact soil (Ball et al., 2000) and may increase the total amount of nutrients in soil. There is an increased the amount of total N in the compacted soil, as total N content in soil is connected with organic C content. Also Lipiec and Stepniewski (1995) found reduced N mineralization in compacted soil and Motavalli et al. (2003) reported N recovery efficiency from 290 to 140 g kg^{-1} by compaction of the soil. However, De Neve and Hofman (2000) concluded that rates of N and C mineralization may or may not be affected by compacted conditions.

3.2 Compaction effect on plant growth

Twenty-eight weed species were identified from the experimental area during four years of the experiment. In the 2001, 24 weed species were described, mostly annual weeds. The amount of perennial weed species increased from 3 species in the first year to 7 species in fourth year due to repeatedly growing barley in monoculture without herbicides and fertilizers. Only 13 weed species emerged on six-times passed soil in 2004. The most widespread weed species were common lambsquarter, field pennycress, common fumitory and common chickweed. After four years without fertilizer use, the dominating annual weed species were field pennycress and common chickweed (Table 2).

Though the changes in soil bulk density (Fig. 1) and penetration resistance (Fig. 2) due to compaction were exiguous in 2001, there was still a decrease of plant shoot dry weight and number of plants (Table 3, 4,). Significant barley dry weight decrease was observed following the six passes treatment and also on barley plants density following the one and six pass treatments, respectively (Table 3, 2001). Compaction had higher effect on annual weeds than perennial weeds. In 2001 weeds formed 6.5% from the total plant shoot dry weight and 23.4% from plants density in non-compacted and 2.8% and 17.2% in six times passed soil, respectively. Compaction decreased common lambsquarters dry weight, but did not affect density (Table 4). Common lambsquarters weed mass was 15.3% to 32.5% of the total weed mass and comprised 25.3% to 44.4% of the weed density, depending on compaction treatment. The most sensitive weed to the soil compaction was common fumitory; its mass decreased 53%–86% and density 20%–76%, depending on compaction level. Compaction affected also other investigated weed species, but the differences were not biologically significant.

In 2002, compaction increased weed shoot mass from 12% in the control up to 52% in the six times pass treatment and decreased significantly barley yield (Table 3). Changes in perennial weed density and weight were not statistically significant. In 2002, one, three and six pass treatments had positive effect on annual weed shoot dry weight. This was mainly caused by changes in weight and density of common lambsquarters, which comprised more than 50% of the weed community (Table 4).

Weed species Common name	Scientific name	2001				2004			
		Number of passes[a]							
		0	1	3	6	0	1	3	6
Annual weeds									
Common lambsquarters	*Chenopodium album* L.	+[b]	+	+	+	+	+	+	+
Field pennycress	*Thlaspi arvense* L.	+	+	+	+	+	+	+	+
Corn bindweed	*Polygonum convolvulus* L.	+	+	+		+		+	+
Hairy tare (vetch)	*Vicia villosa* Roth.	+	+	+	+	+	+	+	+
Common chickweed	*Stellaria media* (L.) Vill.	+	+	+	+	+	+	+	+
Wall speedwell	*Veronica arvensis* L.	+	+	+	+	+	+	+	+
Field pansy	*Viola arvensis* Murr.	+	+	+	+	+	+	+	+
Red dead nettle	*Lamium purpureum* L.	+	+	+	+	+	+	+	+
Common hemp nettle	*Galeopsis tetrahit* L.		+						
Common fumitory	*Fumaria officinalis* L.	+	+	+	+		+	+	
Corn bugloss	*Lycopsis arvensis* L.	+	+		+	+			
Cleavers	*Galium aparine* L.	+	+	+	+	+	+		
Shepherds-purse	*Capsella bursa-pastoris* (L.) Med.	+	+	+	+	+	+		+
Corn spurry	*Spergula arvensis* L. (coll.)	+				+	+		+
Common scorpion grass	*Myosotis arvensis* (L.) Hill	+	+						
Wild mustard	*Sinapis arvensis* L.	+	+	+	+	+		+	+
Corn mayweed	*Matricaria inodora* L.	+	+				+		
Storks-bill	*Erodium cicutarium* (L.) L'Her.	+			+				
Sun spurge	*Euphorbia helioscopia* L.						+		
Peachwort	*Polygonum persicaria* L.	+							
Perennial weeds									
Canadian thistle	*Cirsium arvense* (L.) Scop.					+	+	+	
Coltsfoot	*Tussilago farfara* L.					+		+	
Perennial sowthistle	*Sonchus arvensis* L.				+		+	+	
Corn mint	*Mentha arvensis* L.					+			
Mugwort	*Artemisia vulgaris* L.	+	+	+	+	+	+	+	+
Great (broadleaf) plantain	*Plantago major* L.		+				+	+	+
Field horsetail	*Equisetum arvense* L.				+				
Knotgrass	*Polygonum aviculare* L.					+			

[a] 0 – control plot without special compaction; 1, 3, 6 number of special passes
[b] Presence (+) or absence of weeds

Table 2. Presence of weed species depending on soil compaction and year in earing phase of spring barley (*Hordeum vulgare* L.) in the first (2001) and last (2004) year of the experiment

Treatment	Dry weight (g m⁻²)				LSD$_{0.05}$[a]	Density (plants m⁻²)				LSD$_{0.05}$
	0[b]	1	3	6	(comp.)	0	1	3	6	(comp.)
2001										
Spring barley	797	632	758	551	190	883	778	859	738	104
Annual weeds	54	31	21	15	18	245	236	187	146	86
Perennial weeds	1.4	1.2	1.2	1.0	ns[c]	24	14	13	7	ns
Total biomass	855	674	780	567	215	1151	1028	1059	890	169
2002										
Spring barley	302	262	169	61	78	450	228	266	137	132
Annual weeds	33	39	58	60	22.6	213	148	182	80	73
Perennial weeds	8.3	16	2.2	5.5	ns	6	32	5	7	19
Total biomass	344	317	230	126	68	669	408	453	224	119
2003										
Spring barley	186	167	169	91	21	550	466	463	475	58
Annual weeds	21	17	27	9	7.2	462	350	344	181	107
Perennial weeds	0.3	1.2	2.6	0.8	1.7	15	34	19	19	15
Total biomass	208	185	199	101	19	1028	850	825	791	100
2004										
Spring barley	151	164	132	92	18	573	496	429	462	97
Annual weeds	9	11	12	19	2.4	264	264	210	464	101
Perennial weeds	19	4.5	15	15	ns	88	38	74	22	22
Total biomass	180	179	159	125	20	920	781	669	1043	179
LSD$_{0.05}$ (year)										
Spring barley	161	124	143	55		134	99	112	100	
Annual weeds	44.2	20.2	34.1	24.0		120	72	59	112	
Perennial weeds	8.6	16.5	3.7	6.5		24	19	18	7	
Total biomass	196	134	144	53		175	165	134	184	

[a] Least significant difference at p<0.05
[b] Number of passes
[c] No significant differences between treatments

Table 3. Soil compaction effect on spring barley (*Hordeum vulgare* L.) and weed dry mass and density in community in earing phase of spring barley during experiment

Common fumitory did not emerge following the six pass compaction treatment. However, compaction had no strongly positive or negative effect on pennycress and chickweed.

After three years of soil compaction without fertilizer use, the total shoot dry weight and barley yield was only ¼ from first year shoot dry mass on non-compacted soil (Table 3). Changes in barley shoot dry weight were similar to earlier years, one and three pass treatments decreased barley density; however, six passes increased it. Weeds formed 9.7% to 14.9% from total shoot mass, depending on treatment and year. There was significant increase of perennial plant dry weight and density, mostly great plantain, which composition increased from 0.5% on non-compacted soil to 12.2% on six pass compacted soil in the weed community (Table 4). Again, the dominating weed species was common chickweed, with a shoot mass of 43.2% in the control and 17.6%, 36.1% and 38.8% in one, three and six pass compacted treatments, respectively. Field pennycress increased in dry weight with compaction. Again, common fumitory was not found following the six pass treated soil in the earing phase of barley.

Treatment	Dry weight (g m^{-2})				LSD$_{0.05}$[a]	Density (plants m^{-2})				LSD$_{0.05}$
	0[b]	1	3	6	(comp.)	0	1	3	6	(comp.)
2001										
Chenopodium album L.	8.5	9.8	7.2	5.2	2.1	68	88	71	68	ns[c]
Fumaria officinalis L.	5.8	0.8	2.7	1.4	4.7	25	12	20	6	14
Lamium purpureum L.	2.6	5.5	1.0	0.6	ns	23	10	9	14	9
Thlaspi arvense L.	2.0	1.4	0.3	0.8	ns	16	18	8	14	ns
Stellaria media (L.) Vill.	4.8	4.3	1.5	1.7	ns	43	63	35	24	ns
Plantago major L.	0	0.1	0	0	ns	0	3	0	0	ns
Weeds total	58	42	22	16	19	285	250	200	153	ns
2002										
Chenopodium album L.	21.8	30.6	32.7	36.0	8.4	91	95	68	36	25
Fumaria officinalis L.	1.0	1.0	1.1	0.1	ns	14	8	6	1	7
Lamium purpureum L.	0.5	0.6	0.7	0	ns	9	6	4	0	6
Thlaspi arvense L.	0.7	0.2	1.9	1.5	ns	11	4	7	8	ns
Stellaria media (L.) Vill.	1.0	0.3	5.8	1.7	ns	26	3	19	10	16
Plantago major L.	1.6	0	0.2	1.8	ns	4	0	1	2	ns
Weeds total	42	55	61	65	21.3	219	180	187	87	73
2003										
Chenopodium album L.	1.8	3.3	3.7	1.2	0.7	91	97	109	56	20
Fumaria officinalis L.	0.3	0	0.2	0	0.1	16	0	13	0	5
Lamium purpureum L.	2.4	3.3	1.6	0.1	0.3	66	66	31	3	7
Thlaspi arvense L.	0.6	1.7	0.6	1.1	ns	22	38	9	19	ns
Stellaria media (L.) Vill.	9.2	3.2	10.7	3.8	5.8	109	50	63	41	45
Plantago major L.	0.1	2.4	0.2	1.2	0.1	3	3	6	13	7
Weeds total	22	18	30	10	5	477	384	363	200	95
2004										
Chenopodium album L.	0.4	0.3	0.6	0.4	0.2	25	21	17	33	3
Fumaria officinalis L.	0	0.8	0.3	0	0.3	0	22	9	0	9
Lamium purpureum L.	0.1	0.2	0.1	0.2	0.08	4	20	14	24	5
Thlaspi arvense L.	3.0	2.4	5.1	3.1	1.7	66	34	40	48	11
Stellaria media (L.) Vill.	1.0	1.4	0.8	3.7	1.4	50	25	30	150	86
Plantago major L.	0.02	0.3	0.6	25.2	0.4	1	2	3	6	1
Weeds total	28	16	27	33	4	342	302	284	486	97
LSD$_{0.05}$ (year)										
Chenopodium album L.	9.8	11.4	20.4	18.5		25	31	19	28	
Fumaria officinalis L.	4.3	ns	2.4	1.3		14	11	ns	2	
Lamium purpureum L.	1.8	ns	ns	0.3		7	8	11	11	
Thlaspi arvense L.	ns	ns	2.3	1.7		10	15	12	20	
Stellaria media (L.) Vill.	ns	3.2	6.9	ns		ns	38	ns	81	
Plantago major L.	ns	0.1	0.1	0.6		ns	ns	5	2	
Weeds total	24.7	17.2	32.1	22.3		128	81	69	102	

[a] Least significant difference at p<0.05

[b] Number of passes

[c] No significant differences between treatments

Table 4. Soil compaction effects on the most abundant six weed species dry mass and density in weed community in earing phase of spring barley (*Hordeum vulgare* L.) during experiment

In 2004, due to persistent rain and lower penetration resistance (Fig. 2), the impact of soil compaction on barley dry weight and density was lower than in previous two years (Table 3). The lowest barley weed dry weight and density were recorded following one time passed soil, where their share from total shoot dry weight was 8.6% and from total plant density 37.9%. Following six times passed soil weeds formed 27% total shoot dry weight and 51.2% total plant density. Perennial weed dry weight was 29% to 67% of total, depending on treatment. Also in 2004, annual weed mass increased with increasing soil bulk density; however, this was likely due to higher plant density on compacted soil and not likely due to higher plant shoot mass like in 2002 (Table 4). Compaction had significant effect on all investigated weed species dry weight and density. In six pass treated soil the most widespread weed species was great plantain by comprising 74.1% of the weed community. No common fumitory plants were detected on non-compacted and six times compacted soil in earing phase of barley after four years. Year had less impact on field pennycress shoot mass and common chickweed shoot mass and density (Table 4). Again, the most affected weed species were common lambsquarters and common fumitory.

The visible result of soil compaction on plant growth is plants height reduction. Compaction had more effect on barley stalks length than weed density (Fig. 4). One compaction pass reduced barley height by 0.02 to 0.04 m, three passes 0.03 to 0.06 m and six passes 0.08 to 0.2 m depending on experiment year. In the same level of vegetation mixture with barley we observed wild mustard, corn bindweed and common lambsquarters plants. Most of the observed weed species were shorter than barley, except in 2002, when weeds over-topped barley. Compaction also reduced average weed height, but increased the differences of plant height for only common lambsquarter. Although compaction affected common fumitory mass and density, there was no impact on common fumitory height.

As compaction changes soil properties, the direct effect will be on plant root growth. Our results showed that moderate compaction (one and three compaction passes) might increase root mass in the upper part of soil compare to non-compacted soil (Fig. 5). In the draughty year 2002, the highest root mass in the top 45 cm depth was detected in three times compacted soil, two times higher than in non-compacted soil. One time and six times compaction decreased root mass by 37% and 13%, respectively, especially in deeper soil layers. In 2003, one compaction pass increased root mass in the upper 15 cm soil layer by 50 g m^{-2}, but decreased in deeper layers to 54%. Following three and six compaction passes total root mass decreased 66% and 80% respectively. In 2004, compaction decreased root mass relative to the increase of soil bulk density. Still the highest root mass was detected in three and six pass compacted soil in 15–30 cm depth likely due to high composition of perennial weed roots. In the very rainy year 2004, total root mass was four times higher than in the droughty year 2002, and two times higher than in year 2003.

Changes of dominant weed species in plant community during the four year experiment were likely caused by changes in soil conditions: decrease of available nutrients in soil, higher soil penetration resistance and soil bulk density. Low availability of major nutrients such as N, P and K play an essential role in maintaining species richness in weed communities. Like cultivated plants, weeds have different advantages depending on environmental parameters. In this experiment, decrease of common lambsquarters dry weight was due to nutrient availability and high soil resistance to root growth, conditions where it thrived compared to other species. Common lambsquarters has been shown to grow well in a wide range of climates and soils, especially those with high organic matter content (Mitch, 1988). It possesses a prolific rooting system, which allows it to resist adverse environmental conditions, such as soil compaction. Common lambsquarters emergence rate has been reported to increase with temperature and decreases with increasing soil penetration resistance and depth (Vleeshouwers, 1997).

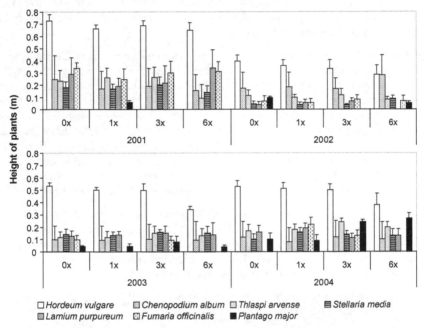

Fig. 4. Effect of soil compaction on the height of the most common seven weeds observed in earing phase of spring barley (*Hordeum vulgare* L.) in years 2001 – 2004; 0 – non-compacted control, 1, 3, 6 – number of passes; bars indicates the standard deviation

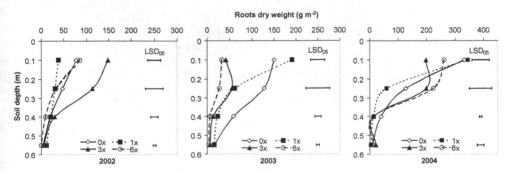

Fig. 5. Effect of soil compaction on plant roots (barley and weeds) dry weight in earing phase of spring barley (*Hordeum vulgare* L.) in years 2002, 2003 and 2004; 0 – non-compacted control, 1, 3, 6 – number of passes; $LSD_{0.05}$ – least significant differences at significance at $p < 0.05$; ns – differences are not significant

Common fumitory favours well-drained soil. The plant is suited to sandy and medium loamy acid, neutral and alkaline soils (Mitch, 1997). In this respect, common fumitory has only limited ability to compete with other weeds and suffers strongly from intraspecific competition. We observed some common fumitory in tillering phase of barley, but they did not survive in competition with other weeds on compacted soil until the earing phase of barley. Field pennycress also thrives on fertile soils but the plant can also tolerate dense soils. If the intensity of tilling is reduced, the field pennycress composition density in weed community increases (Stevenson et al., 1998).

Great plantain and corn mayweed are commonly observed on edges of field and waysides, while corn spurry is observed on soils with low fertility (Trivedi & Tripathi, 1982). Great plantain is characteristic of relatively fertile, disturbed habitats and where its root system is restricted by compaction (Whitfield et al., 1996). Few great plantain plants were observed in the weed community at the beginning of this experiment but were likely out competed by other weed species. Following, compaction, soil strength inhibited establishment of most of other existing weed species and great plantain started to dominate in weed flora (Photo 1). Species which increase in abundance under changed soil properties or low nutrient conditions (Liebman, 1989) may do so due to their intolerance of earlier conditions or high nutrient levels, or they may be suppressed by other species which respond better.

Common chickweed is a cosmopolitan species, common in cereal and broad-leaved cultivated crops (Lutman et al., 2000). Walter et al. (2002) found that chickweed was positively cross-correlated with clay and negatively cross-correlated with pH and potassium content. In our experiment common chickweed tolerated moderately compacted soil more than severely compacted soil in most years of our investigation.

Photo 1. Differences in great plantain (*Plantago major* L.) abundance on compacted and non-compacted soil after two years of continuous compaction (A) and great plantain on field edge likely indicating compaction problems (B)

Heterogeneous occurrence of some weed species, such common hemp nettle, common scorpion grass, storks-bill, spun spurge and peachwort, was caused by their heterogeneous seed distribution in the experimental area soil (Table 2). Increase of perennial weed species after four years was the result of herbicides free management and reduced tillage intensity (no additional tillage or cultivation operations, except ploughing, were made to control weeds). Without herbicides use also weed density increased in all compaction treatments (Table 3). While specific soil conditions have been associated with weed infestations, it should also be recognized that these same soil conditions may reduce the vigour of the crop, making the crop less competitive with weeds. Therefore, the weeds associated with a specific soil condition may be a secondary effect related to crop vigour rather than a weed response to soil conditions (Buhler, 2003). However, the soil physical properties and the position of weed seeds within the soil matrix play an important role in seedling emergence and seed survival. Grundy et al. (2003) found that the weed species with smaller seeds, such corn mayweed and wall speedwell showed a sharp decline in emergence when burial depth exceeded 1 cm, but some species (common chickweed and common lambsquarter) have the physical reserves to emerge from a wider range of burial depths and soil densities than normally observed in the field, suggesting an ability to exploit opportunities when they occur.

3.3 Compaction effect on plant nutrition

Most of the observed weed species had higher nitrogen content in their shoots than barley, especially common chickweed and common lambsquarters (Fig. 6). Nitrogen content in common chickweed and common lambsquarters dry matter reached 27 g kg^{-1}. Only in 2001, barley had higher nitrogen content than common lambsquarters, common fumitory and field pennycress and lower nitrogen content than common chickweed. In droughty 2002, the nitrogen content in weeds was more than 2 times higher than in barley. Also in 2003 and 2004, barley contained the lowest nitrogen content. Plant root (barley and weeds) nitrogen content was similar to barley in 2002 and 2003, while barley had the most roots mass. In 2004, the nitrogen content was higher in barley roots than in observed weed species. The lowest nitrogen contents during experiment were measured in 2004. Compaction did not cause any significant changes in plant nitrogen content after first year of soil compaction (Fig. 6). There was some decrease due to one and six pass compaction in case of barley, common chickweed and common lambsquarters. In 2002 and 2003, soil compaction decreased nitrogen content in most investigated weed species while three the pass compaction and six pass compaction treatments increased nitrogen content again, except in case of common chickweed (Fig. 6, 2002 and 2003). Nitrogen content in barley dry matter increased with increasing of soil bulk density. In 2004, compaction had only negative effect on plant nitrogen content regardless of compaction intensity (Fig. 6). Nitrogen decrease was detected regardless of species. Changes in root nitrogen content due to the compaction were similar to aboveground plant parts in 2002 and 2003. In 2004 increasing amount of perennial weed roots in compacted soil also increased the root nitrogen content.

Phosphorus content in plant dry matter was highest in common chickweed (3 to 4.5 g kg^{-1}) in all years (Fig. 7). Lowest phosphorus content was detected in barley in all years (1.0 to 1.7 g kg^{-1}). Changes in phosphorus content in plants due to compaction were similar to nitrogen changes in 2001. In 2002, soil compaction in most cases decreased phosphorus content. Increase of phosphorus content due to the six pass compaction treatment was observed only for common fumitory in 2002. Compaction had the highest negative effect on common lambsquarters and barley phosphorus content. In 2004, one pass compaction increased while three and six times compaction decreased phosphorus content in barley, common lambsquarters and common chickweed. Compaction had no significant impact on roots phosphorus content. However, the six pass compaction treatment increased roots phosphorus content in all years.

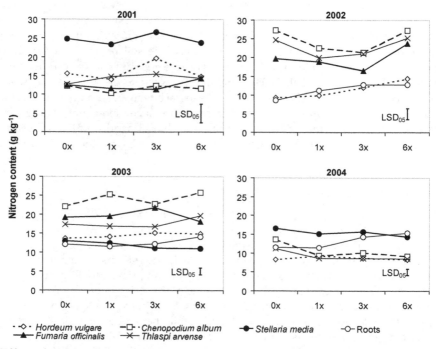

Fig. 6. Effect of soil compaction on plants and roots nitrogen content in earing phase of spring barley (*Hordeum vulgare* L.) in years 2001 – 2004; 0 – non-compacted control, 1, 3, 6 – number of passes; $LSD_{0.05}$ – least significant differences at significance at $p < 0.05$

The weeds highest in potassium were common chickweed and common lambsquarters in all years of experiment (Fig. 8). The potassium content in common chickweed ranged from 55 g kg⁻¹ to 80 g kg⁻¹ and in common lambs quarters from 25 to 62 g kg⁻¹, depending on soil compaction and year. Stabile low potassium content was observed in barley and field pennycress dry matter with 12 to 20 g kg⁻¹ and in roots with 5 to 12 g kg⁻¹, depending on compaction and year. No positive correlation between nutrient content and soil compaction was observed in case of potassium in plants aboveground parts. Compaction inhibited potassium uptake of all investigated species and the differences between compaction treatments were significant in 2001. The six pass compaction treatment caused highest decrease on common fumitory by 50%, on common lambsquarters and common chickweed by 21% in 2003, and on common lambsquarters and common chickweed by 33% and 35%, respectively, in 2004. Significant increase of roots potassium content was observed in 2003 due to the six pass compaction treatment and in 2004 due to one, three and six pass compaction treatments.

Similar to this experiment, our earlier investigations (Reintam & Kuht, 2004) and investigations of other researchers (Salonen, 1992) reported higher nutrient content in weeds compared to barley. The lower nitrogen need of many weed species can give them advantage in competition with cereals (Di Tomaso, 1995) and thus they have a greater ability to compete with barley for nutrients, water and light. Because weeds are more efficient in nutrient uptake, in particular nitrogen, the nitrogen content of a crop decreases with increasing competition with weeds. However, some researchers suggest that competition between weeds and crops is lower on nutrient rich than on nutrient-poor soils (Pyšek et al., 2005) and competition is most intense in plots with lowest resource levels (Wilson & Tilman, 1993). In our experiment, weeds were more able to compete with barley under moderate compaction conditions (3 pass treatment), where in many cases nutrient content in weeds increased, especially in common lambsquarters and common chickweed.

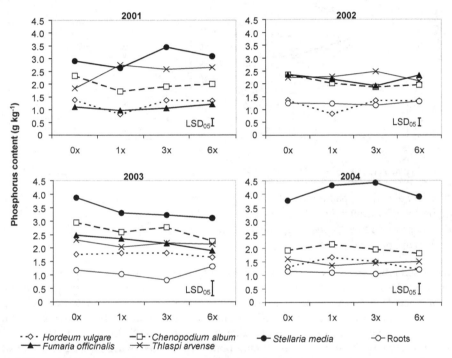

Fig. 7. Effect of soil compaction on plants and roots phosphorus content in earing phase of spring barley (*Hordeum vulgare* L.) in years 2001 – 2004; 0 – non-compacted control, 1, 3, 6 – number of passes; $LSD_{0.05}$ – least significant differences at significance at $p<0.05$

Also Bockholt and Schnittke (1996) observed the high nutrient, especially potassium assimilation of young chickweed plants. Common lambsquarters was especially rich in nitrogen, but also potassium. Common lambsquarters is reported as the highest competitor for the nutrients to cultivated plants because of its high mass and nutrient content (Parylak, 1996) and high ability to compete with cultivated plants. Common lambsquarters taproot makes it more competeveness on dense soil compared to the barley, which have fibrous roots. Thicker roots are better able to penetrate the compacted soil compared to thinner roots (Whitely & Dexter, 1984) and compaction influences less dicotyledonous than monocotyledonous plants (Materachera et al., 1991).

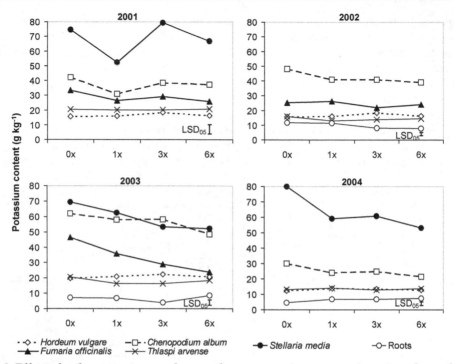

Fig. 8. Effect of soil compaction on plants and roots potassium content in earing phase of spring barley (*Hordeum vulgare* L.) in years 2001–2004; 0 – non-compacted control, 1, 3, 6 – number of passes; $LSD_{0.05}$ – least significant differences at significance at $p<0.05$; ns – differences are not significant

The increased nitrogen content in barley and weed species dry matter with increasing soil bulk density was likely due to better competitive conditions available to the survived plants on the most compacted soil. Plant density was there 1.3 to 3-times lower than on non-compacted soil and nutrient area per one plant was higher. In unsuitable conditions, barley tillering is higher and in dry year new sprouts may grow after adequate rainfall. In 2002, barley grew new sprouts in middle of the summer, and young plant tissues are always richer in nutrients than older. In the better competitive conditions plants are also producing more leaves than under lover radiation and less competitive conditions, and leaves usually containing more nitrogen than in stems. Planting density and ontogenetic processes significantly influence dry matter partitioning between leaves and stems (Röhrig & Stützel, 2001; Causin, 2004). With increasing competition (on non-compacted soil) common lambsquarters, field pennycress and common fumitory allocated relatively more biomass to stems than to leaves. In addition, higher soil moisture content over time is observed in compacted soil in dry seasons compared to less compacted soil. In moist soil, there are more plant available nitrates than in dry soil. In a dry year, due to compacted soil, uptake of elements such as nitrogen, calcium and magnesium, which are moving into the plant with water, might increase. Decrease of nitrogen content in plant dry matter in wet years, especially in 2001 and 2004, was probably connected next to poorer root development also with increased denitrification and decreased mineralization of organic matter in highly compacted soil due to decreased soil aeration. N_2O flux increases with decreasing distance from straw residues and air permeability, and with increasing cone resistance and wet bulk density (Ball et al., 2000).

No increase of phosphorus and potassium content in plants (barley and weeds) due to increasing soil bulk density and penetration resistance were detected during this experiment (Fig. 7 and 8). Due to compacted soil, the plants are in stress and in the stressed conditions (increased cellular pH) plants nutrient acquisition through proton pumping via the H^+-ATPase and transporters from roots to the stems and leaves is reduced (Bucher et al., 2001; Reintam & Kuht, 2003) and results in increased nutrients uptake by roots. These processes likely explain the increase of nutrients in the roots due to the compaction. Liepiec and Stepniewski (1995) found that root growth greatly affects uptake of nutrients transported by diffusion, such as phosphorus.

4. Conclusion

Soil over-compaction inhibits the nutrition of cultivated plants and decreases their ability to compete with weeds. Changing the field conditions also changes the weed composition with which cultivated plants will compete. In compacted soils without fertilizer use, relatively easily controlled weed species will likely be replaced with harder to control weed species due to selection for competitive species. Weeds are serious competitors in agricultural systems; they accumulated free nutrients from soil, especially in dense soil, at the detriment to less competitive cultivated crops. At the same time the nutrient assimilation by weeds may stop their leaching from soil and store the nutrients in organic matter also for the next growing period. However, in severely compacted soil even weeds are not able to flourish and free nutrients may start to pollute the environment. Both, changes in weed community composition and nutrient assimilation deserves further investigations to understand better plant–soil and plant–plant interactions of other cultivated plants and soils under stress conditions, such is soil compaction.

5. Acknowledgment

The study was supported by Estonian Science Foundation grant No 5418, 4991 and 7622.

6. References

Andersson, T.N. & Milberg, P. (1998). Weed flora and the relative importance of site, crop, crop rotation, and nitrogen. *Weed Science*, Vol. 46, pp. 30–38.

Arvidsson, J. (1999). Nutrient uptake and growth of barley as affected by soil compaction. *Plant and Soil*, Vol. 208, pp. 9–19.

Ball, B.C.; Horgan, G.W. & Parker, J.P. (2000). Short-range spatial variation of nitrous oxide fluxes in relation to compaction and straw residues. *European Journal of Soil Science*, Vol. 51, pp. 607–616.

Bischoff, A. & Mahn, E.-G. (2000). The effect of nitrogen and diaspore availability on the regeneration of weed communities following extensification. *Agriculture, Ecosystems and Environment*, Vol. 77, pp. 237–246.

Blackshaw, R.E.; Larney, F.J.; Lindwall, C.W.; Watson, P.R. & Derksen, D.A. 2001. Tillage intensity and crop rotation affect weed community dynamics in a winter wheat cropping system. *Canadian Journal of Plant Sciences*, Vol. 81, pp. 805–813.

Bockholt, R. & Schnittke, C. (1996). Nähr- und Mineralstoffgehalt von Kräutern des Niedermoorgrünlandes unter intensive Bewirtschaftung. *Wirtschaftseigene Futter*, Vol. 42, pp. 209–216.

Bucher, M.; Rausch, C. & Daram, P. (2001). Molecular and biochemical mechanisms of phosphorus uptake into plants. *Journal of Plant Nutrition and Soil Science*, Vol. 164, pp. 209–217.

Buhler, D.D. (2002). Challenges and opportunities for integrated weed management. *Weed Sciences*, Vol. 50, pp. 273–280.

Buhler, D.D. (2003). Weed Biology, Cropping Systems, and Weed Management, *Journal of Crop Production*, Vol. 8, No. 1-2, pp. 245-270.

Carter, M.R. & Ivany, J.A. (2005). Weed seed bank composition under three long-term tillage regimes on a fine sandy loam in Atlantic Canada. *Soil & Tillage Research*, Vol. 90, pp. 29–38.

Causin, H.F. (2004). Responses to shading in *Chenopodium album*: the effect of the maternal environment and the N source supplied. *Canadian Journal of Botany*, Vol. 82, pp. 1371–1381.

Conlin, T.S.S. & Driessche, R. (2000). Response of soil CO_2 and O_2 concentrations to forest soil compaction at the long-term soil productivity sites in central British Columbia. *Canadian Journal of Soil Science*, Vol. 80, pp. 625–632.

De Neve, S. & Hofman, G. (2000). Influence of soil compaction on carbon and nitrogen mineralization of soil organic matter and crop residues. *Biology and Fertility of Soils*, Vol. 30, pp. 544–549.

Derksen, D.A.; Lafond, G.P.; Thomas, A.G.; Loeppky, H.A. & Swanton, C.J. (1993). Impact of agronomic practices on weed communities – tillage systems. *Weed Science*, Vol. 41, pp. 409–417.

Di Tomaso, J. M. 1995. Approaches for improving crop competitiveness through the manipulation of fertilization strategies. *Weed Science*, Vol. 43, pp. 491–497.

Etana, A. & Håkansson, I. (1996). Effect of traffic with a tractor and wheel loader on two ploughed moist soils. *Swedish Journal of Agricultural Research*, Vol. pp. 26, 61–68.

Ferrero, A., Lipiec; J., Loredana, S. & Nosalewicz, A. (2002). Effect of increasing compaction levels on the efficiency of nitrogen topdressing of grasses. In: *Sustainable land management – environmental protection: A soil physical approach. Advances in GeoEcology 35*, Pagliai, M., Jones, R. (Eds.), 351–358.

Fykse, H. & Wærnhus, K. (1999). Weed development in cereals under different growth conditions and control intensities. *Acta Agriculturae Scandinavica, Section B - Plant Soil Science*, Vol. 49, pp. 134–142.

Grundy, A.C.; Mead, A. & Burston, S. (2003). Modelling the emergence response of weed seeds to burial depth: interactions with seed density, weight and shape. *Journal of Applied Ecology*, Vol. 40, pp. 757–770.

Hamza, M.A. & Anderson, W.K. (2005). Soil compaction in cropping systems: A review of the nature, causes and possible solutions. *Soil & Tillage Research*, Vol. 82, pp. 121–45.

Hooker, D.C,; Vyn, T.J. & Swanton, C.J. (1997). Effectiveness of soil-applied herbicides with mechanical weed control for conservation tillage systems in soybean. *Agronomy Journal*, Vol. 89, pp. 579–587.

Jurik, T.W. & Zhang ShuYu. (1999). Tractor wheel traffic effect on weed emergence in central Iowa. *Weed Technology*, Vol. 13, pp. 741–746.

Kuchenbuch, R.O. & Ingram, K.T. (2004). Effect of soil bulk density on seminal and lateral roots of young maize plants (*Zea mays* L.). Journal of Plant Nutrition and Soil Science, Vol. 167, pp. 229–235.

Lampurlanés, J. & Cantero-Martínez, C. (2003). Soil bulk density and penetration resistance under different tillage and crop management systems and their relationship with barley root growth. *Agronomy Journal*, Vol. 95, pp. 526-536.

Liebman, M., (1989). Effects of nitrogen fertilizer, irrigation, and crop genotype on canopy relations and yields on an intercrop/weed mixture. *Field Crops Research*, Vol. 22, pp. 83-100.

Lipiec, J.; Arvidsson, J. & Murer, E. (2003). Review of modelling crop growth, movement of water and chemicals in relation to topsoil and subsoil compaction. *Soil & Tillage Research*, Vol. 73, pp. 15-29.

Lipiec, J. & Hatano, R. (2003). Quantification of compaction effects on soil physical properties and crop growth. *Geoderma*, Vol. 116, pp. 107-136.

Lipiec, J. & Stepniewski, W. (1995). Effects of soil compaction and tillage systems on uptake and losses of nutrients. *Soil & Tillage Research*, Vol. 35, pp. 37-52.

Lutman, P.J.W.; Bowerman, P.; Palmer, G.M. & Whytock, G.P. (2000). Prediction of competition between oilseed rape and *Stellaria media*. *Weed Research*, Vol. 40, pp. 255-269.

Mahn, E.G. & Muslemanie, N. (1989). The influence of increasing nitrogen supply on the development and matter production of *Spergula arvensis* L. under agro-ecosystem conditions. *Angewandte Botanik*, Vol. 63, pp. 347-359.

Materachera, S.A.; Dexter, A.R. & Alston, A.M. (1991). Penetration of very strong soils by seedling roots of different plant species. *Plant and Soil*, Vol. 135, pp. 31-41.

McCloskey, M.; Firbank, L.G.; Watkinson, A.R. & Webb, D.J. (1996). The dynamic of experimental arable weed communities under different management practices. *Journal of Vegetation Sciences*, Vol. 7, pp. 799-808.

Mitch, L.W. (1988). Common lambsquarters. *Weed Technology*, Vol. 2, pp. 550-552.

Mitch, L.W. (1997). Fumitory (*Fumaria officinalis* L.). *Weed Technology*, Vol. 11, pp. 843-845.

Miura, R. & Kusanagi, T. (2001). Variation in the factors determining flowering time in the *Stellaria media* complex. *Weed Research*, Vol. 41, pp. 69-81.

Motavalli, P.P.; Anderson, S.H. & Pengthamkeerati, P. (2003). Surface compaction and poultry litter effects on corn growth, nitrogen availability, and physical properties of a claypan soil. *Field Crop Research*, Vol. 84, pp. 303-318.

Parylak, D. (1996). Competitive uptake of nutrients by spring barley and weeds. *Fragmenta-Agronomica*, Vol. 13, pp. 68-74. (Polish, summary in English)

Pickering, E.W. & Veneman, P.L.M. (1984). Strength characteristics of three indurated horizons in Massachusetts. *Soil Science Society American Journal*, Vol. 48, pp. 133-137.

Pollard, F. & Cussans, G.W. (1981). The influence of tillage on the weed flora in a succession of winter cereal crops on a sandy loam soil. *Weed Research*, Vol. 21, pp. 185-190.

Pyšek, P.; Jarošík, V.; Kropáč, Z.; Chytrý, M.; Wild, J. & Tichý, L. (2005). Effects of abiotic factors on species richness and cover in Central European weed communities. *Agriculture, Ecosystems Environment*, Vol. 109, pp. 1-8.

Raper R.L. (2005). Agricultural traffic impact on soil. *Journal of Terramechanics*, Vol. 42, pp. 259-80.

Reintam, E. & Kuht, J. 2003. Changes in nutrient uptake and cellular fluid pH of spring barley as affected by soil compaction. *Indian Journal of Plant Physiology*. Special Issue, pp. 522-526.

Reintam, E. & Kuht, J. 2004. Soil compaction effect on soil, nutrient cycling and weeds in agroecosystem. In *Integrative Approaches towards sustainability in the Baltic Sea region*, Leal Filho, W., Ubelis, A., (Eds.), Peter Lang Scientific Publishers, pp. 519-527.

Reintam, E. (2006). Changes in soil properties, spring barley (*Hordeum vulgare* L.) and weed nutrition and community due to soil compaction and fertilization on sandy loam Stagnic Luvisol. (PhD thesis) Tartu: Eesti Maaülikool

Reintam, E.; Kuht, J.; Trükmann, K.; Puust, J. (2006). Composition of weed community depending on soil compaction in barley field. In *Sustainable Development in the Baltic Sea and Beyond*, Leal Filho, W.; Ubelis, A. & Berzina, D. (Eds.). 541–552. Peter Lang Verlag

Reintam, E. & Köster, T. (2006). The role of chemical indicators to correlate some Estonian soils with WRB and Soil Taxonomy criteria. *Geoderma*, Vol. 136, pp. 199–209.

Reintam, E.; Trükmann, K.; Kuht, J.; Toomsoo, A.; Teesalu, T.; Köster, T.; Edesi, L. & Nugis, E. (2008). Effect of Cirsium arvense L. on soil physical properties and crop growth. Agricultural and Food Science, Vol. 17, pp. 153–164.

Reintam, E.; Trükmann, K.; Kuht, J.; Nugis, E.; Edesi, L.; Astover, A.; Noormets, M.; Kauer, K.; Krebstein, K. & Rannik, K. (2009). Soil compaction effects on soil bulk density and penetration resistance and growth of spring barley (*Hordeum vulgare* L.). *Acta Agriculturae Scandinavica, Section B - Plant Soil Science*, Vol. 59, No. 3, pp. 265–272.

Röhrig, M. & Stützel, H. (2001). Dry matter production and partiotioning of *Chenopodium album* in contrasting competitive environments. *Weed Research*, Vol. 41, pp. 129–142.

Salonen, J. (1992). Distribution of nitrogen between crop and weeds in spring cereals. *Acta Agriculturae Scandinavica, Section B - Plant Soil Science*, Vol. 42, pp. 218–223.

Shrestha, A.; Knezevic, S.Z.; Roy, R.C.; Ball-Coelho, B.R. & Swanton, C.J. (2002). Effect of tillage, cover crop and crop rotation on the composition of weed flora in a sandy soil. *Weed Research*, Vol. 42, pp. 76–87.

Stevenson, F.C., Légère, A., Simard, R.R., Angers, D.A., Pageau, D., Lafond, J. 1998. Manure, tillage, and crop rotation: Effect on residual weed interference in spring barley cropping systems. Agron. J. 90, 496–504.

Trivedi, S. & Tripathi, R.S. (1982). The effects of soil texture and moisture on reproductive strategies of *Spergula arvensis* L. and *Plantago major* L. *Weed Research*, Vol. 22, pp. 41–49.

Unger, P.W. & Kaspar, T. (1994). Soil compaction and root growth: a review. *Agronomy Journal*, Vol. 86, pp. 759–66.

van den Akker, J.J.H. (2002). Determination of the susceptibility of subsoil to compaction and ways to prevent subsoil compaction. In: *Sustainable land management – environmental protection: A soil physical approach. Advances in GeoEcology 35*, Pagliai, M., Jones, R. (Eds.), 291–304.

van Elsen, T. (2000). Species diversity as a task for organic agriculture in Europe. *Agriculture, Ecosystems and Environment*, Vol. 77, pp. 101–109.

Vleeshouwers, L.M. (1997). Modelling the effect of temperature, soil penetration resistance, burial depth and seed weight on pre-emergence growth of weeds. *Annals of Botany*, Vol. 79, pp. 553–563.

Walter, A.M.; Christensen, S. & Simmelsgaard, S.E. (2002). Spatial correlation between weed species densities and soil properties. *Weed Research*, Vol. 42, pp. 26–38.

Whitely, G. M.& Dexter, A.R. (1984). The behaviour of root encountering cracs in soil. I Experimental methods and results. *Plant and Soil*, Vol. 77, pp. 141–149.

Whitfield, C.P.; Davison, A.W. & Ashenden, T.W. (1996). Interactive effects of ozone and soil volume on *Plantago major*. *New Phytologist*, Vol. 134, pp. 287–294.

Wilson, S.D. & Tilman, D. (1993). Plant competition and resource availability in response to disturbance and fertilization. *Ecology*, Vol. 74, pp. 599–611.

Zanin, G.; Otto, S.; Riello, L. & Borin, M. (1997). Ecological interpretation of weed flora dynamics under different tillage systems. *Agriculture, Ecosystems and Environment*, Vol. 66, pp. 177–188.

12

Interrelationships Among Weed Management in Coffee Plantation and Soil Physical Quality

Cezar Francisco Araujo-Junior[1], Moacir de Souza Dias Junior[2],
Elifas Nunes de Alcântara[3], Paulo Tácito Gontijo Guimarães[3]
and Ayodele Ebenezer Ajayi[4]

[1]*Agronomic Institute of Paraná – IAPAR,*
Rodovia Celso Garcia Cid, Londrina, State of Paraná,
[2]*Department of Soil Science,*
Federal University of Lavras, Lavras, State of Minas Gerais,
[3]*Minas Gerais State Corporation for Agriculture and Livestock Research,*
EPAMIG, CTSM
[4]*Soil Water and Environment Section, Department of Agricultural Engineering,*
Federal University of Technology Akure Ondo State,
[1,2,3]*Brazil*
[4]*Nigeria*

1. Introduction

Coffee bean is one of the most important commodities produced in Brazil. Brazil is responsible for the supply of about 30% of world coffee bean market. Coffee related enterprises are a major economic driver in the regions where it is cultivated in Brazil and elsewhere as it generates jobs, provide income and stimulate development. However, for greater coffee agribusiness competitiveness , it is necessary to meet social-environmental requirements expected by international consumers (Araujo-Junior et al., 2008).

Among several social-environmental expectations met by coffee farmers internationally, biodiversity conservation, sustainable management and subsequent improvement or maintenance of soil structure in order to avoid or minimize additional soil compaction resulting from inadequate management are vital (Brazil Specialty Coffee Association [BSCA], 2005). These requirements help the coffee farmers develop eco-friendly production practices/guidelines: environmentally appropriate, economically viable, socially beneficial and culturally acceptable in their production system. These production guidelines, help in balancing environmental and socio-economic factors in coffee bean production.

Amongst all agronomic practices involved in coffee production, the weed management strategy/system is one of the most intensive in coffee bean production and critical to eco-friendly management ranging from two to five operations per year. The adopted weed management system in coffee plantations can have major effects on the soil environment, affecting physical, chemical and biological conditions, resulting in changes soil compressive behavior and load bearing capacity affecting yield potential in coffee plantations (Araujo-Junior et al., 2008; 2011).

Appropriate weed management systems utilized between coffee rows would help in minimizing soil degradation by erosion (Carvalho et al., 2007), reducing compaction and improving soil workability and machines trafficability (Araujo-Junior et al., 2008, 2011). Weed plants utilized as cover crops residues can be left on the soil surface similar to a cereal stubble mulch to protect against evaporations and erosion (Hillel, 1980; Faria et al., 1998). In a newly developed orchard, Yang et al. (2007) observed that the application of herbicides and tillage favored soil erosion. Yang et al. (2007) pointed out that chemical and mechanical methods are the dominant weed control practices in many production systems due to its effectiveness, but noted on the other hand, that weed presence during the rainy season prevented soil erosion. Studies conducted in tropical conditions showed that mechanical and chemical methods for weed control on coffee plantations had a great influence on the soil compaction state (Kurachi & Silveira, 1984; Alcântara & Ferreira, 2000b; Araujo-Junior et al., 2008, 2011), soil surface crust formation, erosion and coffee yield (Silveira et al., 1985; Alcântara & Ferreira, 2000a).

Soil compaction processes are one of the most important causes of soil degradation and changes on soil structure, affecting soil physical quality. Compaction is a reduction of the volume of a given mass of soil and ceases when the soil structure has become strong enough to withstand the applied stress without further failure, in compacted soils volume of pores is reduced (Dexter, 2004). Soil structure is defined as the arrangement of the solid particles and of the pore space located between them (Marshall, 1962). Also, soil structure may be defined as the combination or arrangement of primary soil particles into secondary units or peds. The secondary units are characterized on the basis of size, shape and grade (Soil Science Society American – SSSA, 2008). Structural changes to the soil could alter their physical quality, thereby altering the soil workability and trafficability, infiltrate rate, drainage, water redistribution and water retention, as a function of pore-size distribution. Due to effects of soil residue coverage on soil, the weed management system has direct influence on soil structure management and physical quality and must therefore be considered from both agronomic and environmental viewpoints.

Structural changes resulting from the traditional bare ground weed management system stand out among the main adverse effects of this practice (Kurachi & Silveira, 1984; Silveira & Kurachi, 1985; Faria et al., 1998; Alcântara & Ferreira, 2000a; Araujo-Junior et al., 2008, 2011). Structural changes due to improper soil management make coffee plants more susceptible to dry conditions by the reduction of infiltration rate and gas flow into the soil profile. Inadequate soil aeration and nutritional deficiency, decreases root growth and enhancing soil erosion, resulting in a compromise of the soil and environmental quality in agro-forestry production (Horn, 1988; Dias Junior et al., 2005; Vogeler et al., 2006).

The water content in the soil profile determines the reaction to tillage, and among the physical properties, soil moisture is the most important for soil-machine interactions, since it controls the consistency of the soil (Hillel, 1980) and governs the amount of soil deformation when subjected to external pressure (Dias Junior & Pierce, 1996). Thus, soil water acts as a lubricant and as a binder between the soils particles, affecting the structural stability and strength of geological materials and soil (Topp & Ferré, 2002).Therefore, knowledge of the interrelationship of weed management and its influence on soil structure is essential to establish sustainable management of the soil in coffee plantations. Mentioned previously, soil structure greatly influences the distribution of the pore size, water and gas movement into the soil, soil strength and soil water retention. Few studies have been investigated the effect of weed management system on soil physical quality. In this book chapter, changes in soil physical attributes (soil bulk density, microporosity, macroporosity, total porosity, soil water retention curve, precompression stress and load bearing capacity) are studied in relation to weed management system in coffee plantation. Load bearing capacity models were developed to assess the influence of the different weed management systems on soil structure.

2. Site description and characterization

The study site was the Experimental Farm of the Minas Gerais State Department for Agriculture and Livestock Research [EPAMIG] (20°55'00'' S, 47°07'10'' W, ≈ 885 m) in the São Sebastião do Paraíso County, State of Minas Gerais, Brazil. The farm has been used for weed control management system experiments since 1977. The average annual temperature of the area is 20.8 °C, (27.6 °C maximum, 14.1 °C, minimum) and the average annual rainfall is 1470 mm (Alcântara & Ferreira, 2000a,b).

The soil in the experimental area is derived from basalt and was classified as a Dystroferric Red Latosol according to the Brazilian Soil Classification System (Brazilian Agricultural Research Council [Embrapa], 2006); Oxisol according to USDA soil taxonomy (Soil Survey Staff, 1998) and Ferralsol (Food and Agriculture Organization [FAO], 2006). Analysis of soil collected close to experimental area under natural forest showed that Dystroferric Red Latosol contains 570 g kg^{-1} clay, 230 g kg^{-1} silt and 200 g kg^{-1} sand, in the top 0 to 30 cm depth and also have a homogeneous structure throughout the profile. The soil has low soil bulk density, high total porosity and macroporosity and exhibit a granular structure like a coffee powder.

2.1 Weed control management systems and conduction of the coffee plantation

Seven weed management systems which had been in use for about 30 years in the coffee plantation were considered in this study (Photo 1; Table 1). The management systems were established in a randomized complete block design with three replicates, each plot 36m in length. The experimental design further included a split-plot with each weed management system in use in three interrows as the main-plot factor, and the soil sampling depths (0–3, 10–13 and 25–28 cm) as a split-plot. In the areas under the coffee canopy, the weeds are managed as needed utilizing manual hoeing or with the application of herbicides. The successful weed management system utilized in the coffee plantation experimental area for the 30 years period prior to treatment establishment influenced the number of operations needed as well as the density and diversity of weeds found in the area at the time of the sampling (Table 1).

Weed management	Operations	Species weed/common name/families
No-Weed Control (NWC)	0	*Marmodica charantia* L., melão-de-são-caetano, Cucurbitaceae; *Ephorbia heterophylla* L., leiteira, Euphorbiaceae; *Digitaria insularis* (L.) Mea ex Ekman, capim-amargoso, Poaceae; *Panicum maximum* Jacq., capim-colonião, Poaceae; *Nicandra physaloides* Gaertn., joá-de-capote, Solanaceae; *Ipomoea acuminata*, corda-de-viola, Convolvulaceae; *Amaranthus viridis*, caruru-de-mancha, Amaranthaceae
Hand Hoeing (HAHO)	8	*Ephorbia heterophylla* L., leiteira, Euphorbiaceae; *Digitaria horizontalis* Willd., capim-colchão, Poaceae; *Cenchrus echinatus* L., timbête, Poaceae.
Rotary Tilling (ROTI)	8	*Cyperus rotundus* L, tiririca, Cyperaceae; *Cynodon dactylon* (L.) Pers., grama-seda, Poaceae; Bidens pilosa L., picão-preto, Compositae.
Post-Emergence Herbicide (POSH)	8	*Amaranthus viridis* (caruru-de-mancha, Amaranthaceae); *Commelina benghalensis* L. (trapoeraba, Commelinaceae).
Mechanical Mowing (MMOW)	9	*Cyperus rotundus* L, tiririca, Cyperaceae; *Cynodon dactylon* (L.) Pers., grama-seda, Poaceae; *Amaranthus viridis*, caruru-de-mancha, Amaranthaceae; *Brachiaria decumbens* Stapf., braquiária, Poaceae.
Disk Harrowing (CTDH)	8	*Cyperus rotundus* L, tiririca, Cyperaceae; *Cynodon dactylon* (L.) Pers., grama-seda, Poaceae; *Brachiaria plantaginea* (Link) Hitchc., marmelada, Poaceae.
Pre-emergence herbicide	6	Without weed plants at the moment of the sampling

Table 1. Weed management system, numbers of operations performed between January 2006 and December 2007, species, common name and genus observed in an experimental area at the time of soil sampling.

1. No-weed control between coffee rows (NWC): the weeds plants were left to grow freely between the coffee rows, thus, high density and diversity of the weed plants were found in the plots at the time of sampling (Table 1).
2. Hand hoeing (HAHO): performed with the aid of a hoe, when the weed reached 45 cm height. These operations were carried out eight times between January 2006 to December 2007 (Table 1).
3. Post-emergence herbicide (POSH): glyphosate, N-(fosfonometil) glicina, was applied with the aid of a knapsack sprayer, at a rate 2.0 L ha-1 of commercial product and 0.72 Kg active ingredient ha-1, soluble concentrate formulation 0,36 Kg L-1, and applied with spray volume of 400 L ha-1, eight applications were performed between January 2006 and December 2007 (Table 1).
4. Mechanical mowing (MMOW): the weed plants were mowed with a mechanical mower Kamaq® model 132 KD, with cutting width of 1.32 m and 340 Kg of static mass
5. Rotary-tilling (ROTI): the axis has five flanges, as two sides with three knives and threes edges with six knives. It's worked at 10 cm depth incorporating the weeds.
6. Coffee tandem disk harrow (CTDH): the equipment is composed by two sections in tandem, each section is equipped with seven flat disks with cut width of 1.3 m and static mass 300 kg. It's worked at 7 cm depth.
7. Pre-emergence herbicide (HPRE): oxyfluorfen (2-cloro-a,a,a-trifluoro-p-tolyl-3-ethoxy-4-nitrophenyl ether), was applied with the aid of a knapsack sprayer, at a rate 2.0 L ha-1 of commercial product and 0.48 Kg active ingredient ha-1 in the soluble concentrate formulation 0.24 Kg L-1, and applied with spray volume of 400 L ha-1 (Rodrigues & Almeida, 2005) six applications were performed from January 2006 to December 2007 (Table 1). For this application, soil surface was free of the vegetation.

A

B

C

Photo 1. Overview of experimental area at the time of the sampling in December 2007. (A) weedy control between coffee rows; (B) pre-emergence herbicide. Note sheet erosion (B) and decreased infiltration due to surface crusting (C) between coffee rows.

The equipment used to apply tillage treatments was mounted on a two-wheel-drive coffee tractor Valmet® model 68. This tractor has engine capacity of 61.9 CV (45 kW), total weight of tractor with equipment was 38.25 kN, front tyres 6-16 (15.24 cm of width x 40.64 cm rim diameter) in inflation pressure 172 kPa and rear tyres 12.4-R28 in inflation pressure 124 kPa. To determine the maximum stress applied by each tyre, the static weight distribution was considered to be 35% for the front tyres and 65% for the rear tyres. The critical volumetric water content for the traffic of the tractor, were considered as those stress that don't exceed the internal strength of the soil expresses in the precompression stress (Araujo-Junior et al., 2011).

2.2 Soil sampling

In each weed management system, 15 undisturbed soil samples (early December, 2007) were collected randomly in the traffic line of the machines and equipments, 80 cm from stems of the coffee trees in the 0-3, 10-13 and 25-28 cm layers, totaling 315 soil samples (15 samples x 3 depths x 7 management system). Additional fifteen samples at each depth were collected in a Dystroferric Red Latosol under natural forest (NAFT) adjacent to coffee cultivation, 45 undisturbed soil samples (15 samples x 3 depths) were collected which served as a reference of soil physical quality. The undisturbed soil samples were collected using a cylindrical Uhland sampler (Uhland, 1949) and aluminum rings, 2.54 cm high by 6.35 cm diameter (Photo 2). The Uhland sampler is pressed into the soil sample in the 0-3 cm depth. To collect the sample at 10-13 cm and 25-28 cm depths, the sampling pit were carefully dug to depths 10 cm and 25 cm.

Photo 2. Uhland undisturbed soil sampler components. 1 – driving assembly; 2 – aluminum cylinder room ; 3- graphite lubricant; 4 – plastic film to cover soil sample; 5 – measuring tape; 6 – mattock for digging soil sampling pit.

2.3 Laboratory analysis

In the laboratory, a knife was used to trim the soil from the ends to the exact size of the rings. This was used to determine the volume of soil and its weight. The scrapped soil materials were later used for physical (particle size distribution, soil particle density) and chemical (total soil organic carbon content) characterization of the soil. The soil particle-size distribution was determined by the pipette method (Day, 1965), by chemical dispersion with a 50 mL 0,1 N sodium hydroxide solution, in contact with the samples for 24 hours. Physical dispersion was accomplished by slowly rotating in a Wiegner mixer that shakes 30 times per minute, adding 20 g coarse sand (Grohmann & Raij, 1977). Soil particle density was determined by the pycnometer method (Blake & Hartge, 1986b). The total soil organic carbon content were determined by wet combustion with carbon oxidation adding 10 mL of digest solution ($Na_2Cr_2O_7$ $2H_2O$ 4 N + H_2SO_4 10 N) (Raij et al., 1987).

Three soil samples for each plot and at the sampled depths were saturated by capillary with distilled water, and equilibrated to a matric potential (Ψm) of - 2 and - 6 kPa, on a suction table (Romano et al., 2002) and - 10, - 33, - 100, - 500 and - 1500 kPa in a ceramic plate inside a pressure chamber (Soilmoisture Equipment Crop., P.O. Box 30025, Santa Barbara, CA 93105) (Dane & Hopmans, 2002). The soil-water retention data were fitted through the van Genuchten (1980) model with Mualen (1976) constraint. The - 6 kPa matric potential was used to separate the pores with effective diameter greater than 50 μm, drained from the cores (macropores). Water retained at this matric potential is considered as a measure of microporosity.

Precompression stresses were determined from the undisturbed soil samples submitted to uniaxial compression tests. The soil samples were kept within the sleeves of the coring cylinder, which were placed in the compression cell, and afterwards subjected pneumatically (Durham Geo Slope Indicator, USA, model S-450 Terraload®) to pressures 25, 50, 100, 200, 400, 800 and 1600 kPa to reach equilibrium (Bowles, 1986). During each test, a normal vertical stress was applied until 90% of the maximum deformation was reached and then the pressure is increased to the next level (Taylor, 1948). After uniaxial compression tests, the undisturbed soil samples were dried in the oven at 105–110 °C for 48 hours to determine the dry soil weight per unit volume, to calculate the soil bulk density (Blake and Hartge, 1986a). The volumetric total porosity (VTP) was estimated using the relationship between bulk density and particle density (Flint & Flint, 2002). Volumetric water content for each sample was also obtained

3. Soil physical properties

3.1 Bulk density and total soil organic carbon

The soils samples from the coffee-cultivated plots subjected to different weed management systems in the traffic line, had a higher bulk density and lower total soil organic carbon at the three layers studied, when compared to the soil samples from natural forest soil (Fig. 1A and 1B). These results indicated that land use with coffee plantation using different mechanical and chemical methods for weed control, increased the packing of the solids particles in soil thereby affecting the soil structural sustainability.

The bulk densities values from soil samples following post-emergence herbicide and mechanical mowing weed management systems at all the depths, and those from the rotary tilling managements (10–13 and 25–28 cm depths), coffee tandem disk harrowing and pre-emergence herbicide (0–3 and 10–13 cm depths) were considered higher than critical values for clay soils (1.2 Mg m^{-3}) in agreement with other studies including Derpsch et al. (1991); Dexter (2004); Severiano et al. (2011) and critical values for coffee root growth in Dystropherric Red Latosol (Araujo-Junior et al., 2011). The disk harrowing and pre-emergence herbicide weed management systems promote the crusting in the soil surface (Photos 1B and 1C) and increase the values of the bulk density (Fig. 1A).

After 30 years of conventional coffee cultivation, the total organic carbon contents were markedly affected by weed control between the coffee rows in the traffic line (Figure 1B). Total organic carbon contents were greater for native forest compared to the coffee plantation at all depths, except at 0–3 cm following mechanical mowing, which had the same total organic carbon (Fig 1A.) this is understandable considering that weed control with mechanical mower cut the weed in all the interrows and concentrate weed near the edge of the equipment increasing the total soil organic carbon in this region, where soil samples were collected.

The next highest contents of total organic carbon were found in the soils samples from hand-hoed (CAPM), post-emergence herbicide (HPOS), rotary tilling (ENRT) followed by no-weed control (SCAP), disk harrow (GRAD), and lowest was found in the soil from pre-emergence herbicide (Figure 1B). This low organic carbon condition was obviously due to the lack weed on the soil surface in the pre-emergence herbicide management system in agreements with other reports from tropical soil environments (Faria et al., 1998; Alcântara & Ferreira, 2000b; Araujo-Junior et al., 2011).

Published results reveal that weedy soil covers between coffee rows had great influences on the dynamics of total organic carbon content. Plant residues may influence the light soil fraction and thus the organic carbon content as reported by Ding et al. (2006) when these authors assessed the effect of cover crop management on chemical and structural composition of soil organic matter. The constant use of the pre-emergence herbicide for weed control in Dystroferric Red Latosol clay decreases significantly the total organic carbon content in the soil surface, because of the prevalence of soil without weed between the coffee rows. The effect of weed control with pre-emergence herbicide on total soil organic carbon was observed also in the 10–13 cm layer due the absence of weed roots (Figure 1B).

The different weed management system applied to coffee interrows influenced the soil bulk density and organic carbon content of the Latosol, in the 25-28 cm layer (Fig. 1A and 1B), when compared with the soil under natural forest (NAFT); however, when the soil samples were collected in center of the interrows, differences were not observed (Araujo-Junior et al., 2011). These authors observed that different weed management systems used in the interrows did not influenced soil bulk density and total organic carbon content of the Latosol, in the 25-28 cm layer, compared to the soil under natural forest. In our study, it is important highlight that the soil samples were collected in the traffic line of machines, and the total soil organic carbon content did not differ among the weed management systems in coffee plantation at the 25-28 cm depth (Figure 1B). However, Latosol samples from natural forest had greater total organic carbon content when compared to the soil under the different weed management system in coffee plantation. It has been proposed that the conservation of soil organic matter is an essential to protection soil against compaction (Etana et al., 1997; Dexter, 2004; Zhang et al., 2005; Araujo-Junior et al., 2011).

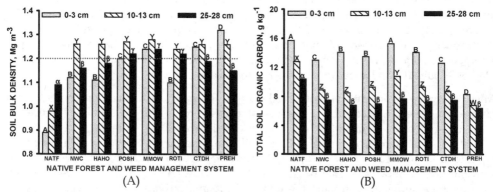

Fig. 1. Soil bulk density (A) and total soil organic carbon (B) of a Dystroferric Red Latosol in 0–3, 10–13 and 25–28 cm layers, affected by different weed management between coffee rows. NATF: natural forest; NWC: no-weed control between coffee rows; POSH: post-emergence herbicide; MMOW: mechanical mower; ROTI: rotary-tilling; CTDH: coffee tandem disk harrow; PREH: pre-emergence herbicide. Mean followed by equal letters compare the layers in the same weed management, and uppercase letters among the managements in the same depth of sampling, were not different, at 5% probability by the Scott-Knott test. Letters A to D compare 0-3 cm, X and Y compare managements at the 10-13 cm and Greek letters 25–28 cm depths. The red horizontal dotted line represents the critical soil bulk density for coffee root growth and soil structure sustainability estimate by Araujo-Junior et al. (2011) based on soil compression curves.

3.2 Total porosity and pore size distribution

Figure 2 shows the total porosity and pore size distribution of the Dystroferric Red Latosol (Oxisol) under native forest compared with the samples from the coffee plantation under different weed management system. We observed that samples taken from natural forest in the 0–3 cm depth have a higher total porosity (0.73 cm³ cm⁻³), macroporosity (0.44 cm³ cm⁻³) and lower microporosity (0.29 cm³ cm⁻³) when compared to the soil in different weed management system in the coffee plantation. For other depths (10–13 cm and 25–28 cm), the Latosol total porosity and pore size distribution were not different under natural forest and coffee plantation in the different weed management systems. Studies have been shown that under native forest the most Latosols found in Brazil with the gibbsite minerals content and high hematite contents on the clay fraction have percentage of macropores higher than 20% (Kemper & Derpsch, 1981; Ferreira et al., 1999; Oliveira et al., 2003a,b; Ajayi et al., 2009; Severiano et al., 2011).

Macropores are the pores in the soil in which water percolates due to gravity and their number is also measure of soil compaction (Kemper & Derpsch, 1981). In addition, macropores facilitates gas movement, thus it relates to the ability of the soil both to store and to transport gas (Stepniewski et al., 1994). These authors concluded that macroporosity of 25% (v/v) provides good aeration while in the 10–25% (v/v) range, there may be a limitation to gas exchange under certain conditions and that air-filled porosities < 10% (v/v) are characteristic of deficient aeration.

The lowest macroporosities (0.08 cm³ cm⁻³) in the 0–3 cm depth (pores with effective diameter greater than 50 µm, drained from cores) were observed for the samples under mechanical mowing and coffee tandem disk harrowing weed management system (Figure 2). The soil compaction process reduces the large pores in size first (Hillel, 1980; Dexter, 2004; Pires et al., 2008; Ajayi et al., 2009; Severiano et al., 2011).

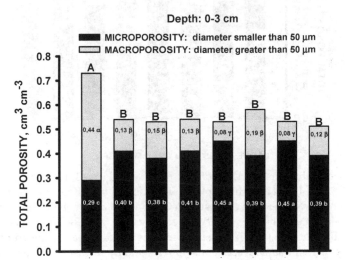

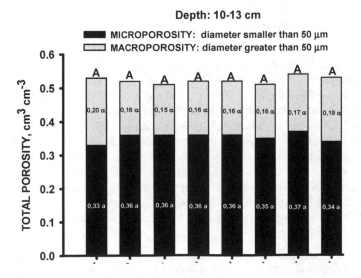

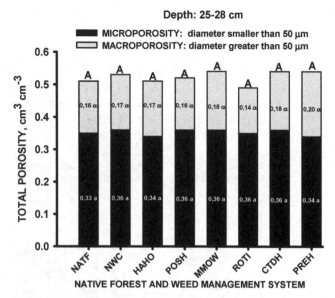

Depth: 25-28 cm

Fig. 2. Pore size distribution for a Dystroferric Red Latosol in 0–3, 10–13 and 25–28 cm layers, under natural forest and coffee plantation affected by different weed management between coffee rows. NATF: native forest; NWC: no-weed control between coffee rows; POSH: post-emergence herbicide; MMOW: mechanical mower; ROTI: rotary-tilling; CTDH: coffee tandem disk harrow; PREH: pre-emergence herbicide. Mean followed by equal letters compare the layers in the same weed management, and uppercase letters among the managements in the same depth of sampling, were not different, at 5% probability by the Scott-Knott test.

3.3 Soil-water retention curve

The soil-water retention curve defines the relationship between the soil matric potential and soil volumetric water content (Figure 3). This relationship may also assess the effect of weed management practices on soil structure. The differences between water retention behaviour for the soil samples collected at the interrows (center of the coffee rows, non-tracked soil) and the traffic line (wheel-tracked soil) at the 0 to 3 cm depth suggests that these curves are influenced by soil structure. The saturated water content (0.57 cm^3 cm^{-3}) for retention curve for traffic line decreased as a consequence of destruction of large pores or structural pores. On the other hand, the non-tracked interrow soil water retention curve revealed higher saturated water content (0.66 cm^3 cm^{-3}). As stated earlier, the large pores can be transformed into smaller pores and thus increase the soil-water holding capacity in low matric potential (- 1500 kPa). In this study, residual water content or water content at permanent wilting point (- 1500 kPa) increased in 0.04 cm^3 cm^{-3} in the traffic line as compared to interrows (Figure 3).

Recently, Dexter (2004) proposed to calculate the soil water retention curve parameters at inflection point (slope at inflection point, S-index) to assess soil physical quality. This author showed that the slope at inflection point governs directly many of the principal soil physical quality and is a measure of soil microstructure that can be used as an index of soil physical quality. According to Pires et al. (2008) soil compaction decreases large pores followed by a rising amount of small pores, that committing soil physical quality decreases the S-index values (Dexter, 2004). They showed that large values for S-index indicating good soil physical quality and presence of structural pores.

Based on soil water retention curve behaviors for a Eutric Nitossol (430 g kg⁻¹ clay) under coffee plantation Pires et al. (2008) assessed the effect of wetting and drying cycles. They found that the wetting and drying treatments did not affect the S-index for this soil. However, they showed that for the other soils S-index were affecting for the wetting and drying cycles.

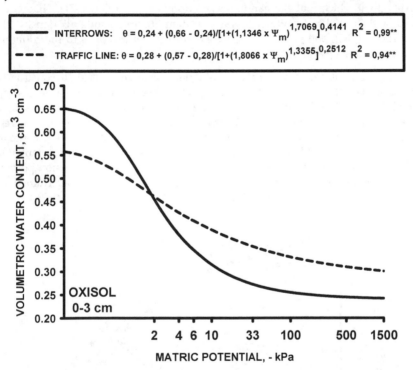

Fig. 3. Soil water retention curves for a Dystroferric Red Latosol in 0–3 cm in two sampling position interrows (no-wheel tracked soil) and traffic line (wheel-tracked soil).

3.4 Soil compressive behavior and load bearing capacity models

The soil compression curve is a conceptual and interpretative tool by which the compressive behaviour of the soil can be understood. The soil compression curve or stress-deformation curve can be described as a measure of soil deformation under given external loads (Holtz & Kovacz, 1981) (Figure 4) and defines the relationship between the logarithm of applied normal stress on the top of the sample and some parameter related to the packing state of soil; for example soil void ratio or soil bulk density (Casagrande, 1936; Larson et al., 1980; Holtz & Kovacs, 1981; Horn, 1988; Dias Junior & Pierce, 1995). This curve is divided into two regions so-called: a region of plastic and unrecoverable deformation called the virgin compression curve, and a region of small, elastic and recoverable deformation called the secondary compression curve (Larson et al., 1980; Holtz & Kovacs, 1981; Dias Junior & Pierce, 1995; Gregory et al., 2006). The point that separates these two regions in a compression curve is the precompression stress or preconsolidation pressure (σp) depending on if air or water is being eliminated from the soil, and can be variously defined.

In this study, we assumed, the precompression stress as indicator of internal strength of soils, which resulted from pedogenetic processes, anthropogenic effects, or hydraulic site-specific conditions (Horn et al., 2004) the maximum vertical overburden stress that particular sample has sustained in the past (Holtz & Kovacs, 1981) or as a predictor of the critical strength at which root elongation ceases (Römkens & Miller, 1971). This parameter is influenced by the initial soil volumetric water content (θ), initial soil bulk density (Bd), total organic carbon (TOC), soil structure and stress history, as it relates to the different weed management in coffee plantation.

The stress in a logarithmic scale versus strain data were then used to construct the soil compression curves (Larson et al., 1980), from which the precompression stress (σp) were determined (Figure 4) following the procedure of Dias Junior & Pierce (1995). In this procedure, precompression stress was estimated as the intersection of two lines: the regression line obtained for the first two (for soil samples with initial volumetric water content higher than matric potential – 100 kPa) or four points (for soil samples with matric potential lower or equal – 100 kPa) of the applied stress sequence in the secondary compression portion of the compression curve and the extension of the virgin compression line determined from the points associated with applied stress of 800 and 1600 kPa (Figure 4).

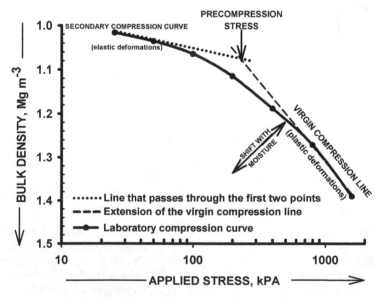

Fig. 4. Soil compression curve illustrating the position of the precompression stress
Source: "From Dias Junior, 1994"

Soil load bearing capacity has been defined as the capability of a soil structure to withstand stresses induced by field traffic without changes in the three-dimensional arrangement of its constituent soil particles (Alakukku et al., 2003). Soil load bearing capacity models (LBC) represents mathematically the relationship between soil volumetric water content (θ) and soil precompression stress (σp) and may be described by the Equation 1 (Dias Junior, 1994). In this model, the precompression stress decreases exponentially with the increases in the volumetric soil water content.

$$\sigma_p = 10^{(a + b\theta)} \tag{1}$$

Where, precompression stress (σp), estimated linear "a" and angular "b" coefficients and θ the initial volumetric soil water content. All the models obtained for the Dystroferric Red Latosol were significant at 1% probability level, for t-Student test and the coefficient of determination (R^2) ranged from 0.75 to 0.96 (Table 2).

The estimated linear "a" and angular "b" coefficients of the load bearing capacity models values varied from 2.57 for the soil under native forest at 0–3 cm depth to 2.89 for the soil samples collected from rotary tiller at 25–28 cm depth, and from -1.60 for the soil samples under pre-emergence herbicide at 25–28 cm depth, to - 0.71, for the soil samples collected from native forest at 0–3 cm depth (Table 2). Others studies done in Brazilian Latosols and Ultisols (Silva & Cabeda, 2006; Oliveira et al., 2003a; Kondo & Dias Junior, 1999) are in agreement with this results, which found lowest linear coefficients for soils under native forest when compared to the soil under different tillage management. The soil samples collected from native forest presented lower soil bulk density, microporosity and higher total organic carbon content, total porosity and macroporosity (Figures 1 and 2) due to the lack of anthropogenic activity and stress history. These findings suggest that the fitted parameter, "a" is interrelated to the packing of the solid particles expressed by soil bulk density and air-filled porosity (macropores) which affect the pore water pressure.

In all the models, the dependence of soil precompression stress on the water content in the soil was displayed. It was observed that the strength of the Latosol soil samples reduces although not linearly, with increases in the water content of the soil. The observation was consistent with results from several studies on the strength of soil samples (Kondo & Dias Junior, 1999; Peng et al., 2004; Dias Junior et al., 2005; Araujo-Junior et al., 2008, 2011).

Reported results from soil samples from three Ultisols under subtropical climate, Peng et al. (2004) also suggested that precompression stress decreases in exponential way with the initial water content. These authors suggest that the parameter "a" indicates the intrinsic strength of dry soil and the parameter "b" influences of soil properties such as soil texture and organic matter on the soil strength.

3.4.1 Influence of weed management system on soil load bearing capacity

To assess the influence of the adoption of different weed management on soil load bearing capacity, undisturbed soil samples collected from native forest and coffee plantation submitted to different weed management system were subjected to uniaxial compression test to obtain the soil compression curves. This load bearing capacity model was used to verify possible effects of different weed management systems on soil structure. This model is based on stress history or either, of the stress and other changes that have occurred during their history, and these changes are preserved in the soil structure (Casagrande, 1932 cited by Holtz & Kovacs, 1932).

Native forest and weed management	a	b	R^2	n
Depth: 0–3 cm				
Native forest	2,57	- 0,71	0,80**	15
No-weed control between coffee rows	2,65	- 1,26	0,96**	15
Hand hoe	2,82	- 1,56	0,84**	15
Post-emergence herbicide	2,72	- 0,92	0,92**	15
Mechanical mower	2,86	- 1,19	0,83**	15
Rotary-tilling	2,74	- 1,14	0,79**	14
Coffee tandem disk harrow	2,73	- 0,84	0,77**	15
Pre-emergence herbicide	2,78	- 1,35	0,86**	15
Depth: 10-13 cm				
Native forest	2,61	- 0,90	0,77**	15
No-weed control between coffee rows	2,77	- 1,26	0,84**	15
Hand hoe	2,77	- 1,05	0,86**	15
Post-emergence herbicide	2,77	- 1,43	0,87**	15
Mechanical mower	2,79	- 1,37	0,82**	15
Rotary-tilling	2,82	- 1,24	0,81**	15
Coffee tandem disk harrow	2,71	- 0,92	0,77**	15
Pre-emergence herbicide	2,83	- 1,49	0,78**	14
Depth: 25-28 cm				
Native forest	2,66	- 1,11	0,90**	14
No-weed control between coffee rows	2,66	- 0,93	0,82**	15
Hand hoe	2,76	- 1,40	0,94**	15
Post-emergence herbicide	2,86	- 1,51	0,86**	15
Mechanical mower	2,80	- 1,26	0,75**	15
Rotary-tilling	2,89	- 1,45	0,83**	15
Coffee tandem disk harrow	2,76	- 1,27	0,84**	14
Pre-emergence herbicide	2,81	- 1,60	0,83**	14

Table 2. Linear (a) and angular (b) coefficients of the load bearing capacity models $[\sigma_p = 10^{(a + b\theta)}]$, with respective coefficients of determination (R^2), and number of undisturbed soil samples (n) collected at 0–3, 10–13 and 25–28 cm depths in the traffic line in a Dystroferric Red Latosol (Oxisol) under native forest and coffee plantation submitted to different weed management systems.

The load bearing capacity models of the sample collected from different land uses (native forest and coffee plantation), but at different depths, and those of the various weed management systems were compared in multiple scatter plots (Fig. 5 - 7) and using the test of homogeneity for comparison of regression lines (Snedecor & Cochran, 1989). In the multiple scatter plots, the entire soil moisture and the corresponding preconsolidation value data in the different sites are pulled together on a single graph. For the homogeneity test, two models are picked and compared together by examining the intercept (a), slope (b) and the homogeneity parameter data (F). To obtain a and b values in each model for comparison, the model equation in the exponential form (Eq. 1) was transformed into a linear model by computing the logarithm of both sides of the equation giving equation of the form (Eq. 2) (Dias Junior et al., 2005; Araujo-Junior et al., 2011).

$$\log \sigma_p = \log 10^{(a + b\theta)} = \log \sigma_p = a + b\theta \tag{2}$$

We observed that soils under natural forest and no-weed control exhibited the lowest load bearing capacities at the 0-3 cm depth when compared with those under the varied weed management system used in coffee plantation (Figure 5 to 7). This observation can be associated with initial soil bulk density and soil organic carbon content (Figure 1A and 1B) and be associated with the absence of stress history and anthropogenic activities on the soil under native forest. On the other hand, the weed control using mechanical mower exhibited the highest load bearing capacity at that depth (Figure 5). The final results are presented in Fig. 5 for the models of the sample collected from different weed management systems at depth 0-3 cm depth. Homogeneity tests of the regression equations (Snedecor & Cochran, 1989) indicated that the soil under hand hoeing and pre-emergence herbicide weed management; post-emergence herbicide and coffee tandem disk harrow weed management had the similar load bearing capacities at the 0-3 cm depth (Table 3). Therefore, the dataset of the homogeneous models were combined and a new equation was fitted to each data set, considering all the values of preconsolidation pressure and volumetric soil water content for these treatments (Figure 5). Generally, it was observed that the load bearing capacity for the Dystroferric Red Latosol under the different weed management systems at the soil surface(0-3 cm depth) decreases in a following order: mechanical mower > post-emergence herbicide = coffee tandem disk harrow > rotary tiller > hand hoeing = pre-emergence herbicide > natural forest > no-weed control (Figure 5). The highest soil load bearing capacity was observed for the Latosol under mechanical mower in 0-3 cm depth (Fig. 5). Others studies, have been shown that high traffic intensity necessary to satisfactory weed control in coffee plantation throughout the year (5 to 6 times) increases the risk of soil compaction (Silveira & Kurachi, 1984; Alcântara & Ferreira, 2000b; Silva et al., 2006) mainly in the rainy season (October to March) when the soils has high soil water content and consequently lower load bearing capacity (Silva et al., 2006) increases the soil susceptibility to compaction. On the other hand, when soil is drier present higher resistance to compression and high load bearing capacity that decreases soil susceptibility to compaction (Dias Junior et al., 2005; Araujo-Junior et al., 2008; 2011).

Our results suggested the mechanical mower had a greater potential for causing soil compaction due to high traffic intensity to satisfactory weed control through the year (5 operations) and this operation must be accomplished when the soil has water content lower than 0.30 cm^3 cm^{-3} to minimize or avoid additional soil compaction.

MANAGEMENT WEED SYSTEM		F	
	F	Angular coefficient, b	Intercept of regression, a
Depth: 0-3 cm			
HAND HOE vs PRE-EMERGENCE HERBICIE	H	ns	ns
HAND HOE and PRE-EMERGENCE HERBICIDE vs ROTARY-TILLING	H	*	ns
HAND HOE and PRE-EMERGENCE HERB. vs MECHANICAL MOWER	H	**	ns
HAND HOE and PRE-EMERGENCE HERBICIDE vs NATIVE FOREST	H	**	ns
HAND HOE and PRE-EMERGENCE HERBICIDE vs NO-WEED CONTROL	H	*	*
POST-EMERGENCE HERBICIDE vs DISK HARROW	H	ns	ns
POST-EMERGENCE HERBICIDE and DISK HARROW vs ROTARY-TILLING	H	ns	**
POST-EMERGENCE HERBICIDE and DISK HARROW vs MECHANICAL MOWER	H	*	ns
POST-EMERGENCE HERBICIDE and DISK HARROW vs HAND HOE and PRE-EMERGENCE HERBICIDE	H	**	**
POST-EMERGENCE HERBICIDE and DISK HARROW vs NATIVE FOREST	H	*	**
POST-EMERGENCE HERBICIDE and DISK HARROW vs NO-WEED CONTROL	H	ns	**
MECHANICAL MOWER vs ROTARY-TILLING	H	**	**
MECHANICAL MOWER vs NO-WEED CONTROL	H	ns	**
NATIVE FOREST vs NO-WEED CONTROL	H	**	ns
NATIVE FOREST vs ROTARY-TILLING	H	*	ns
NATIVE FOREST vs MECHANICAL MOWER	H	**	**

H: homogeneous; ** significant at 1 % probability level; * significant at 5 % probability level; ns: not significant

Table 3. Comparison of the load bearing capacity models for homogeneity of a Dystroferric Red-Latosol at 0-3 cm depth under native forest and in a coffee plantation submitted to different weed management systems

According to Yang et al. (2007) the weed control in an orchard citrus by mowing three times during the growing season could improve soil and mitigate negative effects of weeds on crops. In the study by Zhang et al. (2006), it was observed that the first three passes of the tractor caused the largest increments in the mechanical resistance of the soil in the first 12cm depth. In conservation tillage systems, no - till management promotes higher soil organic carbon content and contribute to aggregate stability under loading, due to improved structural stability (Silva & Cabeda, 2006). Similarly, others authors have shown that increases in the soil organic carbon content reduces the adverse effects of soil compaction

(Etana et al., 1997) while increasing compressibility due to higher soil resilience (Zhang et al., 2005).

The hand hoeing, pre-emergence herbicide and rotary tilling weed management systems load bearing capacities models were intermediate in the behaviour for the studied depth relative to mechanical mowing (highest) and no weed control between coffee rows (lowest). At this depth, our results for the load bearing capacity models were similar to the obtained by Kurachi & Silverira (1984) starting from medium profiles of mechanical resistance of the profile of the soil under different weed management systems. These authors also observed that the mechanical mower was the implement that impact more on the soil strength, followed by the herbicide sprayer and the rotary tilling.

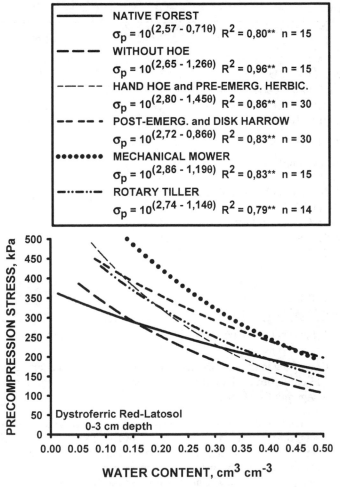

Fig. 5. Load bearing capacity models of a Dystroferric Red Latosol in 0–3 cm layer, cultivated with coffee plants affected by different weed management in interrows of the coffee plantation.

The homogeneity tests of the regression equations for the samples collected in the 10-13 cm depths showed that there were two homogeneous dataset. The mechanical mowing, pre-emergence herbicide, no-weed control and post-emergence herbicide; and rotary-tilling exhibited similarity, while hand hoeing, and coffee tandem disk harrowing were similar (Table 4). Therefore, for each homogeneous dataset, a new equation was fitted, combining all the values of preconsolidation pressure and volumetric soil water content (Figure 6).

MANAGEMENT WEED SYSTEM	F	F	
		Angular coefficient, b	Intercept of regression, a
Depth: 10-13 cm			
MECHANICAL MOWER vs PRE-EMERGENCE HERBICIDE	H	ns	ns
MECHANICAL MOWER and PRE-EMERGENCE HERBICIDE vs NO-WEED CONTROL	H	ns	ns
MECHANICAL MOWER and PRE-EMERGENCE HERBICIDE and NO-WEED CONTROL vs POST-EMERGENCE HERBICIDE	H	ns	ns
ROTARY-TILLING vs HAND HOE	H	ns	ns
ROTARY-TILLING and HAND HOE vs DISK HARROW	H	ns	ns
ROTARY-TILLING and HAND HOE and DISK HARROW vs NATIVE FOREST	H	**	**
MECHANICAL MOWER and PRE-EMERGENCE HERBICIDE and NO-WEED CONTROL vs POST-EMERGENCE HERBICIDE vs NATIVE FOREST	H	**	ns
MECHANICAL MOWER and PRE-EMERGENCE HERBICIDE and NO-WEED CONTROL vs POST-EMERGENCE HERBICIDE vs ROTARY-TILLING and HAND HOE and DISK HARROW	H	*	**

H: homogeneous; ** significant at 1 % probability level; * significant at 5 % probability level; ns: not significant

Table 4. Comparison of the load bearing capacity models for homogeneity of a Dystroferric Red-Latosol at 10-13 cm depth under native forest and in a coffee plantation submitted to different weed management systems.

In general, at 10-13 cm depth the load bearing capacity models for studied area under varying weed management systems were similar and decreased in the following order: hand hoeing = rotary tilling = coffee tandem disk harrow > no-weed control = post-emergence herbicide = mechanical mower = pre-emergence herbicide > natural forest (Figure 6). These responses are associated with lowest soil bulk density value and the greatest soil organic carbon content of the soil under natural forest (Figure 1A and 1B). The lack of anthropogenic activities in the soil under natural forest provides the greater soil organic carbon content and smaller values of soil bulk density, which contribute to smaller values of precompression stress consequently, smaller load bearing capacity at all soil water content. The weed management systems of hand hoeing, rotary tilling and coffee tandem disk harrow had higher soil load bearing capacity at all soil water content (Figure 6). The disturbed soil on soil surface for these weed management favor the stress distribution to 16-21 cm depth (Araujo-Junior et al., 2011), increases the soil load bearing capacity of the samples at the 10-13 cm depth, being the area mainly affected by the distributed stresses (Figure 6).

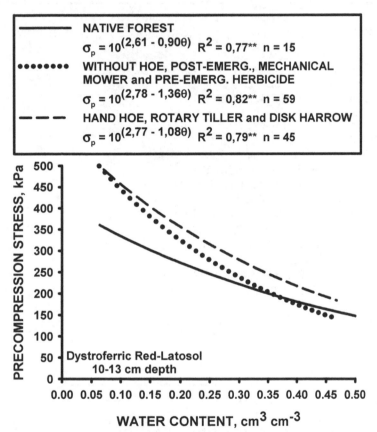

Fig. 6. Load bearing capacity models of a Dystroferric Red Latosol in 10–13 cm layer, cultivated with coffee plants affected by different weed management in interrows of the coffee plantation.

At the 25-28 cm depth, the weed management systems sets consisting of mechanical mowing , post-emergence herbicide and rotary tilling; hand hoeing, pre-emergence herbicide and coffee tandem disk harrow; resulted in homogenous load bearing capacity models (Table 5). Therefore, for each homogeneous set, the data set consisting all the values of preconsolidation pressure and volumetric soil water content were combined and a new equation was fitted (Figure 7). We observed that the load bearing capacity of the soils were similar and decreased in the following order: post-emergence herbicide = mechanical mower = rotary tilling > hand hoeing = pre-emergence herbicide = coffee tandem disk harrow > no-weed control > natural forest (Figure 7).

MANAGEMENT WEED SYSTEM		F	
	F	Angular coefficient, b	Intercept of regression, a
Depth: 25–28 cm			
MECHANICAL MOWER vs POST-EMERGENCE HERBICIDE	H	ns	ns
MECHANICAL MOWER and POST-EMERGENCE HERBICIDE vs ROTARY-TILLING	H	ns	ns
HAND HOE vs PRE-EMERGENCE HERBICIDE	H	ns	ns
HAND HOE and PRE-EMERGENCE HERBICIDE vs DISK HARROW	H	ns	ns
HAND HOE and PRE-EMERGENCE HERBICIDE and DISK HARROW vs NO-WEED CONTROL	NH	**	**
HAND HOE and PRE-EMERGENCE HERBICIDE and DISK HARROW vs NATIVE FOREST	NH	**	ns
HAND HOE and PRE-EMERGENCE HERBICIDE and DISK HARROW vs MECHANICAL MOWER and POST-EMERGENCE HERBICIDE and ROTARY-TILLING	NH	**	**
MECHANICAL MOWER and POST-EMERGENCE HERBICIDE and ROTARY-TILLING vs NO-WEED CONTROL	H	**	*
MECHANICAL MOWER and POST-EMERGENCE HERBICIDE and ROTARY-TILLING vs NATIVE FOREST	H	**	*
NATIVE FOREST vs NO-WEED CONTROL	H	ns	**

H: homogeneous; ** significant at 1 % probability level; * significant at 5 % probability level; ns: not significant

Table 5. Comparison of the load bearing capacity models for homogeneity of a Dystroferric Red-Latosol at 25-28 cm depth under native forest and in a coffee plantation submitted to different weed management systems.

The weed management systems consisting of post-emergence herbicide, mechanical mowing and rotary tilling resulted in most comparisons, higher soil load bearing capacity for the Latosol, indicating that the effect of the traffic of machines in mechanical weed control induced the compaction of the soil in sub-soil region. Kurachi & Silveira (1984) suggest that the weed managements systems that involve the disturbance of the soil had the tendency to increase compaction at the surface; when there is no disturbance, increase compaction is more accentuated starting from the depth of operation of the equipment. However, our result show that the herbicide applicator and mechanical mower as well as rotary tilling increased the soil's mechanical resistance in the moisture levels of 15 cm^3 cm^{-3} and 20 cm^3 cm^{-3}, when compared to hand hoeing. Looking at data presented in Figure 7, it is possible to conclude that, even with the absence of mechanical soil disturbance weed management systems, the soil can still be compacted when wet, when stresses travel up to a depth of 25-28 cm (Figure 7).

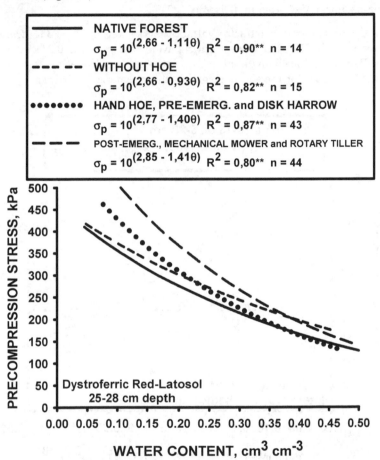

Fig. 7. Load bearing capacity models of a Dystroferric Red Latosol in 25–28 cm layer, cultivated with coffee plants affected by different weed management in interrows of the coffee plantation.

3.4.2 Critical volumetric soil water content for traffic of tractor based on soil load bearing capacity

According to Hillel (1980) soil moisture is the most important soil physical properties to determine soil-machine interactions. This soil physical property also, governs soil deformation when submitted to external loads (Dias Junior, 1994; Dias Junior & Pierce, 1996). To determine the critical volumetric soil water content (θcritical) for traffic of machines and tools, we considered only those stress that can cause additional soil compaction or change the initial state of the soil structure, and are considered that stress do not exceed internal strength expressed by precompression stress (Araujo-Junior et al., 2011). The maximum vertical stress exerted by the tractor and equipments (σ_{max}) and the stress distribution in various wheeled and soil conditions were obtained using the Tyres/Tracks and Soil Compaction-TASC program (Diserens, 2005).

The maximum stress exerted by a tractor Valmet® model 68 was 220 kPa for front tyres 6-16 inflation pressure 172 kPa. The lowest critical water content was 0.27 cm³ cm⁻³ for the Dystroferric Red Latosol in the without hoe no inter-rows control at the 0–3 cm depth and the higher 0.48 cm³ cm⁻³ for the soil managed with pre-emergence herbicide in the 0–3 cm layer.

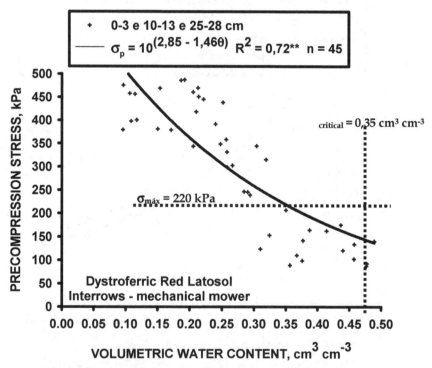

Fig. 8. Soil load bearing capacity models of a Dystroferric Red Latosol in 0–3, 10–13 and 25–28 cm layers, cultivated with coffee plants affected by different weed management in interrows in coffee plantation. ROÇA: mechanical mower. The dotted vertical line represents critical water content (θcritical) for tractor traffic above the soil under mechanical mower management. The dotted horizontal line represents the maximum vertical stress exerted by a tractor (σmax).

Our results show that load bearing capacity models might be useful to assess the effect of the weed management on soil strength or inherent ability of the soil samples to withstand applied pressure without degrading their structure. Also, this soil mechanic approach could be used to define the optimum moisture content for machine traffic without degrading the soil structure.

4. Conclusions

Our results reveal that the weed management system and traffic by machines had a great influence on soil physical quality attributes, mainly on the surface soil (0–3 cm depth) on the inherent strength. The greatest changes in the Latosol structure were observed under mechanical mowing, disk harrowing and pre-emergence herbicide weed management. These observations are related to the applied stress by the machines and direct raindrop impacts to bare soil systems that favored crust formation, thereby increasing the soil strength on the soil surface. In addition, weed control practices that result in the total removal of the soil cover was more prone to compaction due to applied soil stress by machines and equipments.

The soil load bearing capacity and the water content at the time of the traffic machines are the most important soil physical properties; thus these attributes must be considered to minimize additional soil compaction and soil structure damage on coffee plantations under different weed management systems. Recommendations for the sustainable weed management system in coffee plantation must consider the inherent internal strength of the soil expressed by precompression stress.

5. Acknowledgments

The authors are grateful to Brazilian Consortium for Research and Coffee Development (CBP&D - Café) provided financial support for this study and CAPES agency a governmental in scholarship to Dr. Cezar Francisco Araujo Junior.

6. References

Ajayi, A. E.; Dias Junior, M. de S.; Curi, N.; Araujo-Junior, C. F.; Souza, T. T. T. & Inda Junior, A. V. (2009). Strength attributes and compaction susceptibility of Brazilian Latosols. *Soil & Tillage Research*, Vol. 105, No. 1, (September 2009), pp. 122–127, ISSN 0167-1987

Alakukku, L.; Weisskopf, P.; Chamen, W. C. T.; Tijink, F. G. J.; van der Linden, J. P.; Pires, S.; Sommer, C. & Spoor, G. (2003). Prevention strategies for field traffic-induced subsoil compaction: a review Part 1. Machines/soil interactions. *Soil & Tillage Research*, Vol. 73, No. 1/2, (October 2003), pp. 145–160, ISSN 0167-1987

Alcântara, E. N. & Ferreira, M. M. (2000a). Efeito de diferentes métodos de controle de plantas daninhas sobre a produção de cafeeiros instalados em Latossolo Roxo distrófico. (In Portuguese, with English abstract). *Ciência & Agrotecnologia*, Vol.24, No.1, (January 2000), pp. 54–61, ISSN 1413-7054.

Alcântara, E. N. & Ferreira, M. M. (2000b). Efeitos de métodos de controle de plantas daninhas na cultura do cafeeiro (*Coffea arabica* L.) sobre a qualidade física do solo. (In Portuguese, with English abstract). *Revista Brasileira de Ciência do Solo*, Vol.24, No.4, (October to December 2000), pp. 711–721, ISSN 1806-9657

Araujo-Junior, C. F.; Dias Junior, M. de S.; Guimarães, P. T. G. & Pires, B, S. (2008). Resistência à compactação de um Latossolo cultivado com cafeeiro, sob diferentes sistemas de manejos de plantas invasoras. (In Portuguese, with English abstract). *Revista Brasileira de Ciência do Solo*, Vol.32, No.1, (January and February 2008), pp. 25–32, ISSN 0100-0683.

Araujo-Junior, C. F.; Dias Junior, M. de S.; Guimarães, P. T. G. & Alcântara, E. N. (2011). Capacidade de suporte de carga e umidade crítica de um Latossolo induzida por diferentes manejos. (In Portuguese, with English abstract). *Revista Brasileira de Ciência do Solo*, Vol.35, No.1, (January and February 2011), pp. 115–131, ISSN 0100-0683.

Blake, G. R.; Hartge, K. H. (1986a). Bulk density, In: *Methods of soil analysis. Part 1. 2 nd ed*, Klute, A., pp. 363–375, American Society of Agronomy/Soil Science Society of America, Madison, Wisconsin, USA.

Blake, G. R.; Hartge, K. H. (1986b). Partycle density, In: *Methods of soil analysis. Part 1. 2 nd ed.* Klute, A., pp. 377–382, American Society of Agronomy/Soil Science Society of America, Madison, Wisconsin, USA.

Bowles, J. E. (1986). *Engineering properties of soils and their measurements*. Third edition. McGraw-Hill, 1986. 218 p

Brazil Specialty Coffe Association – BSCA. (2005). Associação Brasileira de Cafés Especiais. Lista de Verificação Sistemas de Gestão Sócio-Ambiental. Anexo RA 0552.04 ver. 01. 2005.

Carvalho, R.; Silva, M. L. N.; Avanzi, J. C.; Curi, N. & Souza, F. S. de. (2007). Erosão hídrica em Latossolo Vermelho sob diversos sistemas de manejo do cafeeiro no sul de Minas Gerais. (In Portuguese, with English abstract). *Ciência & Agrotecnologia*, Vol.31, No.6, (November and Dezember 2007), pp. 1679–1687, ISSN 1413-7054.

Casagrande, A. (1936). The determination of pre-consolidation load and its practical significance, *Proceedings International Conference on. Soil Mechanics Foundation*. Cambridge, June 1936.

Dane, J. H. & Hopmans, J. W. (2002). Pressure plate extractor, In: *Methods of soil analysis: physical methods*, Dane, J. H. & Topp, G. C., pp. 688–690, Soil Science Society of America, LCCN 2002109389, Madison, Wisconsin, USA.

Day, P. R. (1965). Particle fractionation and particle-size analysis, In: *Methods of soil analysis*. Black, C. A. et. al. (Ed.). pp. 545–567, n. 1, Part I.. (ASA. Monography of Agronomy, 9), American Society of Agronomy, Madison, Wisconsin, USA.

Derpsch, R.; Roth, C. H.; Sidiras, N. & Köpke, U. (1991). *Controle da erosão no Paraná, Brasil: sistemas de cobertura do solo, plantio direto e preparo conservacionista do solo*. Fundação Instituto Agronômico do Paraná e Deutsche Gesellschaft für Technische Zusammenarbeit (GTZ) GmbH. ISBN 3-88085-433-5, Eschborn, 1991.

Dexter, A. R. (2004). Soil physical quality. Part I. Theory, effects of soil texture, density, and organic matter, and effects on root growth. *Geoderma*, Vol.120, No., (March 2004), pp. 201–214, ISSN 0016-7061

Dias Junior, M. de S. (1994). *Compression of three soils under long-term tillage and wheel traffic*. 1994. 114 p. Tese (Doutorado) - Michigan State University, East Lansing.

Dias Junior, M. de S. & Pierce, F.J. (1995). A simple procedure for estimating preconsolidation pressure from soil compression curves. *Soil Technology*, Vol.8, No 2, (November 1995), pp. 139-151, ISSN 0933-3630.

Dias Junior, M. de S. & Pierce, F. J. (1996). Revisão de Literatura. O processo de compactação do solo e sua modelagem. (In Portuguese, with English abstract). *Revista Brasileira de Ciência do Solo*, Vol.20, No.1, (January to March 1996), pp. 175-182, ISSN 1806-9657.

Dias Junior, M. de S.; Leite, F. P.; Lasmar Júnior, E. & Araujo Junior, C. F. (2005) Traffic effects on the soil preconsolidation pressure due to eucalyptus harvest operations. *Scientia Agricola*, Vol.62, No.3, (May to June 2005), pp. 248-255, ISSN 0103-9016.

Ding, G.; Liu, X.; Herbert, S.; Novak, J.; Amarasiriwardena, D. & Xing, B. (2006). Effect of cover crop management on soil organic matter. *Geoderma*, Vol.130, No.3-4, (March 2006), pp. 229-239, ISSN 0016-7061

Diserens, E. (2005). TASC: tyres/tracks and soil compaction: a pratical tool to prevent soil compaction damage, MS Excel 2000. Zurich: Agroscope FAT Tänikon, 2005. 68 p. Manual.

Empresa Brasileira de Pesquisa Agropecuária - EMBRAPA. (2006). *Sistema brasileiro de classificação de solos*. (2nd Ed.) Embrapa Solos, ISBN, Centro Nacional de Pesquisas de Solos. Rio de Janeiro

Etana, A.; Comia, R. A. & Håkansson, I. (1997). Effects of uniaxial stress on the physical properties of four Swedish soils. *Soil & Tillage Research*, Vol. 44, No. 1/2, (December 1997), pp. 13-21, ISSN 0167-1987

Faria, J. C.; Schaefer, C. E. R.; Ruiz, H. A. & Costa, L. M. (1998). Effects of weed control on physical and micropedological properties of Brazilian Ultisol. *Revista Brasileira de Ciência do Solo*, Vol.22, pp. 731-741, ISSN 0100-0683

Ferreira, M. M.; Fernandes, B. & Curi, N. (1999). Influência da mineralogia da fração argila nas propriedades físicas de Latossolos da região sudeste do Brasil. (In Portuguese, with English abstract). *Revista Brasileira de Ciência do Solo*, Vol.23, No.3, (January to March 1999), pp. 515-524, ISSN 0100-0683

Flint, L. E. & Flint, A. L. (2002). Porosity. In: *Methods of soil analysis: physical methods*, Dane, J. H. & Topp, G. C., pp. 241-254, Soil Science Society of America, LCCN 2002109389, Madison, Wisconsin, USA.

Food and Agriculture Organization - FAO (2006). *World reference for soil resources 2006*: A framework for international classification, correlation and communication. FAO, ISBN, 92-5-105511-4, Rome.

Gregory, A. S.; Whalley, W. R.; Watts, C. W.; Bird, N. R. A.; Hallett, P. D. & Whitmore, A. P. (2006). Calculation of the compression index and precompression stress from soil compression test data. *Soil & Tillage Research*, Vol. 89, No. 1, (August 2006), pp. 45-57, ISSN 0167-1987

Grohmann, F. & van Raij, B. (1977). Dispersão mecânica e pré-tratamento para análise granulométrica de Latossolos argilosos. (In Portuguese, with English abstract). *Revista Brasileira de Ciência do Solo*, Vol.1, No.1, (January to April 1977), pp. 52-53, ISSN 1806-9657.

Hillel, D. (1980). Tillage and soil structure management, In: *Applications of soil physics*, Hillel, D., pp. 234–244, Academic, ISBN 0-12-348580-0, New York

Holtz, R. D. & Kovacz, W. D (1981). *An introduction to geotechnical engineering*. Prentice-Hall, ISBN 0-13-484395-0 Englewood Cliffs, Printed in the United States of America

Horn, R. (1988). Compressibility of arable land. *Catena*, Vol. 11, pp. 53–71, 1988. Supplement, ISSN 0341-8162

Horn, R.; Vossbrink, J.; Becker, S. (2004). Modern forest vehicles and their impacts on soil physical properties. *Soil & Tillage Research*, Vol. 79, No. 2, (December 2004), pp. 207–219, ISSN 0167-1987

Kemper, B. & Derpsch, R. (1981). Soil compaction and root growth in Parana, In: R. Scott Russell, Kozen Igue & Y. R. Mehta, pp. 81–101, *Proceeding of the symposium on the soil/root system*, Instituto Agronômico do Paraná – IAPAR, March, 1980.

Kondo, M. K. & Dias Junior, M. de S. (1999). Compressibilidade de três Latossolos em função da umidade e uso. (In Portuguese, with English abstract). *Revista Brasileira de Ciência do Solo*, Vol.23, No.2, (April to June 1999), pp. 211–218, ISSN 1806-9657

Kurachi, S. A. H. & Silveira, G. M. (1984). Compactação do solo em cafezal provocada por diferentes métodos de cultivo. (In Portuguese, with English abstract). Instituto Agronômico de Campinas, 28 p.

Larson, W. E.; Gupta, S. C. & Useche, R. A. (1980). Compression of agricultural soil from eight soil orders. *Soil Science Society of America Journal*, Vol.44, No. 3, (May to June 1980), pp. 450–457, ISSN 0361-5995

Marshall, T. J. (1962). The nature, development and significance of soil structure. In: Neale, G. J. (Ed.). *Transactions of joint meeting of comissions IV e V (ISSS)*. Palmerston North: New Zealand Society of Soil Science, 1962. p. 243–257.

Mualen, Y. (1976). A new model for predicting the hydraulic conductivity of unsaturated porous media. *Water Resources & Research*, Vol.12, No.4, (August 1976), pp.513–522, ISSN 0043-1397

Oliveira, G. C. de; Dias Junior, M. S. de; Resck, D. V. S. & Curi, N. (2003a). Alterações estruturais e comportamento compressivo de um Latossolo Vermelho distrófico argiloso sob diferentes sistemas de uso e manejo. (In Portuguese, with English abstract). *Pesquisa Agropecuária Brasileira*, Vol. 38, No. 2, (February 2003), pp. 291–299, ISSN 1678-3921

Oliveira, G. C. de; Dias Junior, M. S. de; Resck, D. V. S. & Curi, N. (2003b). Compressibilidade de um Latossolo Vermelho argiloso de acordo com a tensão de água no solo, uso e manejo. (In Portuguese, with English abstract). *Revista Brasileira de Ciência do Solo*, Vol.27, No.4, (September to October 2003), pp. 773–781, ISSN 0100-0683

Peng, X. H.; Horn, R.; Zhang, B. & Zhao, Q. G. (2004). Mechanisms of soil vulnerability to compaction of homogenized and recompacted Ultisols. *Soil & Tillage Research*, Vol. 76, No. 2, (April 2004), pp. 125–137, ISSN 0167-1987

Pires, L. F.; Cássaro, F. A. M.; Reichardt, K. & Bacchi, O. O. S. (2008). Soil porous system changes quantified by analyzing soil water retention curves modifications. *Soil & Tillage Research*, Vol. 100, pp. 72–77, ISSN 0167-1987

Raij, B. V.; Quaggio, J. A.; Cantarela, H.; Ferreira, M. E.; Lopes, A. S. & Bataglia, O. C. (1987). *Analise química do solo*. Fundação Cargil, São Paulo

Rodrigues, B. N. & Almeida, F. S. de. (2005). *Guia de herbicidas* (5th edition), Grafmarke, Londrina, Paraná, Brasil.

Romano, N.; Hopmans, J. W. & Dane, J. H. (2002). Suction table, , In: *Methods of soil analysis: physical methods*, Dane, J. H. & Topp, G. C., pp. 692–698, Soil Science Society of America, LCCN 2002109389, Madison, Wisconsin, USA.

Römkens, M. J. M. & Miller, R. D. (1971). Predicting root size and frequency from one-dimensional consolidation data – A mathematical model. *Plant and Soil*, Vol.35, No.1-3, pp. 237–248, ISSN 1573-5036

Severiano, E. da C.; Oliveira, G. C. de; Dias Junior, M. de S.; Costa, K. A. de P.; Silva, F. G. & Ferreira Filho, S. M. (2011). Structural changes in Latosols of the Cerrado region: I – Relationship between soil physical properties and least limiting water range. *Revista Brasileira de Ciência do Solo*, Vol.35, No.3, (May to June 2011), pp. 773–782, ISSN 0100-0683

Silva, A. J. N. da; & Cabeda, M. S. V. (2006). Compactação e compressibilidade do solo sob sistemas de manejo e níveis de umidade. (In Portuguese, with English abstract). *Revista Brasileira de Ciência do Solo*, Vol.30, No.5, (November to December 2006), pp. 921–930, ISSN 0100-0683

Silva, A. R.; Dias Junior, M. de S.; Guimarães, P. T. G. & Araujo-Junior, C. F. (2006). Modelagem da Capacidade de Suporte de Carga e Quantificação dos Efeitos das Operações Mecanizadas em um Latossolo Amarelo Cultivado com Cafeeiros. (In Portuguese, with English abstract). *Revista Brasileira de Ciência do Solo*, Vol.30, No.2, (March to April) pp. 207–216, ISSN 0100-0683

Silveira, G. M. da; Kurachi, S. A. H. & Fujiwara, M. (1985). Métodos mecânicos e químico no controle de ervas daninhas em cafezal. . (In Portuguese, with English abstract). *Bragantia*, Vol.44, No.1, (June, 1985) pp. 173–178, ISSN 0006-8705

Silveira, G. M. da & Kurachi, S. A. H. (1985). O sistema de cultivo e a estrutura do solo em cafezal. Parte II. (In Portuguese, with English abstract). *Bragantia*, Vol.44, No.1, (June, 1985) pp. 179–185, ISSN 0006-8705

Snedecor, G. W. & Cochran, W. G. (1989). Statical methods. (8th. edition), Ames: Iowa State University, 1989.

Soil Science Society of America.(2008). *Glossary of soil science terms*. ISBN 978-0-89118-851-3, Madison, 2008. 84 p.

Soil Survey Staff. (1998). *Keys to soil taxonomy* (8th ed), USDA-NRCS, ISBN 2-853552-261-X. Washington, DC

Stepniewski, W.; Gliński, J. & Ball, B. C. (1994). Effects of compaction on soil aeration properties. In: *Soil compaction in crop production*, Soane, B. D. & Ouwerkerk, C. van, pp. 45-69, Elsevier, ISBN 0-444-88286-3, Amsterdam

Taylor, D. W. (1948). *Fundamentals of soil mechanics*. John Wiley, ISBN, New York

Topp, G. C. & Ferré, P. A. (2002). Water content, In: *Methods of soil analysis: physical methods*, Dane, J. H. & Topp, G. C., pp. 417–424, Soil Science Society of America, LCCN 2002109389, Madison, Wisconsin, USA.

Uhland, R. E. (1949). Physical properties of soils as modified by crops and management. *Soil Science Society Proceedings*, (August 1949), pp. 361–366

van Genuchten, M. Th. (1980). A closed-form equation for predicting the hydraulic conductivity of unsaturated soils. *Soil Science Society of America Journal*, Vol.44, No. 5, (September 1980), pp. 892–898, ISSN 0361-5995

Vogeler, I.; Horn, R.; Wetzel, H. & Krümmbelbein, J. (2006). Tillage effects on soil strength and solute transport. *Soil & Tillage Research*, Vol. 88, No. 1/2, (July 2006), pp. 193–204, ISSN 0167-1987

Yang, Y.; Wang, H.; Tang, J. & Chen, X. (2007). Effects of weed management practices on orchard soil biological and fertility properties in southeatern China. *Soil & Tillage Research*, Vol. 93, No. 1, (March 2007), pp. 179–185, ISSN 0167-1987

Zhang, B.; Horn, R. & Hallet, P. D. (2005). Mechanical resilience of degraded soil amended with organic matter. *Soil Science Society of America Journal*, Vol.69, No. 3, (May 2005), pp. 450–457, ISSN 0361-5995

Zhang, X. Y.; Cruse, R. M.; Sui, Y. Y.; Jhao, Z. (2006). Soil compaction induced by small tractor traffic in northeast China. *Soil Science Society of America Journal*, Vol.70, No.1, (Feb. 2006), pp. 613–619, ISSN 0361-5995

Permissions

The contributors of this book come from diverse backgrounds, making this book a truly international effort. This book will bring forth new frontiers with its revolutionizing research information and detailed analysis of the nascent developments around the world.

We would like to thank Andrew J. Price, for lending his expertise to make the book truly unique. He has played a crucial role in the development of this book. Without his invaluable contribution this book wouldn't have been possible. He has made vital efforts to compile up to date information on the varied aspects of this subject to make this book a valuable addition to the collection of many professionals and students.

This book was conceptualized with the vision of imparting up-to-date information and advanced data in this field. To ensure the same, a matchless editorial board was set up. Every individual on the board went through rigorous rounds of assessment to prove their worth. After which they invested a large part of their time researching and compiling the most relevant data for our readers. Conferences and sessions were held from time to time between the editorial board and the contributing authors to present the data in the most comprehensible form. The editorial team has worked tirelessly to provide valuable and valid information to help people across the globe.

Every chapter published in this book has been scrutinized by our experts. Their significance has been extensively debated. The topics covered herein carry significant findings which will fuel the growth of the discipline. They may even be implemented as practical applications or may be referred to as a beginning point for another development. Chapters in this book were first published by InTech; hereby published with permission under the Creative Commons Attribution License or equivalent.

The editorial board has been involved in producing this book since its inception. They have spent rigorous hours researching and exploring the diverse topics which have resulted in the successful publishing of this book. They have passed on their knowledge of decades through this book. To expedite this challenging task, the publisher supported the team at every step. A small team of assistant editors was also appointed to further simplify the editing procedure and attain best results for the readers.

Our editorial team has been hand-picked from every corner of the world. Their multi-ethnicity adds dynamic inputs to the discussions which result in innovative outcomes. These outcomes are then further discussed with the researchers and contributors who give their valuable feedback and opinion regarding the same. The feedback is then collaborated with the researches and they are edited in a comprehensive manner to aid the understanding of the subject.

Apart from the editorial board, the designing team has also invested a significant amount of their time in understanding the subject and creating the most relevant covers. They scrutinized every image to scout for the most suitable representation of the subject and create an appropriate cover for the book.

The publishing team has been involved in this book since its early stages. They were actively engaged in every process, be it collecting the data, connecting with the contributors or procuring relevant information. The team has been an ardent support to the editorial, designing and production team. Their endless efforts to recruit the best for this project, has resulted in the accomplishment of this book. They are a veteran in the field of academics and their pool of knowledge is as vast as their experience in printing. Their expertise and guidance has proved useful at every step. Their uncompromising quality standards have made this book an exceptional effort. Their encouragement from time to time has been an inspiration for everyone.

The publisher and the editorial board hope that this book will prove to be a valuable piece of knowledge for researchers, students, practitioners and scholars across the globe.

List of Contributors

Lina Šarūnaitė, Irena Deveikytė and Žydrė Kadžiulienė
Institute of Agriculture, Lithuanian Research Centre for Agriculture and Forestry, Lithuana

Aušra Arlauskienė and Stanislava Maikštėnienė
Joniškėlis Experimental Station of the Lithuanian Research, Centre for Agriculture and Forestry, Lithuania

Timothy Coolong
Department of Horticulture, University of Kentucky, USA

G.R. Mohammadi
Department of Crop Production and Breeding, Faculty of Agriculture and Natural Resources, Razi University, Kermanshah, Iran

Manoj Kumar Yadav
Department of Geophysics, Banaras Hindu University, Varanasi, India

R.S. Singh, Rakesh Kumar, Mahesh Kumar Singh and Amitesh Kumar Singh
Department of Agronomy, Banaras Hindu University, Varanasi, India

Gaurav Mahajan
Department of Agronomy, Jawaharlal Nehru Krishi Vishwavidyalaya, Rewa, India

Subhash Babu
Division of Agronomy, Indian Agricultural Research Institute, New Delhi, India

Sanjay Kumar Yadav
Central Potato Research Station (ICAR), Shillong, India

Amalesh Yadav
Department of Botany, University of Lucknow, Lucknow, India

Jessica Kelton and Jorge Mosjidis
Auburn University, USA

Andrew J. Price
United States Department of Agriculture, USA

Jessica Kelton and Jorge Mosjidis
Auburn University, USA

Andrew J. Price
United States Department of Agriculture, USA

Verica Vasic and Sasa Orlovic
University of Novi Sad, Institute of Lowland Forestry and Environment, Novi Sad, Serbia

Branko Konstantinovic
University of Novi Sad, Faculty of Agriculture, Novi Sad, Serbia

John Carroll
Teagasc Crops Research Centre, Ireland

Nicholas Holden
University College Dublin, Ireland

Hugo De Almeida Dan, Rubem Silvério De Oliveira Junior, Lilian Gomes De Moraes Dan, Jamil Constantin and Guilherme Braga Pereira Braz
Center for Advanced Studies in Weed Research, Agronomy Department, State University of Maringá, Paraná, Brazil

Sergio De Oliveira Procópio
Brazilian Agricultural Research Corporation, Londrina, PR, Brazil

Alberto Leão De Lemos Barroso
University of Rio Verde, Goiás, Brazil

Moses Imo
Department of Forestry and Wood Science, Chepkoilel University College, Moi University, Eldoret, Kenya

Endla Reintam and Jaan Kuht
Estonian University of Life Sciences, Estonia

Elifas Nunes de Alcântara and Paulo Tácito Gontijo Guimarães
Minas Gerais State Corporation for Agriculture and Livestock Research, EPAMIG, CTSM, Brazil

Ayodele Ebenezer Ajayi
Soil Water and Environment Section, Department of Agricultural Engineering, Federal University of Technology Akure Ondo State, Nigeria

Cezar Francisco Araujo-Junior
Agronomic Institute of Paraná – IAPAR, Rodovia Celso Garcia Cid, Londrina, State of Paraná, Brazil

Moacir de Souza Dias Junior
Department of Soil Science, Federal University of Lavras, Lavras, State of Minas Gerais, Brazil

Printed in the USA
CPSIA information can be obtained
at www.ICGtesting.com
JSHW011448221024
72173JS00004B/993